Automatic Control of Aircraft and Missiles

Automatic Control

of Aircraft and Missiles

John H. Blakelock, Colonel, USAF

Associate Professor of Electrical Engineering

Air Force Institute of Technology

John Wiley & Sons, Inc.

New York / London / Sydney

10 9 8 7 6 5 4

ISBN 0 471 07930 8
Library of Congress Catalog Card Number: 65–16402
Printed in the United States of America

Preface

This book was conceived as early as 1958 when I gave a graduate course on the automatic control of aircraft and missiles at the Air Force Institute of Technology, Wright-Patterson Air Force Base, Dayton, Ohio, and I became aware that there was, and still is, with the exception of this book, no work extant which treats the whole subject. It is written for graduate and professional use, but at the same time its contents are presented in such a manner that it also can be used in the senior year of an under-graduate program.

The first four chapters make up the bulk of the book and contain most of the detailed analysis of various automatic control systems. In Chapter 1 the six-degree-of-freedom rigid body equations are derived in detail, followed by the linearization and separation of the equations of motion. The three-degree-of-freedom longitudinal equations are then derived along with the longitudinal stability derivatives, followed by the determination of the longitudinal transfer functions for a jet transport for an elevator input. The transient response of the aircraft as well as the effects of changes in altitude and airspeed on the longitudinal response are demonstrated through the use of analog computer traces. In Chapter 2 various control systems are analyzed by use of the root locus techniques along with the complete analysis and design of a glide slope coupler and automatic flare control, including automatic velocity control, as well as the results from the analog simulation. The same pattern is followed in Chapters 3 and 4 with the three lateral equations. The control systems analyzed in Chapter 4 include the damping of the Dutch roll, various techniques for achieving coordination during a turn (minimizing sideslip), the analysis of the complete lateral autopilot, and the design of a localizer coupler, which when used with the glide slope coupler and automatic flare control would provide an automatic landing system. Chapter 5 follows with an analysis of the effects of inertial cross-coupling including a control system for stabilizing the instability resulting therefrom. The subject of self-adaptive autopilots is discussed in Chapter 6 and includes a discussion of the forerunner of the self-adaptive control system used in the X-15. Chapter 7 follows with the analysis of missile control systems using the transfer function of the Vanguard missile, and Chapter 8 with a study of the effects of flexibility and propellant sloshing, including the derivation of the appropriate transfer functions and a discussion of the methods of compensation.   The book ends with a chapter on the application of statistical design principles.

The final form of the equations of motion for the aircraft, and the aerodynamic terms, are the same as those discussed in a set of unpublished notes prepared by Professor J. Bicknell for a course in Airplane Stability and Control at the Massachusetts Institute of Technology.   I have determined from experience that this is the most useful form for the equations for both numerical analysis and analog simulation.   The aerodynamic stability derivatives used in the equations are the standard NACA non-dimensional stability derivatives.   The various control systems used in this book are not necessarily original with me, but are similar to actual control systems that have been proposed or used for high-performance aircraft.   All the discussion, analysis, computer simulations, and interpretation of the behavior of the various control systems are original with me.

I wish to express my deep appreciation to Mrs. Jane Moone and Miss Charlasine Murph, who typed the original notes.   Their efficiency and ability helped make this book possible.   I also wish to thank the many students who used my material and made many helpful suggestions and corrections.   Particular appreciation is expressed to Dr. C. M. Zieman, Head of the Electrical Engineering Department, and to Professors John J. D'Azzo and Constantine H. Houpis for their continued encouragement, and to Professor H. Phillip Whitaker, whose course in Automatic Control at MIT provided some of the basic concepts on which much of the analysis in this book is based.   Finally, I wish to express my deep appreciation to my wife for her patience, understanding, and encouragement and for doing the necessary typing for the final manuscript.

April, 1965   JOHN H. BLAKELOCK

Contents

Preface   v

Introduction   1

Chapter I   Longitudinal Dynamics   5

1-1   Introduction   5
1-2   The Meaning of Velocities in a Moving Axis System   5
1-3   Development of the Equations of Motion (Controls
      Locked)   6
1-4   Relation of the Aircraft Attitude with Respect to the
      Earth   12
1-5   Linearization and Separation of the Equations of Motion   13
1-6   Longitudinal Equations of Motion   14
1-7   Derivation of Equations for the Longitudinal Stability
      Derivatives   22
1-8   Solution of the Longitudinal Equations (Stick Fixed)   31
1-9   Longitudinal Transfer Function for Elevator Displacement   36
1-10 Transient Response of the Aircraft   46
1-11 Effect of Variation of Stability Derivatives on Aircraft
      Performance   53

Chapter 2   Longitudinal Autopilots   56

2-1   Displacement Autopilot   56
2-2   Pitch Orientational Control System   62
2-3   Acceleration Control System   71
2-4   Glide Slope Coupler and Automatic Flare Control   75
2-5   Flight Path Stabilization   92
2-6   Vertical Gyro as the Basic Attitude Reference   98
2-7   Gyro Stabilized Platform as the Basic Attitude Reference   100
2-8   Effects of Nonlinearities   105
2-9   Summary   106

Chapter 3   Lateral Dynamics   107

   3-1   Lateral Equations of Motion   107
   3-2   Derivation of Equation for the Lateral Stability Derivatives 112
   3-3   Solution of Lateral Equations (Stick Fixed)   116
   3-4   Lateral Transfer Function for Rudder Displacement   118
   3-5   Lateral Transfer Function for Aileron Displacement   122
   3-6   Approximate Transfer Functions   125
   3-7   Transient Response of the Aircraft   128
   3-8   Effect of Stability Derivative Variation   133

Chapter 4   Lateral Autopilots   137

   4-1   Introduction   137
   4-2   Damping of the Dutch Roll   138
   4-3   Methods of Obtaining Coordination   141
   4-4   Discussion of Coordination Techniques   152
   4-5   Yaw Orientational Control System   157
   4-6   Other Lateral Autopilot Configurations   163
   4-7   Turn Compensation   166
   4-8   Automatic Lateral Beam Guidance   167
   4-9   Nonlinear Effects   180
   4-10 Summary   181

Chapter 5   Inertial Cross-Coupling   183

   5-1   Introduction   183
   5-2   Effects of High-Roll Rates   184
   5-3   Determination of the Aircraft Parameters that Affect
         Stability   189
   5-4   System for Controlling an Aircraft Subject to Inertial
         Cross-Coupling   191
   5-5   Improved System for Controlling an Aircraft Subject to
         Inertial Cross-Coupling   194

Chapter 6   Self-Adaptive Autopilots   199

   6-1   Introduction   199
   6-2   General Philosophy of the Self-Adaptive Control System   200
   6-3   Sperry Self-Adaptive Control System   201

6-4 Minneapolis-Honeywell Self-Adaptive Control System   206
6-5 MIT Model-Reference Adaptive Control System for Aircraft 209
6-6 MH-90 Adaptive Control System   215
6-7 Summary   220

Chapter 7   Missile Control Systems   222

7-1 Introduction   222
7-2 Roll Stabilization   223
7-3 Control of Aerodynamic Missiles   225
7-4 Transfer Function for a Ballistic-Type Missile   226
7-5 Vanguard Control System (Rigid Missile)   230
7-6 Alternate Missile Control System (Rigid Missile)   231
7-7 Summary   235

Chapter 8   Structural Flexibility   236

8-1 Introduction   236
8-2 Lagrange's Equation   236
8-3 Lagrange's Equation Applied to a System of Lumped
        Parameters   239
8-4 Mode Shapes and Frequencies   242
8-5 Normal Coordinates   247
8-6 System Transfer Function, Including Body Bending   249
8-7 The "Tail-Wags-Dog" Zero   253
8-8 Effects of Propellant Sloshing   254
8-9 Compensation Required for Body Bending   259
8-10 Summary   266

Chapter 9   Application of Statistical Design Principles 268

9-1 Introduction   268
9-2 Random Processes   269
9-3 Mean-Square Error   273
9-4 Autocorrelation Functions   273
9-5 Cross-Correlation Function   277
9-6 Power Spectral Density   278
9-7 Application of Statistical Design Principles   281
9-8 Additional Applications of Statistical Design Principles 285
9-9 Summary   286

Appendix A   Review of Vector Analysis   289
Appendix B   Some Gyroscopic Theory   298
Appendix C   Basic Servo Theory   306
Appendix D   Fundamental Aerodynamic Principles   328
Appendix E   Matrices   333
Appendix F   F-94A Longitudinal and Lateral Aerodynamic Data   340
              Index   343

Automatic Control of Aircraft and Missiles

Introduction

The seventeenth of December 1903 marked the date of the first successful flight of a powered aeroplane. The Wright brothers, in their efforts to succeed where others had failed, broke with tradition and designed their aeroplane to be unstable but controllable.[1] This break with tradition resulted in a more maneuverable and controllable vehicle that was less susceptible to atmospheric gusts. The approach taken by such pioneers as Lilienthal, Pilcher, Chanute, and Langley in the design of their flying machines was to make them inherently stable, leaving the pilot with no other duty than to steer the vehicle. The price paid for this stability was lack of maneuverability and susceptibility to atmospheric disturbances. The lack of stability introduced by the Wright brothers naturally made the pilot's job more difficult and tiring and more than likely hastened the development of the automatic pilot, often called simply an autopilot.

This inherent instability is still prevalent in aircraft of today in the form of the so-called "spiral divergence" which causes a slow divergence in heading and bank angle as a result of any small disturbance. The main purpose then of the early autopilots was to stabilize the aircraft and return it to the desired flight attitude after any disturbance. Such an autopilot was installed in a Glenn H. Curtis flying boat and first tested in late 1912. The autopilot, using gyros to sense the deviation of the aircraft from the desired attitude and servo motors to activate the elevators and ailerons, was developed and built by the Sperry Gyroscope Company of New York under the direction of Dr. E. A. Sperry. The apparatus, called the Sperry Aeroplane Stabilizer, installed in the Curtis flying boat, won prominence on the eighteenth of June 1914. While piloted by Lawrence Sperry, the son of Dr. Sperry, it won a safety prize of 50,000 francs offered by the Aero Club of France for the most stable aeroplane.[1] It is worthy of note that this event took place only eleven years after the Wright brothers' historic flight. During this demonstration, while flying close to the ground under automatic control, with the pilot Lawrence Sperry standing in the cockpit with both hands over his head, the mechanic was standing and walking back and forth on the wing. In spite of the large yawing and rolling moment generated by the mechanic's presence on the wing, the aircraft maintained its original attitude.[2]

After 1915 the public information on automatic pilots became non-existent due to military security, and after the First World War there was little effort expended on advancing the state of the art. This trend was stopped when in 1933 Wiley Post insisted on the installation of the first prototype of the Sperry pneumatic-hydraulic Gyropilot in his aircraft, the "Winnie May," in which he flew around the world in less than eight days. This flight would have been impossible without the aid of this autopilot that allowed Post to doze for short periods while the aircraft flew on under automatic control. The story goes that Post held a wrench tied to his finger by a piece of string; if he fell sound asleep the wrench would drop when his hand relaxed, the subsequential jerk on his finger would awaken him so that he could check his progress and then doze again.[2] The result of this flight aroused new interest in automatic flight, especially for the so-called navigational autopilots, and in 1947 an Air Force C-47 made a completely automatic trans-Atlantic flight including take-off and landing. During the entire flight the control column was untouched by human hands.

These early autopilots, as already mentioned, were primarily designed to maintain the attitude and heading of the aircraft. With the advent of high-performance jet aircraft new problems have arisen. These problems are in addition to the heading instability already mentioned and are a matter of unsatisfactory dynamic characteristics. The situation may be best illustrated as follows: In general, it may be said that if the period of oscillation inherent in an aircraft is 10 seconds or more, the pilot can adequately control or damp the oscillation, but if the period is 4 seconds or less, the pilot's reaction time is not fast enough to cope with the oscillation; thus, such oscillations should be well damped. The so-called "short period" pitch and "Dutch roll" oscillations inherent in all aircraft fall into the category of a 4-second oscillation. In the more conventional aircraft, the damping of these oscillations is effective enough that the handling characteristics of the aircraft are satisfactory. However, in almost all jet fighter and jet transport-type aircraft artificial damping must be provided by an automatic system. The resulting systems are referred to as pitch and yaw dampers and are analyzed in Chapters 2 and 4.

As the aircraft designers tried to obtain more performance from jet fighters, larger and larger engines were installed along with shorter and thinner wings; this trend resulted in significant changes in the moments of inertia of the aircraft which led to catastrophic results for some aircraft. The culprit was "inertial cross-coupling" that had safely been neglected in the past. This phenomenon, which is discussed in detail in Chapter 5, results when the aircraft rolls at high angular velocities. The normal correction is the installation of a larger and/or more effective vertical stabilizer. This requires a major modification of the airframe that is both costly and time consuming. As shown in Chapter 5 the effects of inertial cross-coupling can be eliminated with a properly designed control system.

Another stability problem that has manifested itself in some jet fighters is the problem of complete loss of longitudinal stability or pitch-up at high angles of attack. This phenomenon is more apt to occur when the horizontal stabilizer is placed on top of the vertical stabilizer to improve lateral stability. The actual cause of pitch-up and a control system that will stabilize the aircraft at these high angles of attack is discussed in Chapter 2. The same control system can be, and is, used to stabilize ballistic-type missiles, thus eliminating the necessity to add stabilizing surfaces at the aft end of the missile. This situation is discussed in Chapter 7.

There are many other problems that face the control engineer, such as the design of approach couplers to provide automatic approaches to landings in bad weather (Chapters 2 and 4), altitude and Mach hold control systems to improve fuel economy during cruise (Chapter 2), compensators to reduce the effects of body bending (Chapter 8), and control systems that automatically perform optimally under variations of air speed and altitude (Chapter 6). All these problems are discussed in the indicated chapters of this book.

Regardless of its role, a particular control system will in general respond faster and more accurately but with less reliability than a human controller; however, the control system in general is unable to exercise judgment. The design engineer must always strive to improve the performance of the system that he is designing in an attempt to meet the desired performance required.

In the study of any airframe-autopilot combination, it is advantageous to represent the aircraft as a "block" in the block diagram of the control system so that the standard methods of analyzing servomechanisms may be employed. To make this representation, the transfer function of the aircraft relating a given input to a given output is required. By definition *the transfer function is the ratio of the Laplace Transform of the output to the Laplace Transform of the input with the initial conditions zero.* The longitudinal and lateral transfer functions are derived in Chapters 1 and 3, respectively.

To obtain the transfer functions of the aircraft it is necessary to define certain quantities referred to as "stability derivatives" which relate the changes in the aerodynamic forces and moments acting on the aircraft caused by changes in its motion, orientation, or control-surface motion. The stability derivatives used are the standard NACA nondimensional stability derivatives with the standard NACA sign conventions.

References

1. C. S. Draper, "Flight Control," *Journal of the Royal Aeronautical Society*, July 1955, Vol. 59.
2. K. I. T. Richardson, *The Gyroscope Applied*, The Philosophical Library, New York, 1954, pp. 261–264.

1

Longitudinal Dynamics

1-1 Introduction

To obtain the transfer function of the aircraft it is first necessary to obtain the equations of motion for the aircraft. The equations of motion are derived by applying Newton's laws of motion which relate the summation of the external forces and moments to the linear and angular accelerations of the system or body. To make this application, certain assumptions must be made and an axis system defined.

The center of the axis system is, by definition, located at the center of gravity of the aircraft. In general, the axis system is fixed to the aircraft and rotates with it. Such a set of axes is referred to as "body axes." It is not necessary to use such an axis system; an axis system could be fixed, for example, to the air mass, and the aircraft could rotate with respect to it. However, for the purposes of this text the axis system is taken as *fixed to the aircraft*.

The axis is taken with OX forward, OY out the right wing, and OZ downward as seen by the pilot to form a right-handed axis system (see Figure 1-2, p. 12). Most aircraft are symmetrical with reference to a vertical plane aligned with the longitudinal axis of the aircraft. Thus, if the OX and OZ axes lie in this plane, the products of inertia J_{xy} and J_{yz} are zero. This result leads to the first assumption, *that the axes OX and OZ lie in the plane of symmetry of the aircraft, and that J_{xy} and J_{yz} are equal to zero.* At this time, the exact direction of OX is not specified, but in general it is not along a principal axis, hence $J_{zx} \neq 0$.

1-2 The Meaning of Velocities in a Moving Axis System

Very often a student has difficulty understanding what is meant by the velocity of a body with respect to an axis system that is moving with the body. How can there be any relative velocity in this situation? Statements about the velocity along the OX axis refer to the *component of velocity with respect to inertial space* taken along the instantaneous

5

direction of the OX axis. At any instant, the aircraft has some resultant velocity vector with respect to inertial space. This vector is resolved into the instantaneous aircraft axes to obtain the velocity components U, V, and W. This resolution also applies to the angular velocity. Resolve the instantaneous angular velocity vector, with respect to inertial space, into the instantaneous direction of the OX, OY, and OZ axes to obtain P, Q, and R, respectively (see Figure 1-2, p. 12). It should be remembered that P, Q, and R are the components of the total angular velocity of the body or aircraft with respect to inertial space. Thus, they are the angular velocities that would be measured by rate gyros fixed to these axes. It should be recalled that inertial space is that space where Newton's laws apply. In general, a set of axes with their origin at the center of the earth but not rotating with the earth may be considered as an inertial coordinate system. Thus, the earth rotates once a day with respect to such an axis system.

1-3   Development of the Equations of Motion (Controls Locked)

The equations of motion for the aircraft can be derived from Newton's Second Law of motion, which states that the summation of all external forces acting on a body must be equal to the time rate of change of the momentum of the body, and the summation of the external moments acting on a body must be equal to the time rate of change of the moment of momentum (angular momentum). The time rates of change are all taken with respect to inertial space. These laws can be expressed by two vector equations.

$$\sum \mathbf{F} = \frac{d}{dt}(m\mathbf{V}_T)]_I \tag{1-1}$$

and

$$\sum \mathbf{M} = \frac{d\mathbf{H}}{dt}\bigg]_I \tag{1-2}$$

where $]_I$ indicates the time rate of change of the vector with respect to inertial space. Now, the external forces and moments consist of equilibrium or steady-state forces and moments and changes in them which cause or result in a disturbance from this steady state or equilibrium condition. Thus,

$$\sum \mathbf{F} = \sum \mathbf{F}_0 + \sum \Delta\mathbf{F}$$

and

$$\sum \mathbf{M} = \sum \mathbf{M}_0 + \sum \Delta\mathbf{M} \tag{1-3}$$

where $\sum \mathbf{F}_0$ and $\sum \mathbf{M}_0$ are the summations of the equilibrium forces and moments. In the dynamic analyses to follow, the aircraft is always considered to be in equilibrium before a disturbance is introduced. Thus, the $\sum \mathbf{F}_0$ and $\sum \mathbf{M}_0$ are identically zero. The equilibrium forces consist of lift, drag, thrust, and gravity, and the equilibrium moments consist of moments resulting from the lift and drag generated by the various portions of the aircraft and the thrust. Therefore, the aircraft is initially in unaccelerated flight and the disturbances in general arise from either the control surface deflection or atmospheric turbulence. Under these conditions, Eqs. 1-1 and 1-2 can be written in the form of

$$\sum \Delta \mathbf{F} = \frac{d}{dt}(m\mathbf{V}_T)]_I \qquad (1\text{-}4)$$

and

$$\sum \Delta \mathbf{M} = \frac{d\mathbf{H}}{dt}\bigg]_I \qquad (1\text{-}5)$$

Before proceeding with the derivation, it is necessary to make some additional assumptions. It is assumed, second, that the *mass of the aircraft remains constant during any particular dynamic analysis.* Actually, there is considerable difference in the mass of an aircraft with and without fuel, but the amount of fuel consumed during the period of the dynamic analysis may be safely neglected. Third, it is assumed *that the aircraft is a rigid body.* Thus, any two points on or within the airframe remain fixed with respect to each other. This assumption greatly simplifies the equations and is quite valid for fighter type aircraft. The effects of aeroelastic deflection of the airframe will be discussed in Chapter 8. Fourth, it is assumed that *the earth is an inertial reference and unless otherwise stated the atmosphere is assumed to be fixed with respect to the earth.* Although this assumption is invalid for the analysis of inertial guidance systems, it is valid for analyzing automatic control systems for both aircraft and missiles and greatly simplifies the final equations. The validity of this assumption is based upon the fact that normally the gyros and accelerometers used for control systems are incapable of sensing the angular velocity of the earth or accelerations resulting from this angular velocity such as the Coriolis acceleration.

It is now time to consider the motion of an aircraft with respect to the earth. Equation 1-4 can be expanded to obtain

$$\sum \Delta \mathbf{F} = \frac{dm}{dt}\mathbf{V}_T + m\frac{d\mathbf{V}_T}{dt}\bigg]_I \qquad (1\text{-}6)$$

but, as the mass is considered constant, and using the fourth assumption Eq. 1-6 reduces to

$$\sum \Delta \mathbf{F} = m\frac{d\mathbf{V}_T}{dt}\bigg]_E \qquad (1\text{-}7)$$

It is necessary to obtain an expression for the time rate of change of the velocity vector with respect to the earth. This process is complicated by the fact that the velocity vector may be rotating while it is changing in magnitude. This fact leads to the expression for the total derivative of a vector given below (see Appendix A)

$$\frac{d\mathbf{V}_T}{dt}\bigg]_E = \mathbf{1}_{V_T} \frac{dV_T}{dt} + \boldsymbol{\omega} \times \mathbf{V}_T \tag{1-8}$$

where $\mathbf{1}_{V_T}(dV_T/dt)$ is the change in the linear velocity, $\boldsymbol{\omega}$ is the total angular velocity of the aircraft with respect to the earth, and $\times$ signifies the cross product. $\mathbf{V}_T$ and $\boldsymbol{\omega}$ can be written in terms of their components so that

$$\mathbf{V}_T = \mathbf{i}U + \mathbf{j}V + \mathbf{k}W \tag{1-9}$$

and

$$\boldsymbol{\omega} = \mathbf{i}P + \mathbf{j}Q + \mathbf{k}R \tag{1-10}$$

where $\mathbf{i}$, $\mathbf{j}$, and $\mathbf{k}$ are unit vectors along the aircraft's X, Y, and Z axes, respectively. Then from Eq. 1-8

$$\mathbf{1}_{V_T} \frac{dV_T}{dt} = \mathbf{i}\dot{U} + \mathbf{j}\dot{V} + \mathbf{k}\dot{W} \tag{1-11}$$

and

$$\boldsymbol{\omega} \times \mathbf{V}_T = \begin{vmatrix} \mathbf{i} & \mathbf{j} & \mathbf{k} \\ P & Q & R \\ U & V & W \end{vmatrix} \tag{1-12}$$

Expanding

$$\boldsymbol{\omega} \times \mathbf{V}_T = \mathbf{i}(WQ - VR) + \mathbf{j}(UR - WP) + \mathbf{k}(VP - UQ) \tag{1-13}$$

$\sum \Delta \mathbf{F}$ can be written in terms of its components as follows

$$\sum \Delta \mathbf{F} = \mathbf{i} \sum \Delta F_x + \mathbf{j} \sum \Delta F_y + \mathbf{k} \sum \Delta F_z \tag{1-14}$$

Equating the components of Eqs. 1-14, 1-11, and 1-13 the equations of linear motion are obtained.

$$\sum \Delta F_x = m(\dot{U} + WQ - VR)$$

$$\sum \Delta F_y = m(\dot{V} + UR - WP)$$

$$\sum \Delta F_z = m(\dot{W} + VP - UQ) \tag{1-15}$$

To obtain the equations of angular motion, it is necessary to return to Eq. 1-5, which is repeated here:

$$\sum \Delta \mathbf{M} = \frac{d\mathbf{H}}{dt}\bigg]_I \tag{1-16}$$

Before proceeding, it is necessary to obtain an expression for **H**. By definition, **H** is the angular momentum or moment of momentum of a revolving body. The momentum of the element of mass dm due to the angular velocity ω will be equal to the tangential velocity of the element of mass times dm about the instantaneous center of rotation. The tangential velocity can be expressed by the vector cross product as follows (see Figure 1-1):

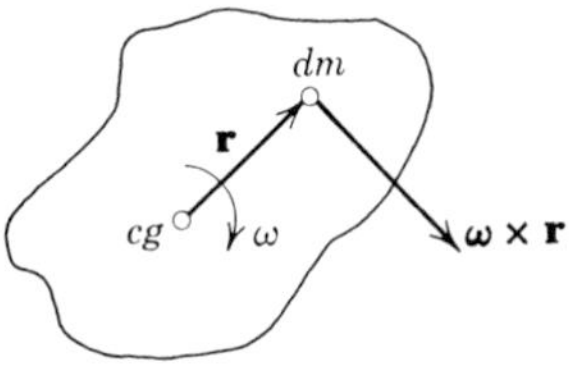

$$\mathbf{V}_{\text{tan}} = \boldsymbol{\omega} \times \mathbf{R} \qquad (1\text{-}17)$$

Figure 1-1 General body with an angular velocity ω about its center of gravity.

Then the incremental momentum resulting from this tangential velocity of the element of mass can be expressed as

$$d\mathbf{M} = (\boldsymbol{\omega} \times \mathbf{r})\, dm \qquad (1\text{-}18)$$

The moment of momentum is the momentum times the lever arm or, as a vector equation,

$$d\mathbf{H} = \mathbf{r} \times (\boldsymbol{\omega} \times \mathbf{r})\, dm \qquad (1\text{-}19)$$

but $\mathbf{H} = \int d\mathbf{H}$ over the entire mass of the aircraft. Thus

$$\mathbf{H} = \int \mathbf{r} \times (\boldsymbol{\omega} \times \mathbf{r})\, dm \qquad (1\text{-}20)$$

In evaluating the triple cross product, if

$$\boldsymbol{\omega} = \mathbf{i}P + \mathbf{j}Q + \mathbf{k}R$$

and

$$\mathbf{r} = \mathbf{i}x + \mathbf{j}y + \mathbf{k}z$$

then

$$\boldsymbol{\omega} \times \mathbf{r} = \begin{vmatrix} \mathbf{i} & \mathbf{j} & \mathbf{k} \\ P & Q & R \\ x & y & z \end{vmatrix} \qquad (1\text{-}21)$$

Expanding,

$$\boldsymbol{\omega} \times \mathbf{r} = \mathbf{i}(zQ - yR) + \mathbf{j}(xR - zP) + \mathbf{k}(yP - xQ) \qquad (1\text{-}22)$$

Then,

$$\mathbf{r} \times (\boldsymbol{\omega} \times \mathbf{r}) = \begin{vmatrix} \mathbf{i} & \mathbf{j} & \mathbf{k} \\ x & y & z \\ zQ - yR & xR - zP & yP - xQ \end{vmatrix} \qquad (1\text{-}23)$$

Expanding,

$$\mathbf{r} \times (\boldsymbol{\omega} \times \mathbf{r}) = \mathbf{i}[(y^2 + z^2)P - xyQ - xzR] + \mathbf{j}[(z^2 + x^2)Q - yzR - xyP] \\ + \mathbf{k}[(x^2 + y^2)R - xzP - yzQ] \qquad (1\text{-}24)$$

Substituting Eq. 1-24 into Eq. 1-20, it becomes

$$\mathbf{H} = \int \mathbf{i}[(y^2 + z^2)P - xyQ - xzR]\, dm$$

$$+ \int \mathbf{j}[(z^2 + x^2)Q - yzR - xyP]\, dm$$

$$+ \int \mathbf{k}[(x^2 + y^2)R - xzP - yzQ]\, dm \qquad (1\text{-}25)$$

But the $\int (y^2 + z^2)\, dm$ is defined as the moment of inertia, I_x, and $\int xy\, dm$ is defined as the product of inertia J_{xy}. The remaining integrals of Eq. 1-25 are similarly defined. By remembering from the first assumption that $J_{xy} = J_{yz} = 0$, Eq. 1-25 can be rewritten in component form as

$$H_x = PI_x - RJ_{xz}$$
$$H_y = QI_y$$
$$H_z = RI_z - PJ_{xz} \qquad (1\text{-}26)$$

However, Eq. 1-16 indicates that the time rate of change of $\mathbf{H}$ is required. As $\mathbf{H}$ can change in magnitude and direction, Eq. 1-16 can be written as

$$\sum \Delta \mathbf{M} = \mathbf{1}_\mathrm{H} \frac{dH}{dt} + \boldsymbol{\omega} \times \mathbf{H} \qquad (1\text{-}27)$$

The components of $\mathbf{1}_\mathrm{H} \dfrac{dH}{dt}$ are

$$\frac{dH_x}{dt} = \dot{P}I_x - \dot{R}J_{xz}$$

$$\frac{dH_y}{dt} = \dot{Q}I_y$$

$$\frac{dH_z}{dt} = \dot{R}I_z - \dot{P}J_{xz} \qquad (1\text{-}28)$$

As the aircraft is assumed to be a rigid body of constant mass, the time rates of change of the moments and products of inertia are zero. Now,

$$\boldsymbol{\omega} \times \mathbf{H} = \begin{vmatrix} \mathbf{i} & \mathbf{j} & \mathbf{k} \\ P & Q & R \\ H_x & H_y & H_z \end{vmatrix} \qquad (1\text{-}29)$$

Expanding,

$$\boldsymbol{\omega} \times \mathbf{H} = \mathbf{i}(QH_z - RH_y) + \mathbf{j}(RH_x - PH_z) + \mathbf{k}(PH_y - QH_x) \quad (1\text{-}30)$$

Also $\sum \Delta \mathbf{M}$ can be written as

$$\sum \Delta \mathbf{M} = \mathbf{i} \sum \Delta \mathscr{L} + \mathbf{j} \sum \Delta \mathscr{M} + \mathbf{k} \sum \Delta \mathscr{N} \quad (1\text{-}31)$$

By equating components of Eqs. 1-28, 1-30, and 1-31 and substituting for H_x, H_y, and H_z from Eq. 1-26, the angular equations of motion are obtained.

$$\sum \Delta \mathscr{L} = \dot{P} I_x - \dot{R} J_{xz} + QR(I_z - I_y) - PQ J_{xz}$$

$$\sum \Delta \mathscr{M} = \dot{Q} I_y + PR(I_x - I_z) + (P^2 - R^2) J_{xz}$$

$$\sum \Delta \mathscr{N} = \dot{R} I_z - \dot{P} J_{xz} + PQ(I_y - I_x) + QR J_{xz} \quad (1\text{-}32)$$

The equations of linear motion from Eq. 1-15 are

$$\sum \Delta F_x = m(\dot{U} + WQ - VR)$$

$$\sum \Delta F_y = m(\dot{V} + UR - WP)$$

$$\sum \Delta F_z = m(\dot{W} + VP - UQ) \quad (1\text{-}33)$$

Equations 1-32 and 1-33 are the complete equations of motion for the aircraft. It will next be necessary to linearize the equations and expand the left-hand sides.

Summary of Nomenclature

Axis	Direction	Name	Linear Velocity	Small Angular Displacement	Angular Velocity
OX	Forward	Roll	U	ϕ	P
OY	Right wing	Pitch	V	θ	Q
OZ	Downward	Yaw	W	ψ	R

Axis	Moment of Inertia	Product of Inertia	Force	Moment
OX	I_x	$J_{xy} = 0$	F_x	$\mathscr{L}$
OY	I_y	$J_{yz} = 0$	F_y	$\mathscr{M}$
OZ	I_z	$J_{zx} \neq 0$	F_z	$\mathscr{N}$

These conclusions are based on the assumptions that

1. OX and OZ are in the plane of symmetry.
2. Mass of the aircraft is constant.
3. The aircraft is a rigid body.
4. The earth is an inertial reference.

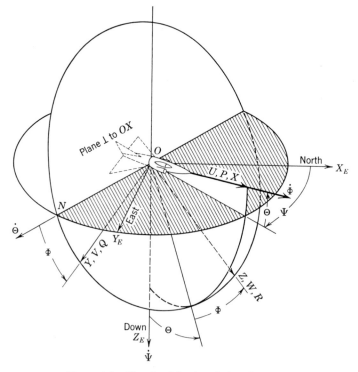

Figure 1-2   Sketch of fixed and aircraft axes.

1-4   Relation of the Aircraft Attitude with Respect to the Earth

In order to describe the motion of the aircraft with respect to the earth or inertial space, it is necessary to be able to specify the orientation of one axis system with respect to another.   This can be done through the use of a set of angles called "Euler angles."   Consider an earth axis system with its origin at the center of gravity of the aircraft and nonrotating with respect to the earth.   Let OX_E and OY_E be in the horizontal plane and OZ_E vertical and down. OX_E may be taken as north or any other fixed direction.   Referring to Figure 1-2, let the following angles indicate the rotation of the XYZ axis from the earth axis.

Ψ the angle between OX_E and the projection of the OX axis on the horizontal plane.

$\dot{\Psi}$ is a vector along OZ_E.

Θ the angle between the horizontal and the OX axis measured in the vertical plane.

$\dot{\Theta}$ is a vector along ON, the line of nodes.

Φ the angle between ON and the OY axis measured in the OYZ plane. Note that this plane is not necessarily vertical.

$\dot{\Phi}$ is a vector along OX.

Thus, the angles Ψ, Θ, and Φ specify the orientation of the aircraft axis system with respect to the earth. The positive direction of these angles is indicated in Figure 1-2.

To transform the components of the angular velocity of the aircraft from the earth axis to the aircraft axis system, take the components $\dot{\Psi}$, $\dot{\Theta}$, and $\dot{\Phi}$ and project them along the OX, OY, and OZ axes to obtain

$$P = \dot{\Phi} - \dot{\Psi} \sin \Theta$$

$$Q = \dot{\Theta} \cos \Phi + \dot{\Psi} \cos \Theta \sin \Phi$$

$$R = -\dot{\Theta} \sin \Phi + \dot{\Psi} \cos \Theta \cos \Phi \qquad (1\text{-}34)$$

These equations can be solved for $\dot{\Phi}$, $\dot{\Theta}$, and $\dot{\Psi}$ to yield

$$\dot{\Theta} = Q \cos \Phi - R \sin \Phi$$

$$\dot{\Phi} = P + Q \sin \Phi \tan \Theta + R \cos \Phi \tan \Theta$$

$$\dot{\Psi} = Q \frac{\sin \Phi}{\cos \Theta} + R \frac{\cos \Phi}{\cos \Theta} \qquad (1\text{-}34a)$$

A similar transformation can be made for linear velocities. It should be noted that $\dot{\Phi}$, $\dot{\Theta}$, and $\dot{\Psi}$ are not orthogonal vectors. Equations 1-34a can be integrated with respect to time and by knowing the initial conditions, Θ, Φ, and Ψ can be determined; however, as the rates of change of these angles are a function of the angles themselves this is best done on a computer.

The components of the gravity force along the aircraft axes are along

$$OX: \quad -mg \sin \Theta$$

$$OY: \quad mg \cos \Theta \sin \Phi$$

$$OZ: \quad mg \cos \Theta \cos \Phi \qquad (1\text{-}35)$$

I-5   Linearization and Separation of the Equations of Motion

A study of Eqs. 1-32 and 1-33 shows that it takes six simultaneous non-linear equations of motion to completely describe the behavior of a rigid aircraft. In this form, a solution can be obtained only by the use of analog or digital computers or by manual numerical integration. In most cases, however, by the use of proper assumptions the equations can be broken down into two sets of three equations each and these linearized to obtain equations amenable to analytic solutions of sufficient accuracy. The six equations are first broken up into two sets of three simultaneous equations. To accomplish this the aircraft is considered to be in straight and level unaccelerated flight and then to be disturbed by deflection of the elevator. This deflection applies a pitching moment about the OY axis, causing a rotation about this axis which eventually causes a change in F_x and F_z, but does not cause a rolling or yawing moment or any change in F_y; thus $P = R = V = 0$ and the $\sum \Delta F_y$, $\sum \Delta \mathscr{L}$, and $\sum \Delta \mathscr{N}$ equations may be eliminated. This leaves

$$\left. \begin{aligned} \sum \Delta F_x &= m(\dot{U} + WQ) \\[1em] \sum \Delta F_z &= m(\dot{W} - UQ) \\[1em] \sum \Delta \mathscr{M} &= \dot{Q} I_y \end{aligned} \right\} \quad \text{longitudinal equations for } P = R = V = 0$$

$$(1\text{-}36)$$

An investigation of the remaining three equations, especially the $\mathscr{L}$ and $\mathscr{N}$ equations, shows that a rolling or yawing moment excites angular velocities about all three axes; thus except for certain cases the equations cannot be decoupled. The assumptions necessary for this decoupling will be discussed in Chapter 3 on the lateral dynamics of the aircraft, and the condition when this separation of the equations is not valid will be discussed in Chapter 5 on inertial cross coupling. The rest of this chapter will be devoted to the expansion of the longitudinal equations of motion.

1-6   Longitudinal Equations of Motion

Previously, the components of the total instantaneous values of the linear and angular velocity resolved into the aircraft axes were designated as U, V, W, P, Q, and R. As these values include an equilibrium value and the change from the steady state, they may be expressed as

$$\begin{aligned} U &= U_0 + u & P &= P_0 + p \\ V &= V_0 + v & Q &= Q_0 + q \\ W &= W_0 + w & R &= R_0 + r \end{aligned}$$

where U_0, V_0, etc., are the equilibrium values and u, v, etc., are the changes in these values resulting from some disturbance. In Section 1-1 a body axis system for the aircraft was discussed and the OX axis was taken forward but the exact orientation was not specified. The OX axis could

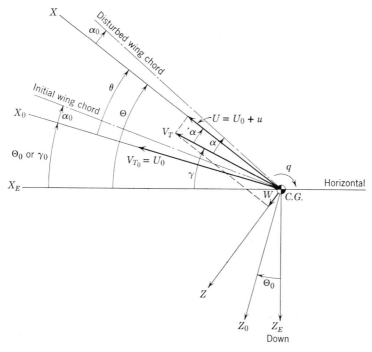

Figure 1-3   Equilibrium and disturbed aircraft stability axes.

$$'\alpha = W/U$$
$$\alpha = \alpha_0 + '\alpha$$
$$\Theta = \Theta_0 + \theta$$
$$\gamma = \Theta - '\alpha$$
$$\text{or}\quad \gamma = \theta - '\alpha \quad \text{for}\quad \Theta_0 = 0$$

be aligned with the longitudinal axis of the aircraft; however, if it is originally aligned with the equilibrium direction of the velocity vector of the aircraft, $W_0 = 0$ (see Figure 1-3). In Figure 1-3 the axes X_E, Y_E, Z_E are earth reference axes; the X_0, Y_0, Z_0 are equilibrium aircraft axes; and X, Y, Z are the disturbed aircraft axes. For any particular problem the aircraft axes, after being aligned with the X axis into the relative wind, remain fixed to the aircraft during the study of perturbations from that initial flight condition. Such a set of aircraft axes are referred to as "stability axes." The stability axes will be used in all the dynamic analysis to follow. As can be seen from Figure 1-3 Θ_0 and γ_0 are measured from the horizontal to the stability X_0 axis. The angle γ is often referred to as the "flight path angle" and is defined as the angle measured, in the vertical plane, between the horizontal and the velocity vector of the aircraft. By using stability axes, Θ_0 and γ_0 are equal. The definition of the angle of attack is standard, that is, the angle between the velocity vector (or the relative wind) and the wing chord.   As the change in Θ, which is equal to θ, is caused by a rotation about the Y axis then $q = \dot{\theta}$.   Under these conditions $U = U_0 + u$, $W = w$ and, as U_0 is a constant, $\dot{U} = \dot{u}$ and $\dot{W} = \dot{w}$.   As the aircraft is initially in unaccelerated flight, Q_0 must be zero, then $Q = q$.   Making these substitutions the force equations of Eq. 1-36 become

$$\sum \Delta F_x = m(\dot{u} + wq)$$

$$\sum \Delta F_z = m(\dot{w} - U_0 q - uq) \tag{1-37}$$

By restricting the disturbances to small perturbations about the equilibrium condition, the product of the variations will be small in comparison with the variations and can be neglected, and the small angle assumptions can be made relative to the angles between the equilibrium and disturbed axes. This fifth assumption somewhat limits the applicability of the equations but reduces them to linear equations.   Thus Eq. 1-37 can be written as follows with the addition of the pitching moment equation from Eq. 1-36.

$$\sum \Delta F_x = m(\dot{u})$$

$$\sum \Delta F_z = m(\dot{w} - U_0 q) = m(\dot{w} - U_0 \dot{\theta})$$

$$\sum \Delta \mathcal{M} = I_y \dot{q} = I_y \ddot{\theta} \tag{1-38}$$

It is now necessary to expand the applied forces and moments and to express them in terms of the changes in the forces and moments that cause or result from these perturbations.   These latter forces are usually of an aerodynamic and gravitational origin.   For instance, the components of gravity along the X and Z axes are a function of the angle Θ, as shown in Figure 1-4.

$$F_{g_x} = -mg \sin \Theta \quad \text{and} \quad F_{g_z} = mg \cos \Theta \tag{1-39}$$

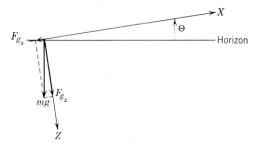

Figure 1-4   Components of gravity resolved into aircraft axes.

The changes in these forces with respect to Θ are

$$\frac{\partial F_{g_x}}{\partial \Theta} = -mg \cos \Theta \quad \text{and} \quad \frac{\partial F_{g_z}}{\partial \Theta} = -mg \sin \Theta \qquad (1\text{-}40)$$

The forces in the X direction are a function of U, W, $\dot{W}$, Θ, and $\dot{\Theta}$. Thus the total differential of F_x can be expressed as

$$\sum dF_x = \frac{\partial F_x}{\partial U}\, dU + \frac{\partial F_x}{\partial W}\, dW + \frac{\partial F_x}{\partial \dot{W}}\, d\dot{W} + \frac{\partial F_x}{\partial \Theta}\, d\Theta + \frac{\partial F_x}{\partial \dot{\Theta}}\, d\dot{\Theta} \qquad (1\text{-}41)$$

The reason for not including the $\partial F_x/\partial \dot{U}$ will be explained in Section 1-6.

If it is assumed that, as the disturbances are small, the partial derivatives are linear, the differentials can be replaced by the actual increments, and Eq. 1-41 becomes

$$\sum \Delta F_x = \frac{\partial F_x}{\partial u}\, \Delta U + \frac{\partial F_x}{\partial w}\, \Delta W + \frac{\partial F_x}{\partial \dot{w}}\, \Delta \dot{W} + \frac{\partial F_x}{\partial \theta}\, \Delta \Theta + \frac{\partial F_x}{\partial \dot{\theta}}\, \Delta \dot{\Theta} \qquad (1\text{-}42)$$

But u, w, etc., are the changes in the parameters, and as the perturbations are small $u = \Delta U$, etc. Therefore Eq. 1-42 becomes

$$\sum \Delta F_x = \frac{\partial F_x}{\partial u}\, u + \frac{\partial F_x}{\partial \dot{w}}\, \dot{w} + \frac{\partial F_x}{\partial w}\, w + \frac{\partial F_x}{\partial \theta}\, \theta + \frac{\partial F_x}{\partial \dot{\theta}}\, \dot{\theta} \qquad (1\text{-}43)$$

Multiplying and dividing the first three terms of Eq. 1-43 by U_0, it becomes

$$\sum \Delta F_x = U_0 \frac{\partial F_x}{\partial u}\, \frac{u}{U_0} + U_0 \frac{\partial F_x}{\partial w}\, \frac{w}{U_0} + U_0 \frac{\partial F_x}{\partial \dot{w}}\, \frac{\dot{w}}{U_0} + \frac{\partial F_x}{\partial \theta}\, \theta + \frac{\partial F_x}{\partial \dot{\theta}}\, \theta$$

$$(1\text{-}44)$$

As the perturbations have been assumed small, then $U \simeq U_0$, hereafter no distinction will be made, and the sub zero will be dropped. However, the value of U that appears explicitly in the equations of motion is the equilibrium value of U. The dimensionless ratios u/U, w/U, and $\dot{w}/U$ are defined as

$$u/U = {}'u$$

$$w/U = {}'\alpha \qquad \text{which is the variation in the angle of attack from equilibrium}$$

and $\quad \dot{w}/U = {}'\dot{\alpha}$

Substituting these values into Eq. 1-44 yields

$$\sum \Delta F_x = U \frac{\partial F_x}{\partial u}\, {}'u + \frac{\partial F_x}{\partial \alpha}\, {}'\alpha + \frac{\partial F_x}{\partial \dot{\alpha}}\, {}'\dot{\alpha} + \frac{\partial F_x}{\partial \theta}\, \theta + \frac{\partial F_x}{\partial \dot{\theta}}\, \dot{\theta} \qquad (1\text{-}45)$$

as $\qquad U \dfrac{\partial F_x}{\partial w} = \dfrac{\partial F_x}{\partial w/U} = \dfrac{\partial F_x}{\partial\, '\alpha} = \dfrac{\partial F_x}{\partial \alpha} \quad$ as $\quad \dfrac{\partial\, '\alpha}{\partial \alpha} \equiv 1$

From Eq. 1-38

$$\sum \Delta F_x = m\dot{u} = m\dot{u}\left(\frac{U}{U}\right) = mU\frac{\dot{u}}{U} = mU'\dot{u}$$

Substituting this expression for $\sum \Delta F_x$ in Eq. 1-45, taking the other terms over to the left-hand side of the equation, and dividing by Sq, the X equation becomes

$$\frac{mU}{Sq}'\dot{u} - \frac{U}{Sq}\frac{\partial F_x}{\partial u}'u - \frac{1}{Sq}\frac{\partial F_x}{\partial \alpha}'\alpha - \frac{1}{Sq}\frac{\partial F_x}{\partial \dot{\alpha}}'\dot{\alpha} - \frac{1}{Sq}\frac{\partial F_x}{\partial \theta}\theta$$

$$- \frac{1}{Sq}\frac{\partial F_x}{\partial \dot{\theta}}\dot{\theta} = \frac{F_{x_a}}{Sq} \quad (1\text{-}46)$$

where F_{x_a} is an applied aerodynamic force of unspecified origin to be explained in Section 1-7. S is the wing area as defined in Appendix D, $q = \frac{1}{2}\rho V^2$, the dynamic pressure in lb/sq ft, and ρ is the air density. Substituting for $\partial F_x/\partial \theta$ from Eq. 1-40 and multiplying and dividing the fourth and sixth terms by $c/2U$, where c is the mean aerodynamic chord (see Appendix D), Eq. 1-46 becomes

$$\frac{mU}{Sq}'\dot{u} - \frac{U}{Sq}\frac{\partial F_x}{\partial u}'u - \frac{1}{Sq}\frac{\partial F_x}{\partial \alpha}'\alpha - \frac{c}{2U}\left(\frac{1}{Sq}\right)\left(\frac{2U}{c}\right)\frac{\partial F_x}{\partial \dot{\alpha}}'\dot{\alpha}$$

$$+ \frac{mg}{Sq}(\cos \Theta)\theta - \frac{c}{2U}\left(\frac{1}{Sq}\right)\left(\frac{2U}{c}\right)\frac{\partial F_x}{\partial \dot{\theta}}\dot{\theta} = \frac{F_{x_a}}{Sq} = C_{F_{x_a}} \quad (1\text{-}47)$$

To show how Eq. 1-47 was nondimensionalized, the $'\dot{\alpha}$ term will be analyzed. First take $\left(\frac{1}{Sq}\right)\left(\frac{2U}{c}\right)\frac{\partial F_x}{\partial \dot{\alpha}}$; as F_x and Sq both have the dimensions of a force (lb), the units of these two terms cancel, leaving $2U/c$ and $1/\dot{\alpha}$. The dimensions of $2U/c$ and $1/\dot{\alpha}$ are 1/sec and sec, respectively; therefore this portion of the $'\dot{\alpha}$ term is nondimensional and can be replaced by a nondimensional coefficient. The remaining portion of the $'\dot{\alpha}$ term, which is $\frac{c}{2U}'\dot{\alpha}$, is also nondimensional. The remaining terms are handled in a similar manner. These nondimensional coefficients are referred to as "stability derivatives," and although there are several forms of these coefficients, the ones used in this text and listed in Table 1-1 are essentially the NACA standard longitudinal stability derivatives. All the terms in Table 1-1 are explained later in Section 1-7, except for C_D and C_L, which are discussed in Appendix D.

Introducing these terms, Eq. 1-47 becomes

$$\frac{mU}{Sq}'\dot{u} - C_{x_u}'u - C_{x_a}'\alpha - \frac{c}{2U}C_{x_{\dot{\alpha}}}'\dot{\alpha}$$

$$+ \frac{mg}{Sq}(\cos \Theta)\theta - \frac{c}{2U}C_{x_q}\dot{\theta} = C_{F_{x_a}} \quad (1\text{-}48)$$

Sometimes mg/Sq is replaced by $-C_w$. If this is done, Eq. 1-48 becomes

$$\frac{mU}{Sq}\,'\dot{u} - C_{x_u}'u - C_{x_\alpha}'\alpha - \frac{c}{2U}C_{x_{\dot\alpha}}'\dot\alpha - C_w(\cos\Theta)\theta - \frac{c}{2U}C_{x_q}\dot\theta = C_{F_{x_a}}$$

$$(1\text{-}49)$$

Table 1-1   Definitions and Equations for Longitudinal Stability Derivatives

Symbol	Definition	Origin	Equation	Typical Values
C_{x_u}	$\dfrac{U}{Sq}\dfrac{\partial F_x}{\partial u}$	Variation of drag and thrust with u	$-2C_D - U\dfrac{\partial C_D}{\partial u}$	-0.05
C_{x_α}	$\dfrac{1}{Sq}\dfrac{\partial F_x}{\partial \alpha}$	Lift and drag variations along the X axis	$C_L - \dfrac{\partial C_D}{\partial \alpha}$	$+0.1$
C_w	—	Gravity	$-\dfrac{mg}{Sq}$	
$C_{x_{\dot\alpha}}$	$\dfrac{1}{Sq}\left(\dfrac{2U}{c}\right)\dfrac{\partial F_x}{\partial \dot\alpha}$	Downwash lag on drag	Neglect	
C_{x_q}	$\dfrac{1}{Sq}\left(\dfrac{2U}{c}\right)\dfrac{\partial F_x}{\partial \dot\theta}$	Effect of pitch rate on drag	Neglect	
C_{z_u}	$\dfrac{U}{Sq}\dfrac{\partial F_z}{\partial u}$	Variation of normal force with u	$-2C_L - U\dfrac{\partial C_L}{\partial u}$	-0.5
C_{z_α}	$\dfrac{1}{Sq}\dfrac{\partial F_z}{\partial \alpha}$	Slope of the normal force curve	$-C_D - \dfrac{\partial C_L}{\partial \alpha} \simeq -\dfrac{\partial C_L}{\partial \alpha}$	-4
C_w	—	Gravity	$-\dfrac{mg}{Sq}$	
$C_{z_{\dot\alpha}}$	$\dfrac{1}{Sq}\left(\dfrac{2U}{c}\right)\dfrac{\partial F_z}{\partial \dot\alpha}$	Downwash lag on lift of tail	$2\left(\dfrac{\partial C_m}{\partial i_t}\right)\left(\dfrac{d\epsilon}{d\alpha}\right)$	-1
C_{z_q}	$\dfrac{1}{Sq}\left(\dfrac{2U}{c}\right)\dfrac{\partial F_z}{\partial \dot\theta}$	Effect of pitch rate on lift	$2K\left(\dfrac{\partial C_m}{\partial i_t}\right)$	-2
C_{m_u}	$\dfrac{U}{Sqc}\dfrac{\partial \mathcal{M}}{\partial u}$	Effects of thrust, slipstream, and flexibility	No simple relation; usually neglected for jets	
C_{m_α}	$\dfrac{1}{Sqc}\dfrac{\partial \mathcal{M}}{\partial \alpha}$	Static longitudinal stability	$(SM)\left(\dfrac{dC_L}{d\alpha}\right)_\delta^a$	-0.3
$C_{m_{\dot\alpha}}$	$\dfrac{1}{Sqc}\left(\dfrac{2U}{c}\right)\dfrac{\partial \mathcal{M}}{\partial \dot\alpha}$	Downwash lag on moment	$2\left(\dfrac{\partial C_m}{\partial i_t}\right)\dfrac{d\epsilon}{d\alpha}\dfrac{l_t}{c}$	-3
C_{m_q}	$\dfrac{1}{Sqc}\left(\dfrac{2U}{c}\right)\dfrac{\partial \mathcal{M}}{\partial \dot\theta}$	Damping in pitch	$2K\left(\dfrac{\partial C_m}{\partial i_t}\right)\dfrac{l_t}{c}$	-8

In like manner the F_z and $\mathcal{M}$ equation can be obtained. The F_z equation can be written

$$\sum \Delta F_z = \frac{\partial F_z}{\partial u} u + \frac{\partial F_z}{\partial w} w + \frac{\partial F_z}{\partial \dot{w}} \dot{w} + \frac{\partial F_z}{\partial \theta} \theta + \frac{\partial F_z}{\partial \dot{\theta}} \dot{\theta} \qquad (1\text{-}50)$$

Multiplying and dividing the first three terms by U and using the definitions of the dimensionless ratios previously stated, following Eq. 1-44, Eq. 1-50 becomes

$$\sum \Delta F_z = U \frac{\partial F_z}{\partial u} \,'u + \frac{\partial F_z}{\partial \alpha} \,'\alpha + \frac{\partial F_z}{\partial \dot{\alpha}} \,'\dot{\alpha} + \frac{\partial F_z}{\partial \theta} \theta + \frac{\partial F_z}{\partial \dot{\theta}} \dot{\theta} \qquad (1\text{-}51)$$

from Eq. 1-38, $\qquad\qquad \sum \Delta F_z = m(\dot{w} - U\dot{\theta})$

Then $\qquad \sum \Delta F_z = m\left(\frac{\dot{w}}{U} U - U\dot{\theta}\right) = mU'\dot{\alpha} - mU\dot{\theta} \qquad (1\text{-}52)$

Substituting this into Eq. 1-51, and dividing by Sq, it becomes

$$-\frac{U}{Sq} \frac{\partial F_z}{\partial u} \,'u + \left(\frac{mU}{Sq} - \frac{1}{Sq} \frac{\partial F_z}{\partial \dot{\alpha}}\right)'\dot{\alpha} - \frac{1}{Sq} \frac{\partial F_z}{\partial \alpha} \,'\alpha$$

$$+ \left(-\frac{mU}{Sq} - \frac{1}{Sq} \frac{\partial F_z}{\partial \dot{\theta}}\right)\dot{\theta} - \frac{1}{Sq} \frac{\partial F_z}{\partial \theta} \theta = \frac{F_{z_a}}{Sq} = C_{F_{z_a}} \qquad (1\text{-}53)$$

Going to the coefficient form of Table 1-1, Eq. 1-53 becomes

$$-C_{z_u}'u + \left(\frac{mU}{Sq} - \frac{c}{2U} C_{z\dot{\alpha}}\right)'\dot{\alpha} - C_{z_a}'\alpha + \left(-\frac{mU}{Sq} - \frac{c}{2U} C_{z_q}\right)\dot{\theta}$$

$$- C_w(\sin \Theta)\theta = C_{F_{z_a}} \qquad (1\text{-}54)$$

The moment equation can be written as

$$\sum \Delta \mathcal{M} = \frac{\partial \mathcal{M}}{\partial u} u + \frac{\partial \mathcal{M}}{\partial w} w + \frac{\partial \mathcal{M}}{\partial \dot{w}} \dot{w} + \frac{\partial \mathcal{M}}{\partial \dot{\theta}} \dot{\theta} \qquad (1\text{-}55)$$

as $\partial \mathcal{M}/\partial \theta = 0$, for there is no change in $\mathcal{M}$ due to a change in θ, if all the other parameters are held constant.

Multiplying and dividing the first three terms by U and using the definitions of the dimensionless ratios previously stated, Eq. 1-55 becomes

$$\sum \Delta \mathcal{M} = U \frac{\partial \mathcal{M}}{\partial u} \,'u + \frac{\partial \mathcal{M}}{\partial \alpha} \,'\alpha + \frac{\partial \mathcal{M}}{\partial \dot{\alpha}} \,'\dot{\alpha} + \frac{\partial \mathcal{M}}{\partial \dot{\theta}} \dot{\theta} \qquad (1\text{-}56)$$

But from Eq. 1-38 $\sum \Delta \mathcal{M} = I_y \ddot{\theta}$ and after dividing by Sqc, Eq. 1-56 becomes

$$-\frac{U}{Sqc}\frac{\partial \mathcal{M}}{\partial u}\,'u - \frac{1}{Sqc}\frac{\partial \mathcal{M}}{\partial \dot{\alpha}}\,'\dot{\alpha} - \frac{1}{Sqc}\frac{\partial \mathcal{M}}{\partial \alpha}\,'\alpha + \frac{I_y}{Sqc}\ddot{\theta}$$

$$-\frac{1}{Sqc}\frac{\partial \mathcal{M}}{\partial \dot{\theta}}\dot{\theta} = \frac{\mathcal{M}_a}{Sqc} = C_{m_a} \quad (1\text{-}57)$$

Going to the coefficient form of Table 1-1, Eq. 1-57 becomes

$$-C_{m_u}\,'u - \frac{c}{2U}C_{m_{\dot{\alpha}}}\,'\dot{\alpha} - C_{m_\alpha}\,'\alpha + \frac{I_y}{Sqc}\ddot{\theta} - \frac{c}{2U}C_{m_q}\dot{\theta} = C_{m_a} \quad (1\text{-}58)$$

Equations 1-49, 1-54, and 1-58 are the longitudinal equations of motion for the aircraft and are rewritten here for reference.

$$\left(\frac{mU}{Sq}\,'\dot{u} - C_{x_u}\,'u\right) + \left(-\frac{c}{2U}C_{x_{\dot{\alpha}}}\,'\dot{\alpha} - C_{x_\alpha}\,'\alpha\right)$$

$$+ \left[-\frac{c}{2U}C_{x_q}\dot{\theta} - C_w(\cos \Theta)\theta\right] = C_{F_{x_a}}$$

$$-(C_{z_u}\,'u) + \left[\left(\frac{mU}{Sq} - \frac{c}{2U}C_{z_{\dot{\alpha}}}\right)'\dot{\alpha} - C_{z_\alpha}\,'\alpha\right]$$

$$+ \left[\left(-\frac{mU}{Sq} - \frac{c}{2U}C_{z_q}\right)\dot{\theta} - C_w(\sin \Theta)\theta\right] = C_{F_{z_a}}$$

$$(-C_{m_u}\,'u) + \left(-\frac{c}{2U}C_{m_{\dot{\alpha}}}\,'\dot{\alpha} - C_{m_\alpha}\,'\alpha\right) + \left(\frac{I_y}{Sqc}\ddot{\theta} - \frac{c}{2U}C_{m_q}\dot{\theta}\right) = C_{m_a}$$

$$(1\text{-}59)$$

These equations assume that:

1. The X and Z axes lie in the plane of symmetry and the origin of the axis system is at the center of gravity of the aircraft.
2. The mass of the aircraft is constant.
3. The aircraft is a rigid body.
4. The earth is an inertial reference.
5. The perturbations from equilibrium are small.
6. The flow is quasisteady (to be explained in Section 1-7).

These equations require that the X axis be aligned with the aircraft velocity vector while the aircraft is in equilibrium flight. The stability derivatives are defined in Table 1-1 and are derived in Section 1-7.

It should be remembered that in these equations $U = U_0$, $q = \frac{1}{2}\rho U_0^2$, $'u = u/U_0$, $'\alpha = w/U_0$, and $'\dot{\alpha} = \dot{w}/U_0$.

These equations are nondimensional; thus all angles and the derivatives thereof must be in radian measure.

In deriving the longitudinal equations of motion certain nondimensional coefficients, referred to as stability derivatives, were introduced. In Section 1-7 the equations for the stability derivatives will be derived and their origin determined.

1-7   Derivation of Equations for the Longitudinal Stability Derivatives

The student probably wondered why in the expansion of the forces and moments the partial derivative with respect to $\dot{u}$ was missing, while the partial derivative with respect to $\dot{w}$ was included. This was based on the assumption of "quasisteady flow." Quasisteady flow assumes that the airflow around the aircraft changes instantaneously when the aircraft is disturbed from equilibrium. This, of course, is not true. If the aircraft is accelerated, it must in turn accelerate a certain mass of air; this effect is called the "apparent mass effect." However, the assumption of quasisteady flow makes the problem simpler, and for low Mach numbers it is completely adequate. For higher Mach numbers (0.8 or higher) this assumption can lead to theoretical results that do not satisfactorily predict the actual performance of the aircraft. For these cases the compressibility effects can be accounted for by use of stability derivatives for the Mach number in question. These, as well as the subsonic stability derivatives are most accurately determined from wind tunnel tests. However, for the examples used in this book it will be assumed sixth, *that quasisteady flow does exist.* As a result of this assumption, all derivatives with respect to the rates of change of velocities are omitted except for those with respect to w, which is retained to account for the effects of downwash on the horizontal stabilizer, to be explained in the next paragraph. When the forces and moments caused by control surface displacement are discussed, the rates of change of the control surface movement will also be neglected according to this same assumption.

The retention of the derivatives with respect to $\dot{w}$ is not in contradiction with the assumption of quasisteady flow; it takes into account the time required for the effect of the downwash produced *by* the wing to reach the horizontal tail. The downwash is caused by the wing tip vortices and is explained as follows: A wing producing lift experiences a low pressure area on the upper surface of the wing and a high pressure area below the wing. At the wing tip the air in the high pressure area below the wing flows up into the area of low pressure on the upper surface of the wing. As the aircraft moves through the air the wing tips leave this circular pattern of air behind in the form of a spiral. These spirals are called "wing tip vortices" and are shown in Figure 1-5. The result of this

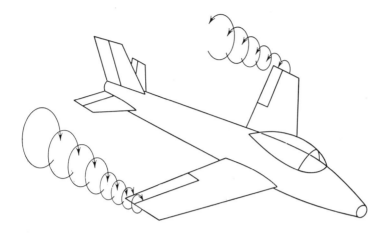

Figure 1-5   An aircraft with wing tip vortices.

wing vortex system is to induce a downward component of velocity to the air flow, which is referred to as downwash. Actually, as shown by Prandtl,[1] in his lifting line theory, the vortex is not restricted to the wing tips but consists of a whole vortex sheet behind the wing. This vortex sheet accounts for the downwash at the wing and at the tail. The downwash velocities are not constant across the span of the wing unless the lift distribution is elliptical. Thus the downwash velocities at a given distance behind the wing vary with the distance from the fuselage. But, as the span of the horizontal tail is small compared to the wing span, the downwash velocity experienced by the tail is almost constant across its span. The effect of the downwash is to reduce the angle of attack of the tail (see Figure 1-6). The angle of attack of the tail can be expressed as

$$\alpha_t = i + \alpha - \epsilon \tag{1-60}$$

where i is the angle of incidence of the stabilizer with respect to the α reference line, α is the wing angle of attack, Dw is the downwash velocity at the tail, and ϵ is the downwash angle $= Dw/U$ in radians.

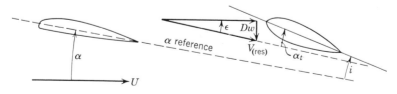

Figure 1-6   Effect of downwash on the angle of attack of the horizontal stabilizer.

The downwash is the result of the lifting action of the wing and increases as the lift and angle of attack of the wing increase. If the vertical velocity of the aircraft is changed, the angle of attack changes, thus the downwash varies with w. Therefore, it is necessary to investigate the effect of a change of w, the vertical velocity, on the angle of attack of the tail. Referring to Figure 1-3, it can be seen that

$$\Delta\alpha \simeq \frac{w}{U} \quad \text{for} \quad W_0 = 0 \tag{1-61}$$

If it is assumed that w is still changing,

$$w_2 = w_1 + \frac{dw}{dt}\Delta t \tag{1-62}$$

where w_1 is the value of w at t_1, w_2 is the value of w at t_2, and $\Delta t = t_2 - t_1$.

As the aircraft is moving at a finite velocity through the air mass it takes a finite time for the effect of this change in w to reach the tail. This time can be expressed by the relation

$$\Delta t = \frac{l_t}{U} \tag{1-63}$$

where l_t is the distance between the quarter chord point of the wing MAC, or some other reference point, and the quarter chord point of the horizontal stabilizer MAC. Let

$$\Delta w = w_2 - w_1 \tag{1-64}$$

Then from Eq. 1-62

$$w_2 - w_1 = \Delta w = \frac{dw}{dt}\Delta t \tag{1-65}$$

Substituting for Δt, Eq. 1-65 becomes

$$\Delta w = \frac{dw}{dt}\frac{l_t}{U} \tag{1-66}$$

As explained following Eq. 1-60 the amount of downwash at the tail is dependent on the angle of attack of the wing and thus dependent on the vertical velocity of the wing; therefore, it can be said that the downwash at the tail is proportional to the vertical velocity of the wing or in equation form

$$(Dw)_t = kw \tag{1-67}$$

Then

$$\Delta(Dw)_t = k\Delta w \tag{1-68}$$

Substituting for Δw from Eq. 1-66, Eq. 1-68 becomes

$$\Delta(Dw)_t = k\frac{l_t}{U}\frac{dw}{dt} = k\frac{l_t}{U}\dot{w} \tag{1-69}$$

But $\dot{w}/U = {}'\dot{\alpha}$; thus Eq. 1-69 becomes

$$\Delta(Dw)_t = kl_t{}'\dot{\alpha} \qquad (1\text{-}70)$$

which is the change in the downwash at the tail due to a rate of change of the angle of attack of the wing. This results in a change of angle of attack of the tail causing a small change in the lift and thus in the Z force and in the pitching moment. Thus partial derivatives with respect to $\dot{w}({}'\dot{\alpha})$ must be included even for quasisteady flow.

It is now time to look at the rest of the stability derivatives. Table 1-1 indicates that $C_{x_{\dot{\alpha}}}$ and C_{x_q} are generally neglected. In the discussion of quasisteady flow the existence of a force due to ${}'\dot{\alpha}$ was explained. In this discussion it is pointed out that this rate of change of angle of attack of the wing results in a change of angle of attack of the horizontal stabilizer, thus changing its lift and drag. The change in drag is the main contribution to a change in the force in the X direction. As the drag of the horizontal stabilizer is usually very small in comparison to the drag of the rest of the aircraft (generally less than 10 per cent of the zero lift drag), any change in this drag can be neglected. In like manner, a pitching velocity $\dot{\theta}$ causes a change in the angle of attack of the horizontal stabilizer, but as in the case of ${}'\dot{\alpha}$ the change in the drag resulting from $\dot{\theta}$ can be neglected. Thus $C_{x_{\dot{\alpha}}}$ and C_{x_q} are usually negligible and are neglected in the following pages. The derivation of the equations for the rest of the stability derivatives follow.

C_{x_u} is the change in the force in the X direction due to a change in the forward velocity. The force in the X direction, when the aircraft is in equilibrium flight with $W_0 = 0$, depends on the thrust and the drag. The thrust vector may not be aligned with the velocity vector of the aircraft, but the angle between these two vectors normally is small. Thus by using small angle assumptions the cosine of the angle can be replaced by 1. Then

$$F_x = T - D = T - C_{Dp}\frac{U^2 S}{2} \qquad (1\text{-}71)$$

Differentiating with respect to u, keeping the other parameters constant, Eq. 1-71 becomes

$$\frac{\partial F_x}{\partial u} = \frac{\partial T}{\partial u} - C_{Dp}\rho US - \frac{\partial C_D}{\partial u}\rho\frac{U^2 S}{2} \qquad (1\text{-}72)$$

It should be noted that all these partial derivatives and C_D must be evaluated at the equilibrium flight condition. Now

$$C_{x_u} = \frac{U}{Sq}\frac{\partial F_x}{\partial u} \quad \text{where} \quad q = \frac{\rho}{2}U^2$$

Multiplying Eq. 1-72 by $\dfrac{U}{Sq}$, it becomes

$$C_{x_u} = \frac{U}{Sq}\frac{\partial T}{\partial u} - 2C_D - U\frac{\partial C_D}{\partial u} \qquad (1\text{-}73)$$

For a jet aircraft the thrust is essentially constant, thus $\partial T/\partial u = 0$. However, for a propeller driven aircraft the thrust decreases as the velocity increases; thus $\partial T/\partial u$ is negative. The variation of drag with velocity is primarily due to Mach number effects, especially in the transonic region where C_D varies with Mach number at a constant angle of attack. Below Mach 0.6, $\partial C_D/\partial u \simeq 0$ (see Appendix D).

C_{z_u} is the change in the force in the Z direction due to a change in the forward velocity. The component of thrust in the Z direction is much less than the lift, thus it may be neglected. Then

$$F_z = -L = -C_L\rho\frac{U^2 S}{2} \qquad (1\text{-}74)$$

Differentiating Eq. 1-74

$$\frac{\partial F_z}{\partial u} = -C_L\rho US - \frac{\partial C_L}{\partial u}\rho\frac{U^2 S}{2} \qquad (1\text{-}75)$$

Multiplying by U/Sq to obtain the nondimensional coefficient, Eq. 1-75 becomes

$$C_{z_u} = \frac{U}{Sq}\frac{\partial F_z}{\partial u} = -2C_L - U\frac{\partial C_L}{\partial u} \qquad (1\text{-}76)$$

The variation of lift coefficient with velocity is primarily due to Mach effects. The lift coefficient normally decreases very rapidly in the transonic region, then rises again. Below about Mach 0.6, $\partial C_L/\partial u \simeq 0$ (see Appendix D).

C_{m_u} is the change in the pitching moment due to a change in forward velocity. This term is primarily a result of slipstream effects. However, aircraft flexibility and compressibility also affect this term. In the transonic region the static stability of the aircraft can vary considerably. Below the transonic region C_{m_u} can be safely neglected for jets. The changes in C_{m_u} are due to the shift, usually rearward, of the center of pressure as the aircraft enters the transonic region making C_{m_u} negative.

C_{x_α} is the change in the force in the X direction due to a change in α caused by a change in w.

Because the aircraft axes are fixed to the aircraft, a disturbance resulting in a vertical velocity produces lift and drag vectors that are no longer perpendicular to and parallel to the X axis (see Figure 1-7). Thus to find the contributions of the lift and drag to the forces in the X direction it is necessary to resolve them into the X axis through the angle $'\alpha$.

Then
$$F_x = L \sin '\alpha - D \cos '\alpha \tag{1-77}$$

Differentiating,

$$\frac{\partial F_x}{\partial '\alpha} = \frac{\partial L}{\partial '\alpha} \sin '\alpha + L \cos '\alpha - \frac{\partial D}{\partial '\alpha} \cos '\alpha + D \sin '\alpha \tag{1-78}$$

As before, the partial derivatives must be evaluated at the equilibrium condition or when $'\alpha = 0$. Thus Eq. 1-78 becomes

$$\left(\frac{\partial F_x}{\partial '\alpha}\right)_{'\alpha = 0} = L - \frac{\partial D}{\partial '\alpha} = L - \frac{\partial D}{\partial \alpha} \quad \text{as} \quad \frac{\partial '\alpha}{\partial \alpha} = 1 \tag{1-79}$$

Multiplying by $1/Sq$ to obtain the nondimensional coefficient, Eq. 1-79 becomes

$$C_{x_\alpha} = \frac{1}{Sq}\left(\frac{\partial F_x}{\partial \alpha}\right) = C_L - \frac{\partial C_D}{\partial \alpha} \tag{1-80}$$

C_{z_α} is the variation of the Z force with angle of attack. This term is very similar to the case of C_{x_α} and referring to Figure 1-7 the new force in the Z direction can be expressed as

$$F_z = -L \cos '\alpha - D \sin '\alpha \tag{1-81}$$

Differentiating,

$$\frac{\partial F_z}{\partial '\alpha} = -\frac{\partial L}{\partial '\alpha} \cos '\alpha + L \sin '\alpha - \frac{\partial D}{\partial '\alpha} \sin '\alpha - D \cos '\alpha \tag{1-82}$$

Evaluating at $'\alpha = 0$, as before, Eq. 1-82 becomes

$$\left(\frac{\partial F_z}{\partial '\alpha}\right)_{'\alpha = 0} = -\frac{\partial L}{\partial '\alpha} - D = -\frac{\partial L}{\partial \alpha} - D \tag{1-83}$$

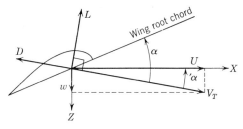

Figure 1-7 Orientation of the lift drag vectors after a disturbance producing a vertical velocity.

Going to coefficient form, Eq. 1-83 becomes

$$C_{z_\alpha} = \frac{1}{Sq}\left(\frac{\partial F_z}{\partial \alpha}\right) = -C_D - \frac{\partial C_L}{\partial \alpha} \simeq -\frac{\partial C_L}{\partial \alpha} \tag{1-84}$$

NOTE. Both $\partial C_D/\partial \alpha$ and $\partial C_L/\partial \alpha$ are the slopes of the drag and lift coefficient curves per radian of angle of attack (see Appendix D).

C_{m_α} is the change in the pitching moment due to a change in angle of attack. This term determines the static longitudinal stability of the aircraft and must be negative for a "statically stable aircraft." A statically stable aircraft is one that tends to return to its equilibrium condition after a disturbance has occurred. A negative C_{m_α} means that as the angle of attack increases positively, the pitching moment becomes more negative tending to decrease the angle of attack. The opposite is true for a positive C_{m_α}. The positive C_{m_α} condition and an automatic system for controlling such an aircraft is discussed in Chapter 2.

Now
$$C_{m_\alpha} = \frac{\partial C_m}{\partial \alpha} = \left(\frac{dC_m}{dC_L}\right)_\delta^a \left(\frac{dC_L}{d\alpha}\right)_\delta^a \tag{1-85}$$

where $\left(\dfrac{dC_m}{dC_L}\right)_\delta^a$ means the change in C_m with respect to C_L for the aircraft

(a) with the elevator fixed ($\delta = 0$). But $\left(\dfrac{dC_m}{dC_L}\right)_\delta^a$ is the same as SM the

static margin, that is, $\left(\dfrac{dC_m}{dC_L}\right)_\delta^a = SM$.[3]

The static margin is equal to x/c, where x is the distance between the fixed control neutral point and the center of gravity of the aircraft and c is the mean aerodynamic chord (MAC). If the center of gravity is ahead of the fixed control neutral point, x is negative and the aircraft is stable. The fixed control neutral point is the location of the center of gravity for which the static margin is zero or the point for which $dC_m/d\alpha$ is equal to zero.[1,3] Therefore

$$C_{m_\alpha} = (SM)_\delta\left(\frac{dC_L}{d\alpha}\right)_\delta^a \tag{1-86}$$

$C_{m_{\dot\alpha}}$ is the effect of the rate of change of angle of attack caused by $\dot w$ on the pitching moment coefficient. As explained in the paragraph preceding Eq. 1-60, this derivative arises from the time lag required for the wing downwash to reach the tail. The change in the angle of attack of the tail due to a rate of change of the downwash is given by the equation

$$\Delta\alpha_t = \frac{d\epsilon}{dt}\Delta t \tag{1-87}$$

where $\Delta t = \dfrac{l_t}{U}$ from Eq. 1-63.

Multiplying Eq. 1-87 by $d\alpha/d\alpha$ and substituting for Δt yields

$$\Delta\alpha_t = \frac{d\epsilon}{d\alpha}\frac{d\alpha}{dt}\frac{l_t}{U} = \frac{d\epsilon}{d\alpha}\frac{l_t}{U}\dot{\alpha} \tag{1-88}$$

Taking the derivative with respect of $\dot{\alpha}$, Eq. 1-88 becomes

$$\frac{d(\Delta\alpha_t)}{d\dot{\alpha}} = \frac{d\epsilon}{d\alpha}\frac{l_t}{U} \tag{1-89}$$

Now
$$\frac{\partial C_m}{\partial\dot{\alpha}} = \left(\frac{dC_m}{di}\right)^t_{\delta,\alpha}\frac{d(\Delta\alpha_t)}{d\dot{\alpha}} \tag{1-90}$$

where $\left(\dfrac{dC_m}{di}\right)^t_{\delta,\alpha}$ is the rate of change of the pitching moment coefficient of the tail with respect to the angle of incidence with α and δ, the elevator deflection, constant.

But
$$C_{m\dot{\alpha}} = \frac{2U}{c}\frac{\partial C_m}{\partial\dot{\alpha}} = \left(\frac{1}{Sqc}\right)\left(\frac{2U}{c}\right)\left(\frac{\partial\mathcal{M}}{\partial\dot{\alpha}}\right) \tag{1-91}$$

as $\mathcal{M} = SqcC_m$. Multiplying Eq. 1-90 by $2U/c$, substituting for $d(\Delta\alpha_t)/d\dot{\alpha}$, and substituting this into Eq. 1-91, it becomes

$$\underline{C_{m\dot{\alpha}} = 2\left(\frac{dC_m}{di}\right)^t_{\delta,\alpha}\frac{d\epsilon}{d\alpha}\frac{l_t}{c}} \tag{1-92}$$

$C_{z\dot{\alpha}}$ is the effect of the rate of change of angle of attack caused by $\dot{w}$ on the Z force. This derivative arises from the same cause as $C_{m\dot{\alpha}}$ and can be derived from $C_{m\dot{\alpha}}$ by dividing by a suitable length to obtain a force coefficient from a moment coefficient. The length used is l_t (see Figure 1-8).

To keep the coefficient nondimensional it is multiplied by c, the mean aerodynamic chord. Thus

$$\underline{C_{z\dot{\alpha}} = \frac{c}{l_t}C_{m\dot{\alpha}} = 2\left(\frac{dC_m}{di}\right)^t_{\delta,\alpha}\left(\frac{d\epsilon}{d\alpha}\right)} \tag{1-93}$$

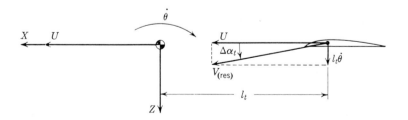

Figure 1-8   Effect of $\dot{\theta}$ on the angle of attack of the tail.

C_{m_q}, the effect on the pitching moment due to a pitch rate, arises from the curvature of the flight path which causes a change in the angle of attack of the tail.   From Figure 1-8 it can be seen that if the aircraft has a pitching velocity, the tail has a vertical velocity component equal to $l_t \dot{\theta}$. This velocity added vectorially to U yields $V_{(res)}$.   Thus the change in the angle of attack resulting from this pitching velocity can be expressed as

$$\Delta \alpha_t = \frac{l_t \dot{\theta}}{U} \tag{1-94}$$

Differentiating,

$$\frac{d(\Delta \alpha_t)}{d \dot{\theta}} = \frac{l_t}{U} \tag{1-95}$$

Now

$$\frac{\partial C_m}{\partial \dot{\theta}} = \left(\frac{dC_m}{di}\right)^t_{\delta,\alpha} \frac{d(\Delta \alpha_t)}{d\dot{\theta}} = \left(\frac{dC_m}{di}\right)^t_{\delta,\alpha} \frac{l_t}{U} \tag{1-96}$$

But

$$C_{m_q} = \frac{2U}{c} \frac{\partial C_m}{\partial \dot{\theta}} = \left(\frac{1}{Sqc}\right)\left(\frac{2U}{c}\right) \frac{\partial \mathcal{M}}{\partial \dot{\theta}} \tag{1-97}$$

Multiplying Eq. 1-96 by $2U/c$ yields

$$C_{m_q} = 2K \left(\frac{dC_m}{di}\right)^t_{\delta,\alpha} \left(\frac{l_t}{c}\right) \tag{1-98}$$

Where the factor K makes a rough allowance for the contribution of the rest of the aircraft to C_{m_q} and is usually about 1.1.   For a swept wing aircraft the factor may be higher.

C_{z_q} is the change in the Z force due to a pitching velocity.   As in the case of $C_{z_{\dot{\alpha}}}$, this stability derivative can be obtained from C_{m_q} by multiplying by c/l_t.

Thus

$$C_{z_q} = \frac{c}{l_t} C_{m_q} = 2K \left(\frac{dC_m}{di}\right)^t_{\delta,\alpha} \tag{1-99}$$

C_{z_q} and $C_{z_{\dot{\alpha}}}$ are often neglected in the final equations of motion.   This results, not from the fact that these terms are negligible in comparison to the other stability derivatives (as in the case of $C_{x_{\dot{\alpha}}}$ and C_{x_q}) but from the fact that they are negligible when combined with the other terms which form the coefficients of $'\dot{\alpha}$ and q, respectively.   Thus

$$\frac{c}{2U} C_{z_{\dot{\alpha}}} \ll \frac{mU}{Sq} \quad \text{and} \quad \frac{c}{2U} C_{z_q} \ll \frac{mU}{Sq}.$$

However, the relative magnitudes of these terms should be checked for any particular set of stability derivatives and flight conditions.

C_{m_a} is the effect on the pitching moment resulting from the movement of some external control surface, in this case and those to follow, the elevator. For an aircraft with conventional elevators

$$C_{m_a} = C_{m_{\delta_e}} \delta_e \qquad (1\text{-}100)$$

where $C_{m_{\delta_e}}$ is the elevator effectiveness. For an aircraft with a so-called "flying tail"

$$C_{m_a} = C_{m_{i_t}} \delta_{i_t} = \left(\frac{\partial C_m}{\partial i}\right)_\alpha^t \delta_{i_t} \qquad (1\text{-}101)$$

$C_{F_{z_a}}$ is the effect on the forces in the Z direction due to the deflection of the elevator.

$$C_{F_{z_a}} = C_{z_{\delta_e}} \delta_e = \frac{c}{l_t} C_{m_{\delta_e}} \delta_e \qquad (1\text{-}102)$$

$C_{F_{x_a}}$ is the effect on the forces in the X direction due to the deflection of the elevator. This is a change in the drag due to the elevator deflection and is usually neglected.

For the longitudinal equations only the effect of the elevator has been considered. The gear, flaps, and dive brakes also cause changes in the forces and moments and would appear on the right hand-side of the equations along with C_{m_a}, $C_{F_{z_a}}$, and $C_{F_{x_a}}$. However, for most cases, if it is necessary to study the dynamics of the aircraft with the gear and/or flaps down, it can be assumed that the aircraft is in equilibrium in the desired configuration. The stability derivatives for that configuration then are used when proceeding with the desired dynamic analysis. If it is necessary to study the effect of the operation of the gear, flaps, or other force or moment producing devices, this can best be done by simulation on an analogue computer.

In Section 1-8 a particular aircraft is studied, and the equations of motions solved.

1-8 Solution of the Longitudinal Equations (Stick Fixed)

In solving the equations of motions it is first necessary to obtain the transient solution, which is obtained from the homogeneous equations, that is, with no external inputs or $C_{m_a} = C_{F_{z_a}} = C_{F_{x_a}} = 0$. Taking the

Laplace transform[4] of Eq. 1-59 with the initial conditions zero and neglecting $C_{x\dot\alpha}$, C_{x_q}, and C_{m_u} yields

$$\left(\frac{mU}{Sq}s - C_{x_u}\right)'u(s) - C_{x_\alpha}'\alpha(s) - C_w(\cos\Theta)\theta(s) = 0$$

$$-C_{z_u}'u(s) + \left[\left(+\frac{mU}{Sq} - \frac{c}{2U}C_{z\dot\alpha}\right)s - C_{z_\alpha}\right]'\alpha(s)$$

$$+ \left[\left(-\frac{mU}{Sq} - \frac{c}{2U}C_{z_q}\right)s - C_w(\sin\Theta)\right]\theta(s) = 0$$

$$\left(-\frac{c}{2U}C_{m\dot\alpha}s - C_{m_\alpha}\right)'\alpha(s) + \left(\frac{I_y}{Sqc}s^2 - \frac{c}{2U}C_{m_q}s\right)\theta(s) = 0 \quad (1\text{-}103)$$

For this example the values of the stability derivatives for a four engine jet transport are used.   The aircraft is flying in straight and level flight at 40,000 ft with a velocity of 600 ft per sec (355 knots), and the compressibility effects will be neglected.   For this aircraft the values are as follows:

$\Theta \quad = 0$

$m \quad = 5800$ slugs

$U \quad = 600$ ft/sec

$S \quad = 2400$ sq ft

$I_y \quad = 2.62 \times 10^6$ slug ft^2

$C_{x_u} = -2C_D = -0.088$

$C_{x_\alpha} = -\dfrac{\partial C_D}{\partial\alpha} + C_L = 0.392$

$C_w \quad = -\dfrac{mg}{Sq} = -C_L = -0.74$

$\dfrac{l_t}{c} \quad = 2.89$

$c \quad = 20.2$ ft

$C_{z_u} = -2C_L = -1.48$

$C_{z\dot\alpha} = \left(\dfrac{dC_m}{di_t}\right)^t_{\delta,\alpha}\left(\dfrac{d\epsilon}{d\alpha}\right)2 = (-1.54)(0.367)(2) = -1.13$

$C_{z_\alpha} = -\dfrac{\partial C_L}{\partial\alpha} - C_D = -4.42 - 0.04 = -4.46$

$C_{z_q} = 2K\left(\dfrac{dC_m}{di}\right)^t_{\delta,\alpha} = 2.56(-1.54) = -3.94$

$$C_{m\dot{\alpha}} = 2\left(\frac{dC_m}{di}\right)^t_{\delta,\alpha} \left(\frac{d\epsilon}{d\alpha}\right) \frac{l_t}{c} = (-1.54)(0.367)(2)(2.89)$$

$$C_{m\dot{\alpha}} = -3.27$$

$$C_{m\alpha} = (SM)_\delta \left(\frac{dC_L}{d\alpha}\right)^a_\delta = (-0.14)(4.42) = -0.619$$

$$C_{m_q} = 2K\left(\frac{dC_m}{di}\right)^t_{\delta,\alpha} \left(\frac{l_t}{c}\right)$$

$$C_{m_q} = (2.56)(-1.54)(2.89) = -11.4$$

Instead of 1.1, 1.28 is used for K, to account for the increased contribution of the swept wing to C_{m_q} and C_{z_q}. Then

$$q \qquad = \frac{\rho}{2} U^2 = \frac{(0.000585)(600)^2}{2} = 105.1 \text{ lb/sq ft}$$

$$\frac{mU}{Sq} \quad = \frac{(5800)(600)}{(2400)(105.1)} = 13.78 \text{ sec}$$

$$\frac{c}{2U} C_{z\dot{\alpha}} = \frac{(20.2)(-1.13)}{(2)(600)} = -0.019 \text{ sec}$$

$$\frac{c}{2U} C_{z_q} = (0.0168)(-3.39) = -0.057 \text{ sec}$$

$$\frac{c}{2U} C_{m\dot{\alpha}} = (0.0168)(-3.27) = -0.0552 \text{ sec}$$

$$\frac{c}{2U} C_{m_q} = (0.0168)(-11.4) = -0.192 \text{ sec}$$

$$\frac{I_y}{Sqc} \quad = \frac{(2.62 \times 10^6)}{(2400)(105.1)(20.2)} = 0.514 \text{ (sec)}^2$$

From the preceding equations it can be seen that $(c/2U)C_{z\dot{\alpha}}$ and $(c/2U)C_{z_q}$ are much smaller than mU/Sq and can be neglected in this case.

Substituting these values into Eq. 1-103 it becomes

$$(13.78s + 0.088)'u(s) - 0.392'\alpha(s) + 0.74\theta(s) = 0$$
$$1.48'u(s) + (13.78s + 4.46)'\alpha(s) - 13.78s\theta(s) = 0$$
$$0 + (0.0552s + 0.619)'\alpha(s) + (0.514s^2 + 0.192s)\theta(s) = 0 \quad (1\text{-}104)$$

The only nonzero solution of these simultaneous equations requires that the determinant of the coefficients be zero.   Thus

$$\begin{vmatrix} 13.78s + 0.088 & -0.392 & 0.74 \\ 1.48 & 13.78s + 4.46 & -13.78s \\ 0 & 0.0552s + 0.619 & 0.514s^2 + 0.192s \end{vmatrix} = 0$$

$$(1\text{-}105)$$

Expanding this determinant, the following quartic equation is obtained (see Appendix E for method).

$$97.5s^4 + 79s^3 + 128.9s^2 + 0.998s + 0.677 = 0 \qquad (1\text{-}106)$$

Dividing through by 97.5, the equation reduces to

$$s^4 + 0.811s^3 + 1.32s^2 + 0.012s + 0.00695 = 0 \qquad (1\text{-}107)$$

At this time it is not necessary to determine the magnitudes of the variations of $'u$, $'\alpha$, and θ. As the transient response of all three variables will be of the same form, their time variation is of greatest interest now. In Section 1-9 the magnitude of these variations will be discussed. The response will be of the form $Ae^{s_1 t} + Be^{s_2 t} + Ce^{s_3 t} + De^{s_4 t}$ where s_1, s_2, etc. are the roots of the characteristic equation. Thus it is necessary to factor Eq. 1-107 to obtain the roots. Since the roots of this quartic lead to two sets of complex roots indicating two damped sinusoidal oscillations, and as a result of the relative magnitudes of the coefficients, Lin's[5] method provides the quickest procedure for obtaining the two quadratic factors. To do this it is necessary to obtain the first trial divisor. This is done by dividing the coefficients of the last three terms by the coefficient of the s^2 term. Then using synthetic division

```
1 + 0.0077 + 0.0053) 1   0.811    1.32      0.0102   0.00695
                     1   0.0077   0.0053
                     ─────────────────────
                         0.8033   1.3147    0.0102
                         0.8033   0.0062    0.0041
                         ─────────────────────────
                                  1.3085    0.0061   0.00695
                                  1.3085    0.0101   0.00694
```

For the next trial divisor take $1.3085 + 0.0061 + 0.00695$ divided by 1.3085 or $1 + 0.00466 + 0.0053$. Then again using synthetic division

```
1 + 0.00466 + 0.0053) 1   0.811     1.32      0.0102   0.00695
                      1   0.00466   0.0053
                      ──────────────────────
                          0.80634   1.3147    0.0102
                          0.80634   0.0037    0.0043
                          ──────────────────────────
                                    1.311     0.0059   0.00695
                                    1.311     0.0061   0.00695
                                    ──────────────────────────
                                              ∼0
```

Therefore the two quadratic factors are

$$(s^2 + 0.00466s + 0.0053)(s^2 + 0.806s + 1.311) = 0 \qquad (1\text{-}108)$$

A common way to write such quadratics is to indicate the natural frequency ω_n and the damping ratio ζ as $(s^2 + 2\zeta\omega_n s + \omega_n{}^2)$. Doing this, Eq. 1-108 becomes

$$(s^2 + 2\zeta_p\omega_{np}s + \omega_{np}^2)(s^2 + 2\zeta_s\omega_{ns}s + \omega_{ns}^2) \tag{1-109}$$

Then
$$\left.\begin{array}{l} \zeta_s = 0.352 \\ \omega_{ns} = 1.145 \text{ rad/sec} \end{array}\right\} \quad \text{Short-period oscillation}$$

$$\left.\begin{array}{l} \zeta_p = 0.032 \\ \omega_{np} = 0.073 \end{array}\right\} \quad \text{Phugoid oscillation}$$

The characteristic modes for nearly all aircraft in most flight conditions are two oscillations: one of short period with relatively heavy damping, the other of long period with very light damping. The periods and the damping of these oscillations vary from aircraft to aircraft and with the flight conditions. The short-period oscillation is called the "short-period mode" and primarily consists of variations in $'\alpha$ and θ with very little change in the forward velocity. The long-period oscillation is called the "phugoid mode" and primarily consists of variations of θ and $'u$ with $'\alpha$ about constant. The phugoid can be thought of as an exchange of potential and kinetic energy. The aircraft tends to fly a sinusoidal flight path in the vertical plane. As the aircraft proceeds from the highest point of the flight path to the lowest point it picks up speed thus increasing the lift of the wing and curving the flight path until the aircraft starts climbing again and the velocity decreases; the lift decreases and the flight path curves downward. This condition continues until the motion is damped out, which generally requires a considerable number of cycles. However, the period is very long and the pilot can damp the phugoid successfully even if it is slightly divergent or unstable.

A good measure of the damping of an oscillation is the time required for the oscillation to damp to one-half amplitude. This measure of the damping can be expressed as

$$T_{1/2} = \frac{0.693}{\zeta\omega_n} \tag{1-110}$$

For the short-period mode

$$T_{1/2} = \frac{0.693}{0.406} = 1.72 \text{ sec} \tag{1-111}$$

For the phugoid mode

$$T_{1/2} = \frac{0.693}{0.00233} = 298 \text{ sec} \quad \text{or} \quad T_{1/2} = 4.96 \text{ min} \tag{1-112}$$

The damping ratio for the short-period mode is fairly representative. Some conventional aircraft will have larger values of damping ratio; however, some of our jet fighters have much lower values making it necessary to add damping by means of a servo system, which is discussed in Chapter 2.

1-9   Longitudinal Transfer Function for Elevator Displacement

To obtain the transfer function it is necessary to obtain values for $C_{m_{\delta e}}$ and $C_{z_{\delta e}}$, these values are $C_{m_{\delta e}} = -0.710$, $C_{z_{\delta e}} = \frac{c}{l_t} C_{m_{\delta e}} = (0.346)(-0.710) = -0.246$. $C_{x_{\delta e}}$ will be neglected.

It is now necessary to define the positive deflection of the elevator. Down elevator (stick forward) is defined as "positive elevator" by NACA convention. Thus a positive elevator deflection produces a negative θ, which means that the transfer function for elevator deflection input to θ output is negative. For this reason some references reverse the elevator sign convention; however, this is not done in this book.

Taking the Laplace Transform of Eq. 1-59 with the initial conditions zero, and after the substitution of the appropriate values, yields

$$(13.78s + 0.088)'u(s) - 0.392'\alpha(s) + 0.74\theta(s) = 0$$

$$1.48'u(s) + (13.78s + 4.46)'\alpha(s) - 13.78s\theta(s) = -0.246\delta_e(s)$$

$$(0.0552s + 0.619)'\alpha(s) + (0.514s^2 + 0.192s)\theta(s) = -0.710\delta_e(s) \quad (1\text{-}113)$$

where δ_e is the elevator deflection in radians. The transfer function for δ_e input to $'u$ output using determinants is (see Appendix E for method).

$$\frac{'u(s)}{\delta_e(s)} = \frac{\begin{vmatrix} 0 & -0.392 & 0.74 \\ -0.246 & 13.78s + 4.46 & -13.78s \\ -0.710 & 0.0552s + 0.619 & 0.514s^2 + 0.192s \end{vmatrix}}{\nabla} \quad (1\text{-}114)$$

where ∇ is the determinant of the homogeneous equation.

$$\nabla = 97.5(s^2 + 0.00466s + 0.0053)(s^2 + 0.806s + 1.311) \quad (1\text{-}115)$$

Expanding the numerator determinant, Eq. 1-114 becomes

$$\frac{'u(s)}{\delta_e(s)} = \frac{-0.0494(s^2 - 68.2s - 45)}{\nabla} \quad (1\text{-}116)$$

or

$$\frac{'u(s)}{\delta_e(s)} = \frac{-0.000506(s - 68.8)(s + 0.6)}{(s^2 + 0.00466s + 0.0053)(s^2 + 0.806s + 1.311)} \quad (1\text{-}117)$$

Putting this expression in the alternate form of the transfer function, Eq. 1-117 becomes (see Appendix C)

$$\frac{'u(s)}{\delta_e(s)} = \frac{-3\left(\frac{s}{68.8} - 1\right)\left(\frac{s}{0.6} + 1\right)}{\left[\left(\frac{s}{0.073}\right)^2 + \frac{2(0.032)}{0.073}s + 1\right]\left[\left(\frac{s}{1.145}\right)^2 + \frac{2(0.352)}{1.145}s + 1\right]} \quad (1\text{-}118)$$

To obtain the steady-state value of u per unit elevator step input, apply the final value theorem to Eq. 1-118 (see Appendix C).

$$\frac{u}{\delta_e} = \frac{'uU}{\delta_e}\frac{\text{ft/sec}}{\text{rad}} = \frac{'uU}{57.3\delta_e}\frac{\text{ft/sec}}{°\delta_e} \quad (1\text{-}119)$$

but $U = 600$ ft sec.   Therefore

$$\frac{u}{\delta_e} = \frac{3(600)}{57.3} = 31.4\frac{\text{ft/sec}}{°\delta_e} \quad (1\text{-}120)$$

The transfer function for δ_e input to $'\alpha$ output in determinant form is

$$\frac{'\alpha(s)}{\delta_e(s)} = \frac{\begin{vmatrix} 13.78s + 0.088 & 0 & 0.74 \\ 1.48 & -0.246 & -13.78s \\ 0 & -0.710 & 0.514s^2 + 0.192s \end{vmatrix}}{\nabla} \quad (1\text{-}121)$$

Expanding, Eq. 1-121 becomes

$$\frac{'\alpha(s)}{\delta_e(s)} = \frac{-0.01785(s^3 + 77.8s^2 + 0.496s + 0.446)}{(s^2 + 0.00466s + 0.0053)(s^2 + 0.806s + 1.311)} \quad (1\text{-}122)$$

Factoring the numerator

$$\frac{'\alpha(s)}{\delta_e(s)} = \frac{-0.01785(s + 77.79)(s^2 + 0.0063s + 0.0057)}{(s^2 + 0.00466s + 0.0053)(s^2 + 0.806s + 1.311)} \quad (1\text{-}123)$$

Going to the alternate form, Eq. 1-123 becomes

$$\frac{'\alpha(s)}{\delta_e(s)} = \frac{-1.14\left(\frac{s}{77.79} + 1\right)\left[\left(\frac{s}{0.0755}\right)^2 + \frac{2(0.041)}{0.0755}s + 1\right]}{\left[\left(\frac{s}{0.073}\right)^2 + \frac{2(0.032)}{0.073}s + 1\right]\left[\left(\frac{s}{1.145}\right)^2 + \frac{2(0.352)}{1.145}s + 1\right]} \quad (1\text{-}124)$$

For δ_e input to θ output

$$\frac{\theta(s)}{\delta_e(s)} = \frac{\begin{vmatrix} 13.78s + 0.088 & -0.392 & 0 \\ 1.48 & 13.78s + 4.46 & -0.246 \\ 0 & 0.0552s + 0.619 & -0.710 \end{vmatrix}}{\nabla} \quad (1\text{-}125)$$

Expanding and factoring,

$$\frac{\theta(s)}{\delta_e(s)} = \frac{-1.31(s + 0.016)(s + 0.3)}{(s^2 + 0.00466s + 0.0053)(s^2 + 0.806s + 1.311)} \quad (1\text{-}126)$$

or the alternate form

$$\frac{\theta(s)}{\delta_e(s)} = \frac{-0.95\left(\dfrac{s}{0.016} + 1\right)\left(\dfrac{s}{0.3} + 1\right)}{\left[\left(\dfrac{s}{0.073}\right)^2 + \dfrac{2(0.032)}{0.073}s + 1\right]\left[\left(\dfrac{s}{1.145}\right)^2 + \dfrac{2(0.352)}{1.145}s + 1\right]} \quad (1\text{-}127)$$

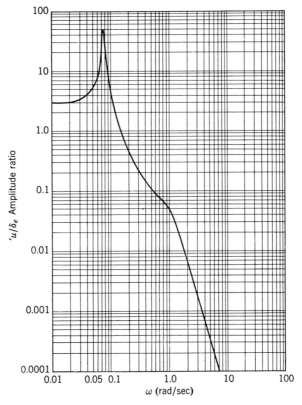

Figure 1-9   Magnitude plot for $'u/\delta_e$ transfer function versus ω for $s = j\omega$.

The steady-state value for $\dfrac{'\alpha(s)}{\delta_e(s)}$ and $\dfrac{\theta(s)}{\delta_e(s)}$ for a unit step input can be obtained directly from Eqs. 1-124 and 1-127, respectively; as both of these equations represent the ratios of two angles, the units may be in either radians or degrees.

Figures 1-9, 1-10, and 1-11 are amplitude ratio plots of Eqs. 1-118, 1-124, and 1-127 against ω for $s = j\omega$. Figure 1-9 shows that the amplitude of the $'u/\delta_e$ response is very small at the natural frequency of the short period oscillation. This statement substantiates the one made in Section 1-8 that there is very little change in the forward velocity during the short period oscillation.

An examination of Eq. 1-124 and Figure 1-10 indicates that the numerator quadratic effectively cancels the phugoid quadratic, thus substantiating the earlier statement that the phugoid oscillation takes place at almost constant angle of attack.

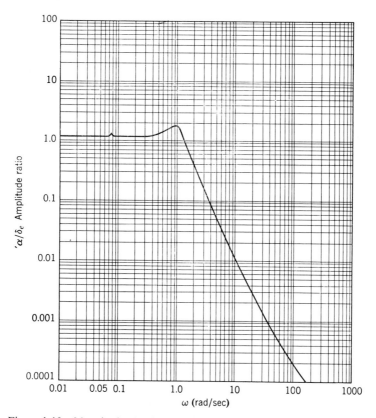

Figure 1-10 Magnitude plot for $'\alpha/\delta_e$ transfer function versus ω for $s = j\omega$.

Figure 1-11 shows that a considerable variation in θ occurs at both the phugoid and short-period frequencies. These observations lead to the following approximations of the phugoid- and short-period oscillations:

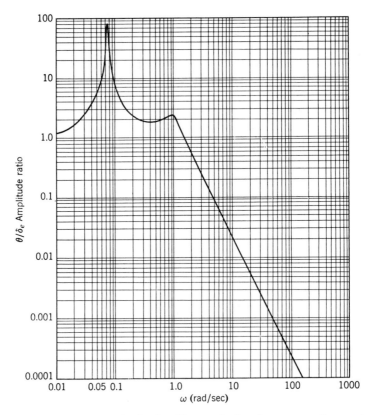

Figure 1-11   Magnitude plot for θ/δ_e transfer function versus ω for $s = j\omega$.

Short-Period Approximation.   As mentioned earlier, the short-period oscillation occurs at almost constant forward speed; therefore, let $'u = 0$ in the equations of motion.   The X equation can be neglected as it does not contribute much to the short-period oscillation, as forces in the X direction contribute mostly to changes in forward speed.   With these assumptions and by neglecting $C_{z\dot\alpha}$ and C_{z_q} and inserting $C_{z_{\delta_e}}$ and $C_{m_{\delta_e}}$, Eq. 1-103 becomes

$$\left(\frac{mU}{Sq} s - C_{z_\alpha}\right)'\alpha(s) + \left[-\frac{mU}{Sq} s - C_w(\sin\Theta)\right]\theta(s) = C_{z_{\delta_e}}\delta_e(s)$$

$$\left(-\frac{c}{2U} C_{m\dot\alpha}s - C_{m_\alpha}\right)'\alpha(s) + \left(\frac{I_y}{Sqc} s^2 - \frac{c}{2U} C_{m_q}s\right)\theta(s) = C_{m_{\delta_e}}\delta_e(s) \quad (1\text{-}128)$$

Before evaluating the transfer functions to compare them with those obtained from the complete equations, the determinant of the homogeneous equation is first expanded in general terms.

Thus with $\Theta = 0$ Eq. 1-128 becomes

$$\begin{vmatrix} \left(\dfrac{mU}{Sq}s - C_{z_\alpha}\right) & \dfrac{-mU}{Sq}s \\[2ex] \left(\dfrac{-c}{2U}C_{m_{\dot\alpha}}s - C_{m_\alpha}\right) & \left(\dfrac{I_y}{Sqc}s^2 - \dfrac{c}{2U}C_{m_q}s\right) \end{vmatrix} = 0 \qquad (1\text{-}129)$$

The expansion of this expression can be written in the form of

$$s(As^2 + Bs + C) = 0$$

where $A = \left(\dfrac{I_y}{Sqc}\right)\left(\dfrac{mU}{Sq}\right)$

$$B = \left(\dfrac{-c}{2U}C_{m_q}\right)\left(\dfrac{mU}{Sq}\right) - \dfrac{I_y}{Sqc}C_{z_\alpha} - \left(\dfrac{c}{2U}C_{m_{\dot\alpha}}\right)\left(\dfrac{mU}{Sq}\right)$$

$$C = \dfrac{c}{2U}C_{m_q}C_{z_\alpha} - \dfrac{mU}{Sq}C_{m_\alpha} \qquad (1\text{-}130)$$

If Eq. 1-130 is divided by A and written in the standard form of the quadratic with ζ and ω_n, then

$$2\zeta\omega_n = \dfrac{B}{A} \qquad (1\text{-}131)$$

and

$$\omega_n{}^2 = \dfrac{C}{A} \qquad (1\text{-}132)$$

Substituting for C/A and taking the square root, Eq. 1-132 becomes

$$\omega_n = \left(\dfrac{\dfrac{c}{2U}C_{m_q}C_{z_\alpha} - \dfrac{mU}{Sq}C_{m_\alpha}}{\left(\dfrac{I_y}{Sqc}\right)\left(\dfrac{mU}{Sq}\right)}\right)^{\!\! \frac{1}{2}} \qquad (1\text{-}133)$$

Substituting $\frac{\rho}{2}U^2$ for q and simplifying, Eq. 1-133 becomes

$$\omega_n = \dfrac{U\rho Sc}{2}\left(\dfrac{\dfrac{C_{m_q}C_{z_\alpha}}{2} - \dfrac{2m}{\rho Sc}C_{m_\alpha}}{I_y m}\right)^{\!\! \frac{1}{2}} \qquad (1\text{-}134)$$

Substituting for B/A and solving for ζ, Eq. 1-131 becomes

$$\zeta = \dfrac{-\dfrac{c}{2U}\dfrac{mU}{Sq}(C_{m_{\dot\alpha}} + C_{m_q}) - \dfrac{I_y}{Sqc}C_{z_\alpha}}{\left(\dfrac{I_y}{Sqc}\right)\left(\dfrac{mU}{Sq}\right)(U\rho Sc)\left(\dfrac{\dfrac{C_{m_q}C_{z_\alpha}}{2} - \dfrac{2m}{\rho Sc}C_{m_\alpha}}{I_y m}\right)^{\!\! \frac{1}{2}}} \qquad (1\text{-}135)$$

Substituting for q and simplifying, Eq. 1-135 becomes

$$\zeta = -\tfrac{1}{4}\left(C_{m_q} + C_{m\dot{\alpha}} + \frac{2I_y}{mc^2}C_{z\alpha}\right)\left(\frac{mc^2}{I_y\left(\frac{C_{m_q}C_{z\alpha}}{2} - \frac{2mC_{m\alpha}}{\rho Sc}\right)}\right)^{1/2} \tag{1-136}$$

From an investigation of Eqs. 1-134 and 1-136 we can see that, for a given altitude, the natural frequency of the short-period oscillation is proportional to U while the damping ratio is constant. However, both the natural frequency and the damping ratio vary with altitude. They are roughly proportional to $\sqrt{\rho}$ and decrease as the altitude increases (see Ref. 3 for a table of ρ versus altitude).

Let us calculate the transfer functions for the approximation to the short-period oscillation. The values of ζ and ω_n can be calculated from Eqs. 1-134 and 1-136, but the numerator must still be evaluated so that it is just as easy to return to the basic equation. Substituting the proper values into Eq. 1-128, the determinant form for the $'\alpha(s)/\delta_e(s)$ transfer function is

$$\frac{'\alpha(s)}{\delta_e(s)} = \frac{\begin{vmatrix} -0.246 & -13.78s \\ -0.710 & 0.514s^2 + 0.192s \end{vmatrix}}{\begin{vmatrix} 13.78s + 4.46 & -13.78s \\ 0.0552s + 0.619 & 0.514s^2 + 0.192s \end{vmatrix}} \tag{1-137}$$

Expanding,

$$\frac{'\alpha(s)}{\delta_e(s)} = \frac{-0.01782(s + 77.8)}{s^2 + 0.805s + 1.325} \tag{1-138}$$

Going to the alternate form,

$$\frac{'\alpha(s)}{\delta_e(s)} = \frac{-1.05\left(\frac{s}{77.8} + 1\right)}{\left(\frac{s}{1.15}\right)^2 + \frac{2(0.35)}{1.15}s + 1} \tag{1-139}$$

A comparison of Eqs. 1-139 and 1-124 shows excellent agreement, again substantiating the original assumption. The $\theta(s)/\delta_e(s)$ transfer function will now be evaluated. From Eq. 1-128

$$\frac{\theta(s)}{\delta_e(s)} = \frac{\begin{vmatrix} 13.78s + 4.46 & -0.246 \\ 0.0552s + 0.619 & -0.710 \end{vmatrix}}{\begin{vmatrix} 13.78s + 4.46 & -13.78s \\ 0.0552s + 0.619 & 0.514s^2 + 0.192s \end{vmatrix}} \tag{1-140}$$

Expanding,

$$\frac{\theta(s)}{\delta_e(s)} = \frac{-1.39(s + 0.306)}{s(s^2 + 0.805s + 1.325)} \tag{1-141}$$

Going to the alternate form,

$$\frac{\theta(s)}{\delta_e(s)} = \frac{-0.321\left(\dfrac{s}{0.306} + 1\right)}{s\left[\left(\dfrac{s}{1.15}\right)^2 + \dfrac{2(0.35)}{1.15}s + 1\right]} \tag{1-142}$$

The magnitude plot of Eq. 1-142 is shown in Figure 1-12. A comparison of Figures 1-11 and 1-12 shows very good agreement in the vicinity of the natural frequency of the short-period mode. It should be noted that Eq. 1-142 has a pole at $s = 0$, that is, a root of the denominator of this

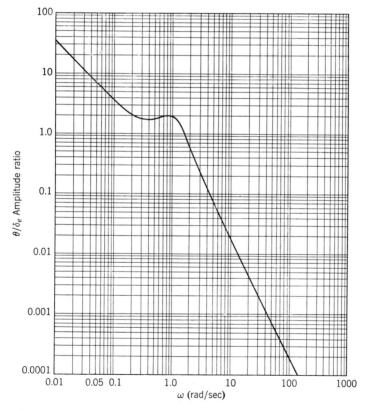

Figure 1-12 Magnitude plot for θ/δ_e transfer function versus ω for $s = j\omega$, short-period approximation.

equation is $s = 0$. This s in the denominator mathematically represents a pure integration. As a result, if δ_e is a step, that is a constant, the output θ will be the integral of this constant and will approach infinity as time approaches infinity. However, $\dot{\theta}$ will reach a constant value for a step input of δ_e. Such a constant value results from the fact that multiplying by s is equivalent to differentiating, thus $\dot{\theta}(s)/\delta_e(s) = s\theta(s)/\delta_e(s)$. The s in the denominator of Eq. 1-142 results from the fact that Θ was taken as zero in Eq. 1-128, which effectively eliminates the effects of gravity from the short-period approximation. However, this does not invalidate the use of the short-period approximation for studying the short-period mode of oscillation. For these studies the frequency of oscillation and the damping after the aircraft has been disturbed from equilibrium is of primary interest. This information can be obtained from the short-period approximation; thus it will be used in the study of the longitudinal autopilots.

Phugoid Approximation. As stated in Section 1-8 the phugoid oscillation takes place at almost constant angle of attack, thus $'\alpha$ can be set to zero in Eq. 1-103. The next problem is to determine which two of the three equations to retain. As the phugoid oscillation is of long period, θ is varying quite slowly; therefore, the inertia forces can be neglected, leaving for the $\mathcal{M}$ equation

$$\frac{-c}{2U} C_{m_q} s\theta(s) = C_{m_{\delta_e}} \delta_e(s) \tag{1-143}$$

If this equation is combined with either the X or Z equations with $'\alpha = 0$, the resulting homogeneous equation is not oscillatory. This could hardly represent the phugoid mode. Thus it is necessary to use the X and Z equations.

Then for, $\Theta = 0$ and neglecting C_{z_q}, Eq. 1-103 becomes

$$\left(\frac{mU}{Sq} s - C_{x_u}\right)'u(s) - C_w\theta(s) = 0$$

$$-C_{z_u}'u(s) - \left(\frac{mU}{Sq} s\right)\theta(s) = 0 \tag{1-144}$$

In determinant form,

$$\begin{vmatrix} \left(\dfrac{mU}{Sq} s - C_{x_u}\right) & -C_w \\ -C_{z_u} & -\dfrac{mU}{Sq} s \end{vmatrix} = 0$$

Expanding,

$$-\left(\frac{mU}{Sq}\right)^2 s^2 + \frac{mU}{Sq} C_{x_u} s - C_w C_{z_u} = 0 \tag{1-145}$$

But $C_w = -C_L$. Substituting for C_w and dividing by $-\left(\dfrac{mU}{Sq}\right)^2$, Eq. 1-145 becomes

$$s^2 - \frac{C_{x_u}}{\dfrac{mU}{Sq}} s - \frac{C_{z_u} C_L}{\left(\dfrac{mU}{Sq}\right)^2} = 0 \qquad (1\text{-}146)$$

Then $\qquad \omega_{np} = \dfrac{\sqrt{-C_{z_u} C_L}}{mU/Sq}$ but $\quad \dfrac{mU}{Sq} = \dfrac{2mU}{S\rho U^2} = \dfrac{2m}{\rho SU}$

Substituting for mU/Sq,

$$\omega_{np} = \frac{\rho SU \sqrt{-C_{z_u} C_L}}{2m} \qquad (1\text{-}147)$$

But from Table 1-1, $C_{z_u} \simeq 2C_L = 2(2mg/\rho SU^2)$. Substituting, Eq. 1-147 becomes

$$\omega_{np} = \frac{g}{U}\sqrt{2} \quad \text{or} \quad T_p = 0.138U \qquad (1\text{-}148)$$

The damping ratio can also be obtained from Eq. 1-146, thus

$$\zeta_p = \frac{-C_{x_u}}{2\,\dfrac{mU}{Sq}\,\dfrac{\sqrt{-C_{z_u} C_L}}{mU/Sq}} = \frac{-C_{x_u}}{2\sqrt{-C_{z_u} C_L}} \qquad (1\text{-}149)$$

From Table 1-1, $C_{x_u} \simeq -2C_D$ and $C_{z_u} \simeq -2C_L$. Substituting this, Eq. 1-149 becomes

$$\zeta_p = \frac{2C_D}{2\sqrt{2C_L^2}} = \frac{C_D}{C_L\sqrt{2}} \qquad (1\text{-}150)$$

This operation assumes that the $\partial C_D/\partial u = \partial C_L/\partial u \simeq 0$, which is valid for low Mach numbers as explained in Section 1-7. If this is not true, the damping of the phugoid is also dependent upon the forward velocity.

With this assumption it can be seen from Eq. 1-150 that the damping of the phugoid is dependent upon the drag. Thus for present-day jet aircraft, where the emphasis is on low drag, the damping of the phugoid is very low. An examination of Eq. 1-148 indicates that the natural frequency of the phugoid is inversely proportional to the forward speed and is independent of ρ. The validity of these conclusions is checked in Section 1-10, where the effects of changes in airspeed and altitude on the short-period and phugoid modes are studied in detail.

The two transfer functions for the phugoid mode can be evaluated from Eq. 1-113 with $'\alpha(s) = 0$ and by taking the X and Z equations. The $'u(s)/\delta_e(s)$ transfer function is

$$\frac{'u(s)}{\delta_e(s)} = \frac{-0.0009}{s^2 + 0.00645s + 0.00582} \tag{1-151}$$

or

$$\frac{'u(s)}{\delta_e(s)} = \frac{-0.165}{\left(\dfrac{s}{0.0765}\right)^2 + \dfrac{2(0.042)}{0.0765}s + 1} \tag{1-152}$$

A comparison of Eqs. 1-152 and 1-118 shows good agreement for the natural frequencies and damping ratios, but here the agreement ends. The error coefficient K' is 0.165 instead of 3 and, as discussed in Section 1-10, there is a $180°$ phase shift between the two equations. For the complete equation a positive elevator deflection gives a positive value of $'u$, while from the approximation the initial value of $'u$ will be negative. This means that for a down elevator, which should cause the aircraft to accelerate, the aircraft slows. This condition results from the elimination of the $\mathcal{M}$ equation, that is, $C_{z_{\delta_e}}$ is the only input. For a positive elevator deflection this gives a force in the $-Z$ direction causing the $180°$ phase shift from the true picture. The $\theta(s)/\delta_e(s)$ transfer function is

$$\frac{\theta(s)}{\delta_e(s)} = \frac{0.018(s + 0.00637)}{s^2 + 0.00645s + 0.00582} \tag{1-153}$$

or

$$\frac{\theta(s)}{\delta_e(s)} = \frac{0.0196\left(\dfrac{s}{0.00637} + 1\right)}{\left(\dfrac{s}{0.0765}\right)^2 + \dfrac{2(0.042)}{0.0765}s + 1} \tag{1-154}$$

When comparing Eqs. 1-127 and 1-154 it can be seen that the only real comparison is in the natural frequency and the damping ratio. Again K' is way off, and there is the $180°$ phase difference. From this it can be concluded that the only real usefulness of the phugoid approximation is to obtain good approximate values of the natural frequency and damping ratio. Thus, as discussed in Section 1-10, the phugoid approximation is not satisfactory for simulation purposes.

1-10   Transient Response of the Aircraft

In the preceding section the transfer functions for the longitudinal equations were obtained and analyzed. In this section the effects of changes in airspeed and altitude on the transient response of the aircraft, as obtained by use of an analog computer, are studied. The usefulness of the short-period approximation and the shortcomings of the phugoid approximation are illustrated. In the computer simulation, the slope of the lift curve slope was assumed constant as were C_{m_α}, $C_{m_{\dot\alpha}}$, and C_{m_q}. In

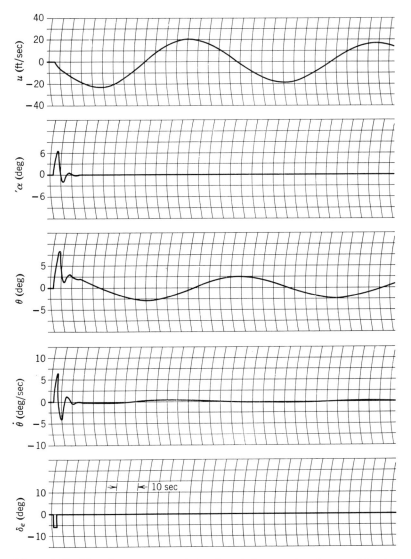

Figure 1-13 Transient response of the aircraft for a pulse elevator deflection (complete longitudinal equations for 600 ft/sec at 40,000 ft).

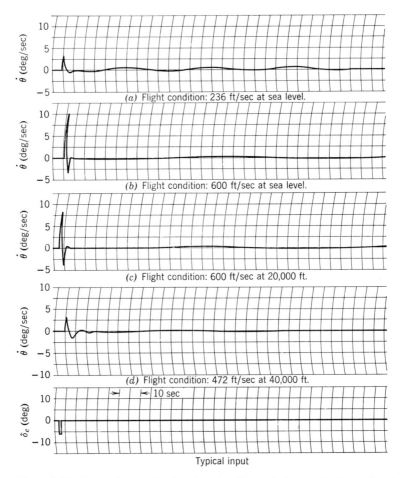

Figure 1-14   Computer results showing the effects of changes in airspeed and altitude on the short-period mode (complete longitudinal equations).

general, this is a valid assumption at the lower Mach numbers. In addition the mass and moment of inertia were held constant. The other stability derivatives were allowed to vary and were calculated using the equations given in Table 1-1. To displace the aircraft from equilibrium a 6° up-elevator deflection of approximately 1-second duration was applied. Three different altitudes and airspeeds were studied. The results are presented in Figures 1-13, 1-14, 1-15 and are summarized in Table 1-2.

In the discussion of the short-period approximation certain predictions were made concerning the effects on the short-period dynamics of changes

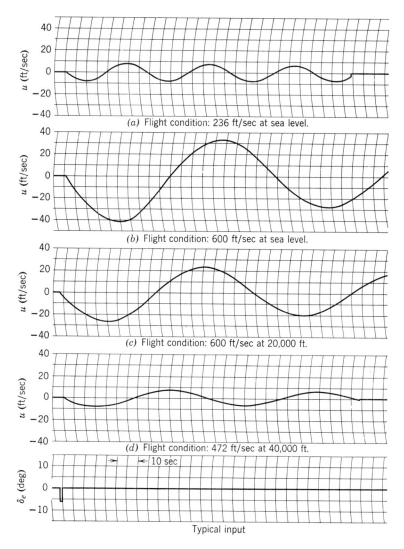

(a) Flight condition: 236 ft/sec at sea level.

(b) Flight condition: 600 ft/sec at sea level.

(c) Flight condition: 600 ft/sec at 20,000 ft.

(d) Flight condition: 472 ft/sec at 40,000 ft.

Typical input

Figure 1-15   Computer results showing the effects of changes in airspeed and altitude on the phugoid mode (complete longitudinal equations).

in airspeed and altitude.   Table 1-2(*a*) shows that these predictions are very well substantiated by the results of the three-degree-of-freedom simulation.   The discussion of the phugoid mode shows that the period of the phugoid (T_p) was proportional to the airspeed and independent of air density or altitude.   The three-degree-of-freedom simulation, as summarized in Table 1-2(*b*), verifies this proportionality; however, the proportionality constant is not independent of altitude nor equal to the theoretical value. Perkins and Hage in their book[1] state that this constant is nearly 0.178. The dependence on altitude and variation from the theoretical value probably results from the second-order effects of the terms neglected in obtaining the phugoid approximation.

Table 1-2 Comparison of the Predicted and Actual Effects of the Variation of Airspeed and Altitude on the Longitudinal Dynamic Response

(*a*) Short-Period Mode

Flight Condition		Change in ζ $\zeta = K_1\sqrt{\rho}$		Change in ω_n $\omega_n = K_2 U\sqrt{\rho}$	
Altitude (ft)	U_0 (ft/sec)	Predicted	Actual	Predicted	Actual
Sea level	236	None	Negligible	2.54	~2
	600				
40,000	472	None	None	1.27	1.3
	600				
Sea level to 40,000	600	0.496	~0.5	0.496	~0.5

(*b*) Phugoid Mode

Flight Condition		Change in ζ $\zeta = D/mg\,\sqrt{2}$		Value of K $T_p = KU$	
Altitude (ft)	U_0 (ft/sec)	Predicted	Actual	Predicted	Actual
Sea level	236	Increases	Increases	0.138	0.161
	600				0.162
40,000	472	Increases	Increases	0.138	0.148
	600				0.145
Sea level to 40,000	600	Less at 40,000 ft	Less at 40,000 ft	0.138	0.162
					0.145

From the phugoid approximation the damping ratio is given by the formula

$$\zeta_p = \frac{C_D}{C_L\sqrt{2}} \tag{1-155}$$

Multiplying the numerator and denominator of Eq. 1-155 by qS yields

$$\zeta_p = \frac{qSC_D}{qSC_L\sqrt{2}} = \frac{D}{L\sqrt{2}} = \frac{D}{mg\sqrt{2}} \qquad (1\text{-}156)$$

Thus the damping of the phugoid is proportional to the ratio of the drag to the lift. If the aircraft is in straight and level flight, the lift is equal to the weight and thus constant. Under these conditions the damping of the phugoid is proportional to the total drag. As the airspeed is increased at a given altitude, the total drag increases and thus the phugoid damping should increase. Similarly, if the airspeed remains constant but the

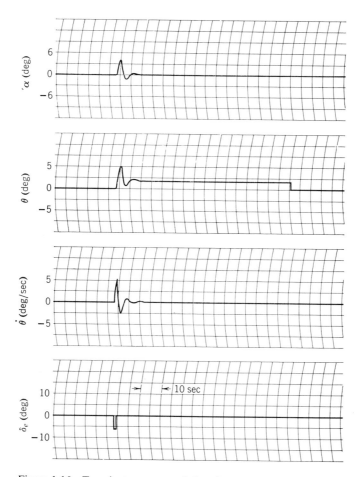

Figure 1-16   Transient response of the aircraft for a pulse elevator deflection. Short-period approximation for 600 ft/sec at 40,000 ft.

altitude is increased, the damping should decrease due to the decrease in drag resulting from the lower air density at the higher altitude. These two observations are verified by the computer runs and are summarized in Table 1-2(*b*).

In deriving the short-period approximate equations it is assumed that the forward velocity remains constant during this oscillation. The computer simulation does not completely verify this, especially at the lower dynamic pressures where the amplitude of the phugoid is small and the period relatively short. However, this result does not invalidate the use of the short-period approximation. The assumption that the angle

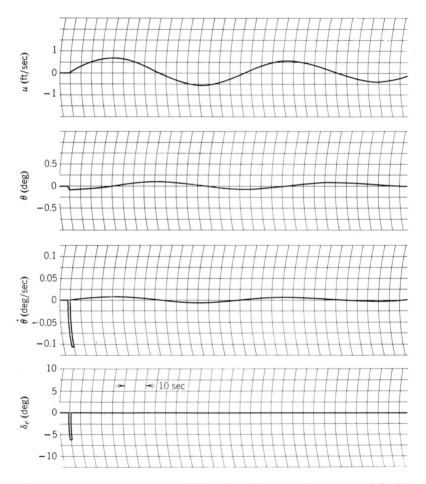

Figure 1-17   Transient response of the aircraft for a pulse elevator deflection. Phugoid approximation for 600 ft/sec at 40,000 ft.

of attack is constant and that the inertia forces are negligible, that is that $\ddot{\theta}$ is very small, during the phugoid oscillation, is confirmed by Figures 1-13 and 1-14 ($\ddot{\theta}$ would be much smaller than $\dot{\theta}$).

Examination of Figure 1-16 shows that the short-period approximation is a perfect duplication of the short-period oscillation in Figure 1-13. The fact that θ does not return to zero was explained in Section 1-9, following Eq. 1-142 and results from dropping the X equation. The fact that the short-period approximation does give such a true picture of the short-period oscillation makes it very useful for simulation purposes. On the other hand, as can be seen from Figure 1-17, the phugoid approximation, except for fairly accurately yielding the natural frequency and damping ratio, does not accurately reproduce the phugoid oscillation. The magnitudes of the oscillations are very much smaller than those obtained from the complete longitudinal equations, and the 180° phase shift is evident.

1-11   Effect of Variation of Stability Derivatives on Aircraft Performance

This section presents the effects of variation of the stability derivatives on the damping ratios and natural frequencies of the longitudinal modes of oscillation of the aircraft. The data for Table 1-3 was obtained for an airspeed of 600 ft/sec at 40,000 ft, through the use of the analog computer, by varying each stability derivative individually and then determining the damping ratios and the natural frequencies from the time recordings. This is an artificial situation; in general, any one of the stability derivatives cannot be changed without changing at least some of the other stability derivatives. However, in performing a dynamic simulation of an aircraft in connection with the design of an autopilot, the engineer must know, or be able to estimate, the values of the stability derivatives that he needs. By knowing which ones are most likely to affect the damping ratios and natural frequencies, and by how much, the engineer knows which stability derivatives should be known most accurately for his particular design problem. The data in Table 1-3 could also aid the engineer in locating errors in his simulation if the damping ratio and/or natural frequency as determined by the simulation differs from the calculated value. For example, if in simulating the short-period mode it is found that the damping is too heavy but that the natural frequency has only increased slightly, the first guess might be that an error in C_{m_q} is the cause. A check of the value of the potentiometer setting that determines this stability derivative, the potentiometer setting itself, or the patch board wiring just ahead or behind this potentiometer might disclose the error and thus save a considerable amount of time that might otherwise be spent trouble-shooting the whole circuit. For these reasons this section is included in this book.

In Table 1-3 the maximum amount that the stability derivatives are increased or decreased is by one order of magnitude; however, if a smaller change gives extremely large variations of the damping ratio or natural frequency, some are not changed by this much. It should be remembered that the data in Table 1-3 is calculated from computer traces similar to those in Figure 1-14; thus extremely small changes in the damping ratios or natural frequencies cannot be detected. But as this data is primarily to aid in analog simulation of the aircraft, the accuracy obtained should be satisfactory. In addition, the main purpose here is to show the trend, for

Table 1-3 *Effects of Variation of the Stability Derivatives on the Damping Ratios and Natural Frequencies of the Longitudinal Modes of Oscillation*

Stability Derivative	Basic Value	Change	Short Period ζ	Short Period ω_n	Phugoid ζ	Phugoid ω_n
C_{x_u}	-0.088	$\times 10$	0.35	1.15	0.4	0.08
		$\times 2$			0.07	0.077
		$\times \frac{1}{2}$			0.007	0.073
		$\times \frac{1}{10}$			$*-0.007$	0.073
C_{x_α}	0.392	$\times 10$			0.075	0.073
		$\times \frac{1}{10}$			0.025	0.073
C_{z_u}	-1.48	$\times 10$			-0.018	0.228
		$\times 2$			0.007	0.103
		$\times \frac{1}{2}$			0.058	0.051
		$\times \frac{1}{10}$			0.1	0.0226
C_{z_α}	-4.46	$\times 4$	~ 0.8	1.53	0.025	0.064
		$\times 2$	0.47	1.21	0.035	0.069
		$\times \frac{1}{2}$	0.27	1.1	0.05	0.075
		$\times \frac{1}{4}$	0.25	1.05	0.13	0.0757
$C_{m_{\dot{\alpha}}}$	-3.27	$\times 10$	~ 1	1.38	0.035	0.073
		$\times 2$	0.4	1.18		
		$\times \frac{1}{2}$	0.32	1.13		
		$\times \frac{1}{10}$	0.31	1.11		
C_{m_α}	-0.619	$\times 2$	0.23	1.62		
		$\times \frac{1}{2}$	0.45	0.92		0.07
C_{m_q}	-11.4	$\times 4$	~ 1	1.85		0.064
		$\times 2$	0.52	1.2		0.069
		$\times \frac{1}{2}$	0.27	1.12		0.075
		$\times \frac{1}{10}$	0.2	1.1		0.076

$*$ — Indicates negative damping or instability.

the amount of change will vary with aircraft. For reference, the basic values of the damping ratios and natural frequencies, as measured from the computer traces, are

Short period: $\zeta_s = 0.35$

$\omega_{ns} = 1.15$ rad/sec

Phugoid: $\zeta_p = 0.035$

$\omega_{np} = 0.073$ rad/sec

From Table 1-3 it can be concluded that the stability derivatives shown in Table 1-4 have the most effect on the damping ratios and/or natural frequencies:

Table 1-4   List of Stability Derivatives That Have the Largest Effects on the Damping Ratios and Natural Frequencies of the Longitudinal Modes of Oscillation

Stability Derivative	Quantity Most Affected	How Affected
C_{m_q}	Damping of the short period	Increase C_{m_q} to increase the damping
C_{m_α}	Natural frequency of the short period	Increase C_{m_α} to increase the frequency
C_{x_u}	Damping of the phugoid	Increase C_{x_u} to increase the damping
C_{z_u}	Natural frequency of the phugoid	Increase C_{z_u} to increase the frequency

References

1. C. D. Perkins and R. E. Hage, *Airplane Performance, Stability, and Control*, John Wiley and Sons, New York, 1949.
2. B. Etkin, *Dynamics of Flight: Stability and Control*, John Wiley and Sons, New York, 1959.
3. D. O. Dommash, S. S. Sherby, and T. F. Connolly, *Airplane Aerodynamics*, Pitman Publishing Corp., New York, 1951.
4. M. F. Gardner and J. L. Barnes, *Transients in Linear Systems*, Vol. 1, John Wiley and Sons, New York, 1942.
5. Shih-Nge Lin, "A Method of Successive Approximations of Evaluating the Real and Complex Roots of Cubic and Higher Order Equations," *Journal of Mathematics and Physics*, Vol. XX, No. 3, Aug. 1941.

Longitudinal Autopilots

2-1 Displacement Autopilot

The simplest form of autopilot, which is the type that first appeared in aircraft and is still being used in some of the older transport aircraft, is the displacement-type autopilot. This autopilot was designed to hold the aircraft in straight and level flight with little or no maneuvering capability. A block diagram for such an autopilot is shown in Figure 2-1.

For this type of autopilot the aircraft is initially trimmed to straight and level flight, the reference aligned, and then the autopilot engaged. If the pitch attitude varies from the reference, a voltage e_g is produced by the signal generator on the vertical gyro. This voltage is then amplified and fed to the elevator servo. The elevator servo can be electromechanical or hydraulic with an electrically operated valve. The servo then positions the elevator causing the aircraft to pitch about the Y axis returning the aircraft to the desired pitch attitude. The elevator servo is, in general, at least a second-order system; but if properly designed, its natural frequency is higher than that of the aircraft. If the damping ratio is high enough, the elevator servo can be represented by a sensitivity (gain) multiplied by a first-order time lag. Representative characteristic times vary from 0.1 to 0.03 sec. The transfer function used to represent the aircraft can be the complete three-degree-of-freedom longitudinal transfer function or the transfer function of the short-period approximation. The choice depends on what characteristic of the aircraft behavior is being studied.

There is one disadvantage of this system, from a servo point of view; that is, it is a Type 0 system (see Appendix C). This means that there are

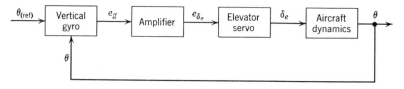

Figure 2-1   Displacement autopilot.

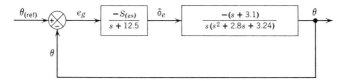

Figure 2-2   Block diagram for the conventional transport and autopilot.

no integrations in the forward loop.   Thus if some fixed external moment were applied, resulting from a change in trim, for example, there would have to be an error in θ to generate a voltage to drive the elevator servo, thus producing an elevator deflection to balance the interfering moment. This, then, would require the pilot to make occasional adjustments of the reference to maintain the desired flight path.

Another disadvantage of this system can be illustrated by studying the root locus of Figure 2-1, for two different aircraft.   The short-period approximation for both aircraft is used here.   The first aircraft is a conventional transport flying at 150 mph at sea level, and the second aircraft is the jet transport studied in Chapter 1, flying at 600 ft/sec at 40,000 ft. Figure 2-2 is the block diagram for the first aircraft obtained by inserting the transfer functions for the components represented by the blocks in

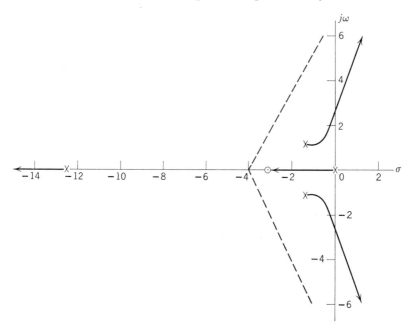

Figure 2-3   Root locus for conventional transport and autopilot.

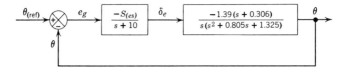

Figure 2-4   Block diagram for the jet transport and autopilot.

Figure 2-1. The amplifier and the elevator servo are combined into one block and the vertical gyro replaced by a summer. As the aircraft transfer function is negative, the sign of the elevator servo transfer function is made negative so that the forward transfer function is positive. This is done so that a positive $\theta_{(ref)}$ causes a positive change in θ, which will be helpful when analyzing the glide slope control system. If the sign of the transfer function of the elevator servo had been left positive, the sign at the summer for the feedback signal would have been positive so that the sign of the open loop transfer function would have been negative (see Appendix C). In either case the analysis would have been the same. Figure 2-3 is the root locus for the conventional transport; Figure 2-4 is the block diagram for the jet aircraft; and Figure 2-5 the corresponding root locus. As can be seen in Figure 2-3, the short-period oscillation for the first

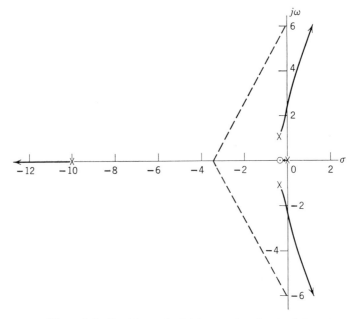

Figure 2-5   Root locus for jet transport and autopilot.

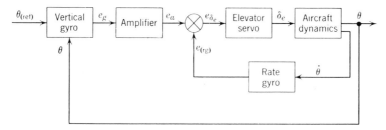

Figure 2-6   Displacement autopilot with pitch-rate feedback for damping.

aircraft is well damped, the damping ratio being 0.78.   As $S_{(es)}$—the gain of the elevator servo—is increased from zero, the damping decreases, and when $S_{(es)}$ is increased to 16.8 deg/volt, the damping ratio decreases to 0.6, which is still sufficient.   In contrast, Figure 2-5 shows that for the jet transport the damping ratio, which is already too low for satisfactory dynamic response, decreases rapidly as the $S_{(es)}$ is increased.   Another consideration is the value of $S_{(es)}$ for instability.   For the first aircraft this value of $S_{(es)}$ is 77.5 deg/volt while for the second aircraft it is only 38.4 deg/volt.   From this example it can be seen that the displacement autopilot, while satisfactory for conventional transports, is not satisfactory for jet transports similar to the one studied in Chapter 1.

The s appearing in the denominator of the aircraft transfer function results from dropping the gravity term ($C_w \sin \Theta$) in Eq. 1-129, that is, $\Theta = 0$.   If Θ is not zero, which would be the case after a disturbance or if Θ_0 were not zero, the s in the denominator becomes a real pole near the origin.   For this reason, although the systems shown in Figures 2-2 and 2-4 are technically Type 1 systems (see Appendix C), they will be considered here as Type 0 systems.   This in no way affects the validity of the analysis to follow, for even if Θ_0 is not zero, the real pole introduced is so close to the origin as to be indistinguishable from a pole at the origin when plotting the root locus.

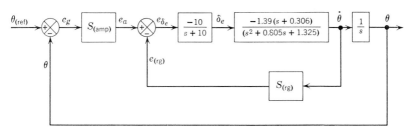

Figure 2-7   Block diagram for the jet transport and displacement autopilot with pitch-rate feedback added for damping.

If, in the case of the jet aircraft, it is desired to increase the damping of the short-period oscillation, this can be accomplished by adding an inner feedback loop utilizing a rate gyro. Figure 2-6 is a block diagram of such a system. A root locus analysis is presented here to show the effect of the rate gyro feedback. First, Figure 2-6 is redrawn indicating the transfer functions for the various blocks which yields Figure 2-7. Figure 2-8 shows the root locus for the inner loop as the rate gyro sensitivity is increased from zero. As seen in Figure 2-8, there is a range of rate gyro sensitivities that yields all real roots. The question that now arises is

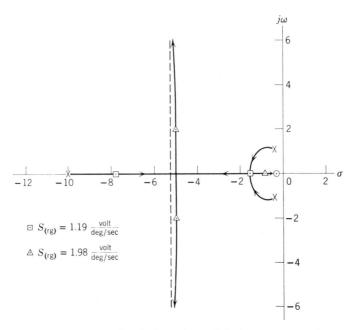

Figure 2-8   Root locus for the inner loop of the jet transport and auto-pilot with pitch-rate feedback.

what value of rate gyro sensitivity should be used to obtain the best overall system response after closing the outer loop. There is no simple rule to aid the engineer in selecting the final sensitivity. The final choice usually results in a compromise between the desirability of rapid response and the desire to reduce excessive overshoot. For illustration, two rate gyro sensitivities are chosen, as indicated in Figure 2-8. Using these closed loop poles a root locus can now be drawn for each rate gyro sensitivity for the outer loop (see Figures 2-9 and 2-10). For each root locus the value of the amplifier sensitivity required to reduce the damping

Table 2-1 Amplifier Sensitivities Required for the Conditions Indicated for Displacement Autopilot for Two Values of Rate Gyro Sensitivity

$S_{(rg)}\left(\dfrac{\text{volt}}{\text{deg/sec}}\right)$	$S_{(amp)}$(volt/volt)	
	For $\zeta = 0.6$	For $\zeta = 0$
1.19	1.41	15.2
1.98	3.3	24.6

ratio to 0.6 and to zero is tabulated in Table 2-1. The result of this root locus study illustrates an important point; that is, the higher the gain of

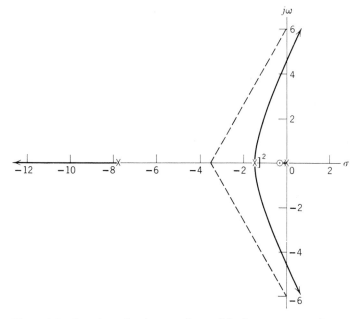

Figure 2-9 Root locus for the outer loop of the jet transport and autopilot for $S_{(rg)} = 1.19$ volt/deg/sec.

the inner loop, the higher the allowable gain for neutral stability for the outer loop. However, too high an inner loop gain results in higher natural frequencies and lower damping as the inner loop poles move out along the 90° asymptotes.

Although the displacement autopilot with rate feedback adequately controls a jet transport; it is desirable to obtain a Type 1 system to eliminate the steady-state error for a step input. This fact plus the desirability for more maneuverability, especially in fighter aircraft, has led to the pitch orientational control system with control stick steering.

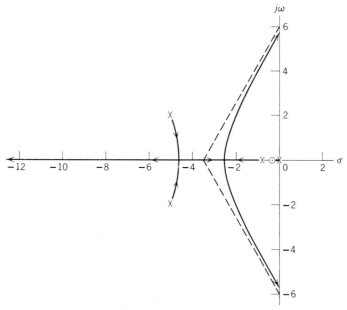

Figure 2-10   Root locus for the outer loop of the jet transport and autopilot for $S_{(rg)} = 1.98$ volt/deg/sec.

2-2   Pitch Orientational Control System

The pitch orientational control system has as its input a desired pitch
rate; such an input, plus the desire to obtain a Type 1 system, which calls
for an integration in the forward loop, makes it advisable to use an integrating gyro as the command receiver.   Figure 2-11 is a block diagram of
a pitch orientational control system.   The necessity for the rate gyro
depends upon the characteristics of the system.   If this system is to be

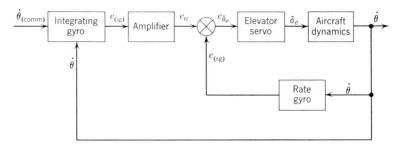

Figure 2-11   Pitch orientational control system.

used to control the attitude of the aircraft, an altitude or Mach hold control system can be used for the third and outer loop (see Section 2-4) and thus provide the $\dot{\theta}_{(comm)}$ signal. The command signal can also come from the pilot.

The pilot's input might come from a so-called "maneuver stick," a small control stick which, when moved, positions a potentiometer and thus produces a voltage. However, the maneuver stick does not provide any feel to the pilot, that is the pilot does not have to exert any force to move the handle. The alternative method is to use a force stick. For this a portion of the stick is necked down so that the stick will flex slightly when a force is applied. The stick is so installed that with the system on, the force exerted by the pilot on the stick is resisted by the servo actuator, which is locked by the fluid trapped when the actuating valve is in the off position (see Figure 2-12). The result of the applied force is to cause a deflection of the control stick at the necked-down area. Strain gauges on this portion of the stick sense this deflection producing a voltage which is the command signal. For the case in question, this voltage is amplified and fed as torquing current to the torque generator of the integrating gyro. As a result of this torquing current a signal is produced from the integrating gyro signal generator. This signal is amplified and used to position

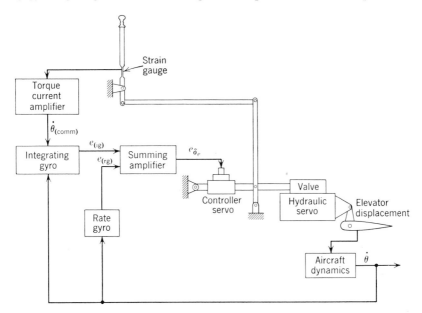

Figure 2-12 Functional diagram of a pitch orientational control system with control stick steering.[1]

the selector valve of the elevator servo positioning the elevator. As the elevator is positioned the stick follows this movement, and the aircraft acquires the desired pitch rate. To maintain a steady pitch rate the pilot must continue to apply pressure to the stick. The system can be adjusted to provide any stick force gradient per "g" of acceleration desired, thus providing the desired feel to the pilot. In the future discussion of the pitch orientational control system it will be assumed that the command to the integrating gyro comes from the force stick; this is often referred to as "control stick steering." This autopilot configuration can be used to provide an all attitude autopilot; however, one very important application is the control of an aircraft that is longitudinally unstable, that is, C_{m_α} is positive.

Although all aircraft are designed to be statically stable, C_{m_α} negative, certain flight conditions can result in large changes in the longitudinal stability. Such changes occur in some high-performance aircraft as they enter the transonic region. However, an even more severe shift in the longitudinal stability results in some high performance aircraft at high angles of attack, a phenomenon referred to as "pitch-up." An aircraft that is subject to pitch-up might have a curve of C_m versus α, similar to the one shown in Figure 2-13. As long as the slope of the curve in Figure 2-13 is negative the aircraft is stable but as the angle of attack is increased, the slope changes sign and the aircraft becomes unstable. If corrective action is not taken, the angle of attack increases until the aircraft stalls. This usually happens so rapidly that the pilot is unable to control or stop the pitch-up. Before proceeding with the study of the pitch orientational control system, the aerodynamic causes of pitch-up will be discussed.

Pitch-up is most likely to occur in aircraft that have the horizontal stabilizer mounted well above the wing of the aircraft; a common place is the top of the vertical stabilizer. This is sometimes done to obtain the end-plate effect on the vertical stabilizer and thus increase the effectiveness of the vertical stabilizer. Another factor that contributes to this unstable flight condition is a wing with a low aspect ratio. Such a wing has a

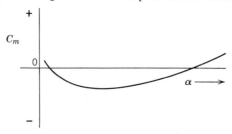

Figure 2-13 C_m versus α for an aircraft
subject to pitch-up.

large downwash velocity that increases rapidly as the angle of attack of the wing is increased. As the high horizontal tail moves down into this wing wake, pitch-up occurs if the downwash velocity becomes high enough. Pitch-up may occur in straight-wing as well as swept-wing aircraft. For the swept-wing aircraft the forward shift of the center of pressure of the wing at high angles of attack is also a contributing factor. A practical solution of the pitch-up problem is to limit the aircraft to angles of attack below the critical angle of attack; however, this also limits the performance of the aircraft. An aircraft that is subject to pitch-up generally will fly at these higher angles of attack; thus an automatic control system that makes the aircraft flyable at angles of attack greater than the critical angle of attack increases the performance capabilities of the aircraft. The pitch orientational control system, if properly designed, provides this control. The rest of this section will be devoted to a root locus study of this control system to show how the sensitivities of the rate gyro and the integrating gyro should be determined to obtain the desired control. Included also will be the results of some computer runs showing the effectiveness of this control. The transfer function used to represent the aircraft is representative of modern high performance fighter aircraft, although not the transfer function of any actual aircraft. Figure 2-14 is the block diagram of the system. The amplifier that was shown in Figure 2-11 is combined with the elevator servo. A root locus study for both the stable and unstable condition demonstrates the factors that influence the selection of the rate gyro and integrating gyro sensitivities.

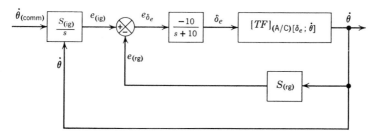

Figure 2-14   Block diagram of pitch orientational control system for the control of pitch-up.

For low angles of attack (stable condition):

$$[TF]_{(A/C)[\delta_e; \dot\theta]} = \frac{-15(s + 0.4)}{s^2 + 0.9s + 8} \frac{\text{deg/sec}}{\text{deg}}$$

For high angles of attack (unstable condition):

$$[TF]_{(A/C)[\delta_e; \dot\theta]} = \frac{-9(s + 0.3)}{(s + 3.8)(s - 2.9)} \frac{\text{deg/sec}}{\text{deg}}$$

Figure 2-15 is the root locus for the inner loop for low angles of attack as the rate gyro sensitivity is varied. The location of the closed loop poles for two values of rate gyro sensitivity are illustrated. One value results in the maximum damping; the reason for the higher rate gyro

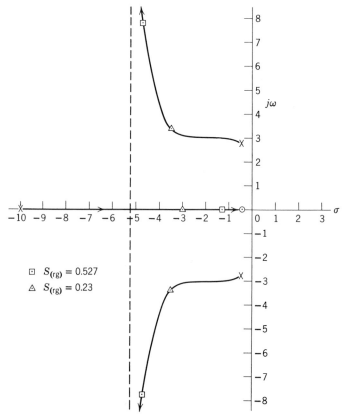

Figure 2-15   Root locus for the inner loop of the control system, shown in Figure 2-14, for low angles of attack.

sensitivity will be shown later. Figure 2-16 is the root locus for the inner loop for high angles of attack. From this root locus it can be seen that as the rate gyro sensitivity is increased, the pole in the right-half plane moves toward the zero in the left-half plane. The locations of the closed loop poles for the same rate gyro sensitivities as were used in Figure 2-15 are shown. These closed loop poles are the open loop poles for the outer loop. The root locus for the outer loop for the unstable condition is seen in Figure 2-17. Note the additional pole at the origin, which results from the integrating gyro. Now the reason for selecting the higher rate gyro sensitivity is evident. If the rate gyro sensitivity is not high enough, that is, the pole in the right-half plane is too far to the right, the system is only stable for a limited range of integrating gyro sensitivities. As is

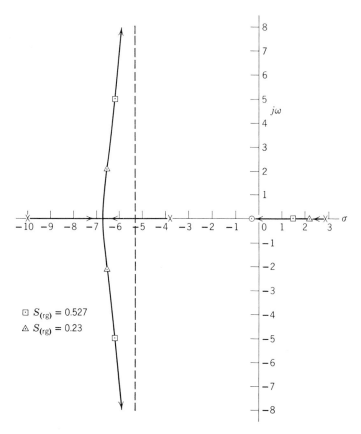

☐ $S_{(rg)} = 0.527$
△ $S_{(rg)} = 0.23$

Figure 2-16 Root locus for the inner loop of the control system shown in Figure 2-14, for high angles of attack.

Table 2-2 Maximum and Minimum Integrating Gyro Sensitivities for System Stability for Two Rate Gyro Sensitivities (High Angle of Attack)

Rate Gyro Sensitivities (volt/deg/sec)	Integrating Gyro Sensitivities (volt/deg/sec)	
	Minimum	Maximum
0.23	1.56	2.63
0.527	1.12	5.78

shown in Table 2-2, the higher the rate gyro sensitivity, the wider the allowable range of integrating gyro sensitivities.   In fact, if the pole in the right-half plane is too far to the right, regardless of the integrating gyro sensitivity the system is unstable.   The locations of the final closed loop poles for the whole system which result in the fastest response time are

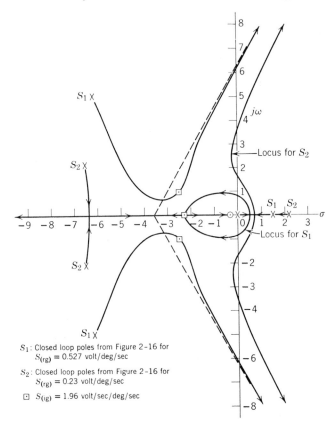

S_1: Closed loop poles from Figure 2-16 for
   $S_{(rg)} = 0.527$ volt/deg/sec

S_2: Closed loop poles from Figure 2-16 for
   $S_{(rg)} = 0.23$ volt/deg/sec

⊡ $S_{(ig)} = 1.96$ volt/sec/deg/sec

Figure 2-17   Root locus for the complete pitch orientational control
system for high angles of attack.

indicated by Figure 2-17.   Figure 2-18 is the root locus for the outer loop at low angles of attack.   The locations of the closed loop poles for the same integrating gyro sensitivity used at large angles are indicated in Figure 2-18.   The integrating gyro sensitivities to produce instability are listed in Table 2-3.   From this table it can be seen that stability at high angles of attack can be obtained without producing instability at low angles of attack.

To check the ability of this system to control an aircraft that is subject to pitch-up at high angles of attack, the aircraft and control systems were simulated on an analog computer. To add realism, a mock-up of an elevator servo with force stick input was "tied" into the computer. The angle of attack was displayed on an oscilloscope. Several runs were

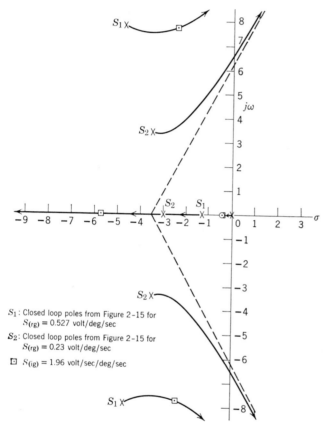

S_1: Closed loop poles from Figure 2-15 for
    $S_{(rg)} = 0.527$ volt/deg/sec
S_2: Closed loop poles from Figure 2-15 for
    $S_{(rg)} = 0.23$ volt/deg/sec
□ $S_{(ig)} = 1.96$ volt/sec/deg/sec

Figure 2-18   Root locus for the complete pitch orientational control system for low angles of attack.

Table 2-3   *Integrating Gyro Sensitivities for Instability for Two Rate Gyro Sensitivities (Low Angle of Attack)*

Rate Gyro Sensitivity (volt/deg/sec)	Integrating Gyro Sensitivity (volt/deg/sec)
0.23	2.36
0.527	5.64

made during which the angle of attack was varied from the trim angle of attack for straight and level flight up to about 23° angle of attack.   Figure 2-19 is a recording of one of these runs.   Neutral stability occurred at about 13° angle of attack at which point the elevator angle started decreasing, and as the angle of attack increased still further, the elevator angle

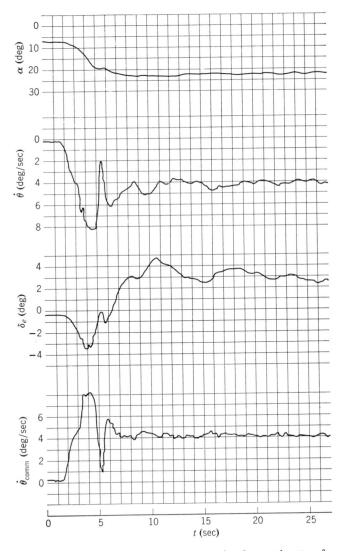

Figure 2-19   Response of the pitch orientational control system for force stick input for Mach of 0.8 at 40,000 ft.

went through zero and became positive. As the force stick was used the stick moved forward as the elevator angle went positive; however, it was still necessary to apply "back pressure" to the stick. By simply releasing the back pressure the commanded pitch rate became zero and the angle of attack returned to the trim angle of attack. Besides being able to control an aircraft that is subject to pitch-up, this control system would also take care of trim changes encountered when an aircraft passes through the transonic region. The characteristics of the integrating gyro as discussed in Section 2-7 make this control possible.

Until now, we have studied longitudinal autopilots in which either the pitch angle or the pitch rate were used to obtain control or to add damping. It is sometimes desirable or necessary to control the acceleration along the aircraft's Z axis. This type of autopilot is referred to as an acceleration autopilot and is discussed in Section 2-3.

2-3   Acceleration Control System

Figure 2-20 is a block diagram for an acceleration control system. A pure integration is introduced in the forward loop to provide a Type 1 system. To understand the requirement for the rate gyro feedback loop and the reason for the signs used at the summers, we must examine the transfer function for the aircraft for an elevator input to an acceleration output. The required transfer function can be derived from Eq. 1-38, which is repeated here.

$$a_z = (\dot{w} - U_0\dot{\theta}) \tag{2-1}$$

Factoring U_0 out of the right-hand side of Eq. 2-1 yields

$$a_z = U_0\left(\frac{\dot{w}}{U_0} - \dot{\theta}\right) = U_0('\dot{\alpha} - \dot{\theta}) \tag{2-2}$$

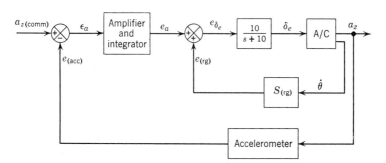

Figure 2-20   Block diagram of an acceleration control system.
$S_{(acc)} = 1$ volt/g

as $\dot{w}/U_0 = {}'\dot{\alpha}$. The transfer function can be obtained by dividing Eq. 2-2 by δ_e which yields

$$\frac{a_z}{\delta_e} = U_0\left(\frac{'\dot{\alpha}}{\delta_e} - \frac{\dot{\theta}}{\delta_e}\right) \tag{2-3}$$

Thus the acceleration transfer function is simply the forward velocity, times the $'\dot{\alpha}$ transfer function, minus the $\dot{\theta}$ transfer function. The question that now arises is whether to use the three-degree-of-freedom equations or the short-period approximation. In Section 1-10 it was shown that during the phugoid oscillation $'\dot{\alpha}$ and $\dot{\theta}$ were almost zero. From this it can be concluded that the short-period approximation can be used for the acceleration transfer function.

As an acceleration control system is used generally in fighter type aircraft, the transfer functions used are typical of a high performance fighter. For the aircraft used in Section 2-2, the $'\dot{\alpha}$ and $\dot{\theta}$ transfer functions are:

$$\frac{'\dot{\alpha}(s)}{\delta_e(s)} = \frac{-(0.1s^2 + 15s)}{(s^2 + 0.9s + 8)} \tag{2-4}$$

$$\frac{\dot{\theta}(s)}{\delta_e(s)} = \frac{-(15s + 6)}{(s^2 + 0.9s + 8)} \tag{2-5}$$

Subtracting Eq. 2-5 from Eq. 2-4, multiplying by U_0, and substituting into Eq. 2-3 yields

$$\frac{a_z(s)}{\delta_e(s)} = \frac{-77.7(s^2 - 60)}{s^2 + 0.9s + 8} \tag{2-6}$$

Factoring,

$$\frac{a_z(s)}{\delta_e(s)} = \frac{-77.7(s + 7.75)(s - 7.75)}{s^2 + 0.9s + 8}\frac{\text{ft/sec}^2}{\text{rad}} \tag{2-7}$$

But the units for the transfer function of the elevator servo are deg/volt, and for the accelerometer they are volt/g. The units of Eq. 2-7 can be changed to g's/deg by dividing by 32.2 ft/sec^2/g and 57.3 deg/rad. Thus Eq. 2-7 becomes

$$\frac{a_z(s)}{\delta_e(s)} = \frac{-0.042(s + 7.75)(s - 7.75)}{s^2 + 0.9s + 8}\frac{\text{g's}}{\text{deg}} \tag{2-8}$$

An examination of Eq. 2-8 indicates that there is a zero in the right-half plane, thus indicating a so-called nonminimum phase angle transfer function. This means that for a positive step input of δ_e the steady-state sign of a a_z will be positive which is consistent with the sign convention already established. Thus the sign at the summer for the acceleration feedback must be negative for negative feedback. The sign for the elevator

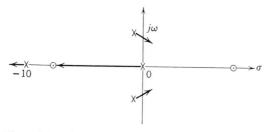

Figure 2-21   Angle of departure of the complex poles for the acceleration control system without rate feedback (zero-angle root locus).

servo remains positive so that a positive $a_{z(\text{comm})}$ yields a positive a_z output.

Having settled on the sign of the feedback element, the sign of the open loop transfer function can be determined (see Appendix C). The sign of the open loop transfer function is determined by the sign of each transfer function and the sign at the summer for the feedback element. Thus the sign of the open loop transfer function is positive (the zero in the right-half plane has no effect on the sign of the open loop transfer function). The denominator of the closed loop transfer function is $1-[TF]_{OL}$ which, as the sign of $[TF]_{OL}$ is plus, means that $G(s)H(s) = +1$, and the zero angle root locus must be plotted. If the rate gyro loop were not included, the complex poles of the aircraft would move toward the imaginary axis as shown in Figure 2-21. The complex poles of the aircraft must therefore be moved further to the left before closing the accelerometer feedback loop. This can be accomplished by using pitch-rate feedback.

The block diagram for the rate feedback loop is shown in Figure 2-22. The system depicted in Figure 2-22 is the same as the inner loop of the pitch orientational control system for low angles of attack, the root locus of which was seen in Figure 2-15. The closed loop transfer function for the inner loop of the acceleration control system for $S_{(\text{rg})} = 0.23$ is

$$\frac{\theta(s)}{e_a(s)} = \frac{-150(s + 0.4)}{(s + 3)(s^2 + 7s + 24)} \frac{\text{deg/sec}}{\text{volt}} \tag{2-9}$$

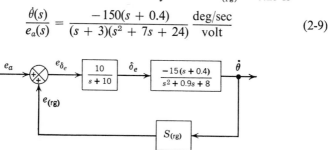

Figure 2-22   Block diagram for the inner loop of the acceleration control system.

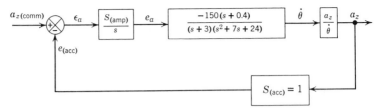

Figure 2-23   Block diagram for the outer loop for the acceleration control system.

The block diagram for the outer loop is shown in Figure 2-23 for $S_{(rg)} = 0.23$. The $a_z/\dot\theta$ block is required to change the output of the inner loop $\dot\theta$ to the required output for the outer loop a_z. The transfer function for this block can be obtained by taking the ratio of the $a_z(s)/\delta_e(s)$ and $\dot\theta(s)/\delta_e(s)$ transfer functions, thus

$$\frac{a_z(s)}{\dot\theta(s)} = \frac{-0.042(s + 7.75)(s - 7.75)}{-15(s + 0.4)} \quad \frac{\text{g's}}{\text{deg/sec}} \tag{2-10}$$

Then the $a_z(s)/e_a(s)$ transfer function is the product of Eqs. 2-9 and 2-10, or

$$\frac{a_z(s)}{e_a(s)} = \frac{-0.42(s + 7.75)(s - 7.75)}{(s + 3)(s^2 + 7s + 24)} \quad \frac{\text{g's}}{\text{volt}} \tag{2-11}$$

This operation changes only the numerator of the forward transfer function by replacing the zero of the $\dot\theta(s)/\delta_e(s)$ transfer function with the zeros of the $a_z(s)/\delta_e(s)$ transfer function. The denominator remains the same.

Using Eq. 2-11 with the transfer function of the amplifier and integrator, the root locus for the outer loop can now be drawn and is shown in Figure 2-24. The closed loop poles for $S_{(amp)} = 2.22$ are illustrated in Figure 2-24, and the closed loop transfer function is:

$$\frac{a_z(s)}{a_{z(comm)}(s)} = \frac{-0.93(s + 7.75)(s - 7.75)}{(s^2 + 2.2s + 2.4)(s^2 + 7.8s + 24)} \tag{2-12}$$

Although the accelerometer control system provides good operation, there are some practical problems involved. One of these is the fact that the accelerometer cannot distinguish between the acceleration due to gravity and accelerations caused by aircraft motion. The acceleration of gravity can be balanced out so that in straight and level flight at normal cruise air speed and altitude the output of the accelerometer would be zero. However, at different angles of attack the accelerometer output would not be zero. For example, if the angle of attack changed by 10° from that angle of attack for which the accelerometer was nulled, the output would be proportional to $\pm 0.5 \text{ ft/sec}^2$. The accelerometer can be adjusted so that it is insensitive to accelerations that are less than 1 ft/sec², thus eliminating this problem. Another problem which would probably be

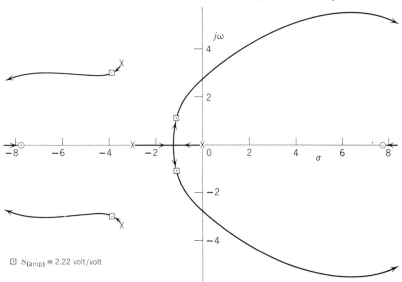

Figure 2-24   Root locus of the outer loop of the acceleration control system (zero angle root locus).

harder to overcome is the unwanted accelerations arising from turbulence. This shows up as noise and has to be filtered out.   As a result of these problems, and due to the fact that there are not many requirements that call for an aircraft maneuvering at constant acceleration, the acceleration autopilot is not often employed.   However, there are some requirements that make the acceleration autopilot ideal.   An example is the necessity to perform a maximum performance pull-up in connection with a partic- ular tactical maneuver.   The system can be adjusted so that the maximum allowable g's for the aircraft are experienced during pull-up.

Now that we have looked at some of the basic longitudinal control systems, their integration into various automatic flight control systems are discussed in Sections 2-4 and 2-5.

2-4   Glide Slope Coupler and Automatic Flare Control

The ultimate goal for both military and commercial aviation is all- weather operation.   To achieve this goal, it must be possible to land the aircraft without visual reference to the runway.   This can be accomplished by an automatic landing system which would guide the aircraft down a predetermined glide slope and then at a preselected altitude reduce the rate of descent and cause the aircraft to "flare" out and touch down with an

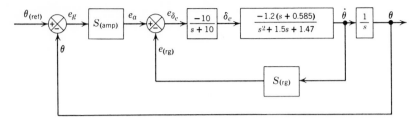

Figure 2-25   Basic autopilot for the jet transport, landing configuration.

acceptably low rate of descent. This section covers the longitudinal control portion of such a system; lateral control is covered in Chapter 4. The design of a glide slope coupler, an automatic flare control system, and an altitude hold control system (to be studied in Section 2-5) have one thing in common, that is, the control of the flight path angle γ. The automatic control of the flight path angle without simultaneous control of the airspeed, either manual or automatic, is impossible. This point is illustrated by studying the response of the attitude control system, without velocity control, that will later be coupled to the glide slope. For this analysis the jet transport in the landing configuration is used. The appropriate longitudinal equations for sea level at 280 ft/sec are

$$(7.4s + 0.15)'u(s) - 0.25'\alpha(s) + 0.85\theta(s) = C_{x_T}\delta_{(\text{rpm})}(s)$$

$$1.7'u(s) + (7.4s + 4.5)'\alpha(s) - 7.4s\theta(s) = -0.246\delta_e(s)$$

$$(0.118s + 0.619)'\alpha(s) + (0.59s^2 + 0.41s)\theta(s) = -0.718_e(s) \qquad (2\text{-}13)$$

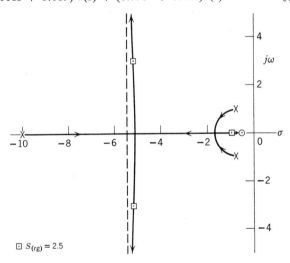

$\square \ S_{(\text{rg})} = 2.5$

Figure 2-26   Root locus for the inner loop of the basic autopilot for the jet transport.

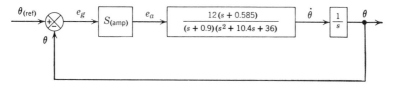

Figure 2-27   Block diagram for the root locus of the outer loop of the basic autopilot.

where $C_{x_T} = \dfrac{1}{Sq} \dfrac{\partial T}{\partial \delta_{(rpm)}}$ (rpm)$^{-1}$.   For the basic autopilot, $\delta_{(rpm)}$ is considered zero.

Figure 2-25 is a block diagram of the basic autopilot used in this analysis.   The root locus for the inner loop is shown in Figure 2-26.   The rate gyro sensitivity was selected so the complex poles would lie on the branches approaching the 90° asymptotes with a damping ratio greater than 0.7, in this case 0.866.   The higher damping ratio was chosen because the outer loop tends to decrease the damping of the complex roots.   For the closed

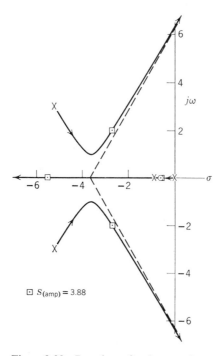

Figure 2-28   Root locus for the outer loop of the basic autopilot for the jet transport.

loop poles indicated in Figure 2-26, $S_{(rg)}$ = 2.5 volt/rad/sec. Using the closed loop poles from the root locus shown in Figure 2-26, the block diagram for the outer loop can be drawn and is seen in Figure 2-27. The root locus for the outer loop is shown in Figure 2-28, with the location

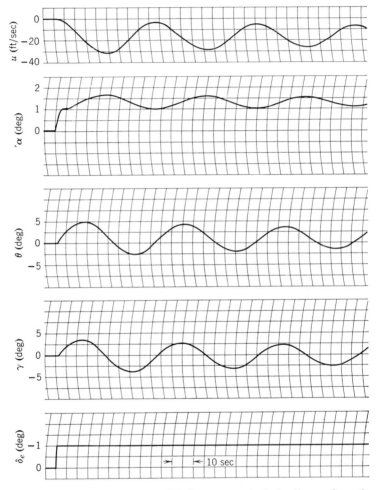

Figure 2-29   Transient response of the jet transport in the landing configuration at sea level at 280 ft/sec for an elevator step input.

of the closed loop poles for $S_{(amp)}$) = 3.88 volt/volt, which yields a damping ratio of 0.8.   The reason for this selection will be more obvious when the root locus for the final outer loop through the glide slope coupler is examined later in this Section.

After selection of the gains for the two loops of the basic autopilot, the aircraft and the basic autopilot were simulated on the analog computer. For the aircraft simulation the complete three-degree-of-freedom longitudinal equations were used. The response of the aircraft alone, for a

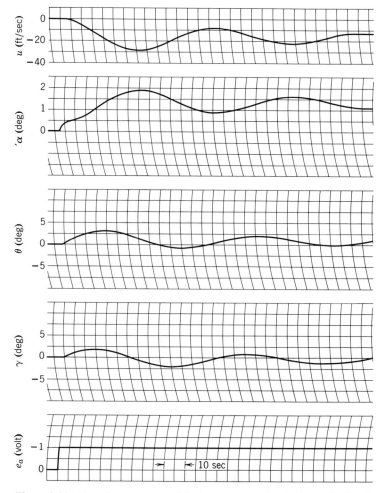

Figure 2-30 Transient response of the jet transport with pitch-rate feedback for $S_{(rg)} = 2.5$ volt/deg/sec, for step input of e_a.

step elevator input with no velocity control is depicted in Figure 2-29. The effect of adding rate gyro feedback is shown in Figure 2-30. It is evident that the rate gyro has little effect on the damping of the phugoid mode. The response of the aircraft, including the basic autopilot for a

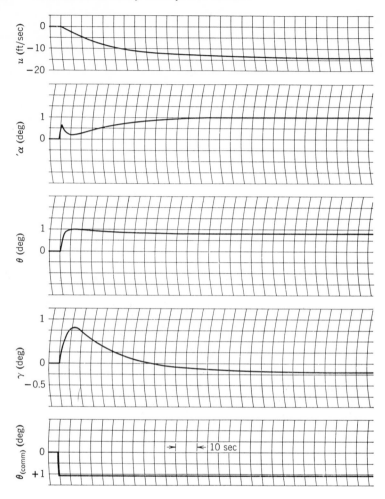

Figure 2-31   Transient response of the jet transport and basic autopilot for
$S_{(amp)} = 3.88$ volt/volt, for a θ step command.

step $\theta_{(comm)}$ can be seen in Figure 2-31.   The necessity for velocity control
is now evident, since the steady-state value of γ is $-0.2°$.   This value
results from a positive θ command, which it is hoped will cause the aircraft to climb (positive γ).   However, as the nose of the aircraft rises and
the aircraft starts to climb, the airspeed decreases requiring an increase in
the angle of attack to maintain the required lift.   In the steady state this
increase in the angle of attack is greater than the change in θ and as $\gamma =$
$\theta - {'\alpha}$, the result is that the aircraft starts a very shallow glide.   In order

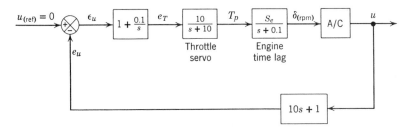

Figure 2-32   Velocity control system using throttle.

for γ to have the same steady-state sign as $\theta_{(comm)}$, some form of velocity control must be employed.

An automatic velocity control system, sometimes referred to as "phugoid damping," is shown in Figure 2-32. The time constant for the jet engine is taken as 10 sec and may be excessive; however, if the system can be made to operate with this lag, its performance would be even better if the actual engine time constant were less. This time lag represents the time required for the thrust of the jet engine to build up after a movement of the throttle T_p. Since u is the change in airspeed, the input to the aircraft block is the change in the engine rpm from that required for straight and level unaccelerated flight. The first block in the forward loop of Figure 2-32 is a proportional plus integral network which makes the system Type 1 and also provides compensation by adding a zero at $s = -0.1$. Velocity and the rate of change of velocity are used in the feedback path. Both the velocity error (actually airspeed error) and the velocity rate signal can be supplied by a single instrument developed by the Avion Instrument Corp., New York 10, N.Y.[2] The aircraft transfer function can be obtained from Eq. 2-13 and is

$$\frac{u(s)}{\delta_{(rpm)}(s)} = \frac{37.9 C_{x_T} s(s^2 + 1.5s + 1.47)}{(s^2 + 1.51s + 1.48)(s^2 + 0.009s + 0.0186)} \frac{\text{ft/sec}}{\text{rpm}} \quad (2\text{-}14)$$

The existence of the quadratic, in the numerator of Eq. 2-14, that almost exactly cancels the short-period quadratic in the denominator is not a coincidence. The reason for this can be seen by examining the determinant of the numerator used to determine Eq. 2-14. This determinant is

$$\begin{vmatrix} C_{x_T} & -0.25 & 0.85 \\ 0 & 7.4s + 4.5 & -7.4s \\ 0 & 0.118s + 0.619 & 0.59s^2 + 0.41s \end{vmatrix}$$

The result of expanding the foregoing determinant yields C_{x_T} times its minor, but the minor of C_{x_T} is identical to the determinant of the denominator for any of the transfer functions obtained by using the short-period

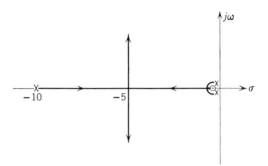

Figure 2-33   Sketch of root locus for the velocity
control loop (not to scale).

approximation.   As the short-period poles are effectively canceled by the
complex zeros for the $u(s)/\delta_{(rpm)}(s)$ transfer function, the velocity control
loop has no effect on the short-period mode.   This characteristic of the
velocity control loop is used to simplify the remaining analysis.

In Figure 2-32 the open loop transfer function using Eq. 2-14 is

$$[TF]_{OL} = \frac{-3790 C_{x_T} S_e(s + 0.1)}{(s + 10)(s^2 + 0.009s + 0.0186)} \qquad (2\text{-}15)$$

Figure 2-33 is a sketch of the root locus for the velocity control loop.
The effectiveness of the velocity control loop for damping the phugoid is
very apparent.   The value of the product of $C_{x_T} S_e$ selected is 0.2.   This
value is determined experimentally from the results of the computer
simulation.   For this value, the complex poles on the vertical branches of
Figure 2-33 have a damping ratio of 0.182 and a natural frequency of
27.5 rad/sec.   The real pole near the real zero is thus the dominant root.
The effect of the velocity control loop for the basic aircraft with no other
feedback is illustrated in Figure 2-34 (*a*), and its effect on the aircraft with
the basic autopilot is shown in Figure 2-34 (*b*).   As can be seen from
Figure 2-34 (*b*), the steady-state sign of γ is not only positive for a positive
$\theta_{(comm)}$ but also is equal to the commanded pitch angle.   It should also
be noted that the velocity control loop has not affected the initial portion
of the transient response of $'\alpha$, θ, and the initial signs of the γ trace,
Figure 2-34 (*b*).   From this, and the fact that the short-period quadratic
is canceled by the numerator quadratic in Eq. 2-14, it can be concluded
that the velocity control loop does not affect the short-period dynamics of
the aircraft.   Therefore if velocity control is assumed, either manual or
automatic, the short-period transfer functions can be used for any analysis
of any control system employed for the control of the flight path angle.

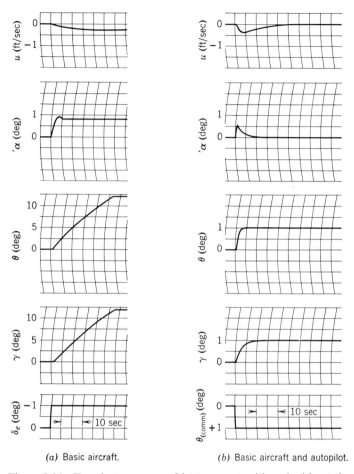

(a) Basic aircraft.   (b) Basic aircraft and autopilot.

Figure 2-34   Transient response of jet transport, with and without the basic autopilot, showing the effects of velocity control.

The effect of velocity control has been illustrated, and the design and analysis of a glide slope coupler may now be continued.

The geometry associated with the glide slope problem is shown in Figure 2-35. If the aircraft is below the center line of the glide slope (Figure 2-35), d is considered negative, as is γ, when the velocity vector is below the horizon, that is, the aircraft descending. The component of U perpendicular to the glide slope center line is $\dot{d}$ and is given in Eq 2-16

$$\dot{d} = U \sin (\gamma + 2\tfrac{1}{2})° \simeq \frac{U}{57.3} (\gamma + 2\tfrac{1}{2})° \qquad (2\text{-}16)$$

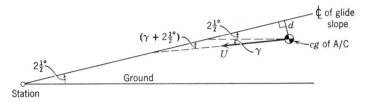

Figure 2-35   Geometry of the glide slope problem.   *Note: d* and *γ* are negative.

for small angles.   For the condition shown as $\gamma < 2\frac{1}{2}°$ in Figure 2-35, then $\gamma + 2\frac{1}{2}$ is positive; therefore d is positive and as d initially was negative, the aircraft is approaching the glide path from below.   Integrating Eq. 2-16 yields

$$d \simeq \frac{U}{57.3s}(\gamma + 2\frac{1}{2})° \qquad (2\text{-}17)$$

One other factor must be accounted for, and that is that the glide slope receiver does not measure the perpendicular distance to the glide slope center line but the angular error resulting therefrom.   Thus for a given value of d the angular error increases as the aircraft nears the station, which has the effect of increasing the system gain as the range to the station decreases as shown in Figure 2-36.   From Figure 2-36, tan $\Gamma = d/R$ or for small angles and Γ in degrees

$$\Gamma \simeq 57.3d/R \text{ deg} \qquad (2\text{-}18)$$

Through the use of Eqs 2-17 and 2-18 the flight path angle γ can be related to the angular error of the aircraft from the glide slope centerline, and the block diagram for the glide slope control system can be drawn, including the geometry (see Figure 2-37).   Figure 2-37 shows that the coupler contains a proportional, plus integral circuit, plus a lead compensator. The zero of the lead network is selected so that it cancels the closed loop pole at $s = -0.5$ from the basic autopilot (see Figure 2-28).   The factor of 10 is included to provide a steady-state gain of 1 for the lead circuit.

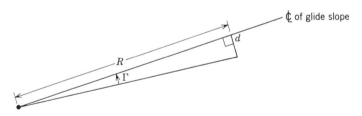

Figure 2-36   Effect of beam narrowing.

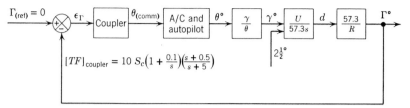

Figure 2-37   Block diagram of automatic glide slope control system.

The $\gamma(s)/\theta(s)$ transfer function is required to convert the θ output from the aircraft and autopilot transfer function to γ, and can be derived from the relation $\gamma = \theta - {}'\alpha$ (see Figure 1-3).

Then

$$\frac{\gamma(s)}{\theta(s)} = 1 - \frac{{}'\alpha(s)}{\theta(s)}$$

But

$$\frac{{}'\alpha(s)}{\theta(s)} = \left[\frac{{}'\alpha(s)}{\delta_e(s)}\right]\left[\frac{\delta_e(s)}{\theta(s)}\right] = \frac{0.0276s(s + 36.7)}{s + 0.585}$$

As

$$\frac{\theta(s)}{\delta_e(s)} = \frac{-1.2(s + 0.585)}{s(s^2 + 1.5s + 1.47)}$$

And

$$\frac{{}'\alpha(s)}{\delta_e(s)} = \frac{-0.0332(s + 36.7)}{s^2 + 1.5s + 1.47}$$

Substituting and simplifying

$$\frac{\gamma(s)}{\theta(s)} = -\frac{0.0276(s + 4.85)(s - 4.35)}{s + 0.585} \qquad (2\text{-}19)$$

Equation 2-19 is the required transfer function; however, most control system designers use a simplified form of Eq. 2-19 which is

$$\frac{\gamma(s)}{\theta(s)} = \frac{0.585}{s + 0.585} \qquad (2\text{-}20)$$

Equation 2-20 can be obtained by neglecting the s terms in the numerator of Eq. 2-19. Although a root locus analysis using Eq. 2-20 gives good results, Eq. 2-19 is used in the analysis to follow. For the closed loop poles shown in Figure 2-28

$$\frac{\theta(s)}{\theta_{(comm)}(s)} = \frac{46.5(s + 0.585)}{(s + 0.5)(s + 5.5)(s^2 + 5.4s + 11.4)} \qquad (2\text{-}21)$$

Using the transfer functions shown in Figure 2-37 and Eqs. 2-19 and 2-21, the forward transfer function for the glide slope control system can be

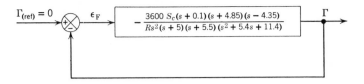

$\Gamma_{(ref)} = 0$ ϵ_F

$$- \frac{3600\, S_c(s + 0.1)(s + 4.85)(s - 4.35)}{Rs^2(s + 5)(s + 5.5)(s^2 + 5.4s + 11.4)}$$

Γ

Figure 2-38   Simplified block diagram of automatic glide slope control system.

obtained and is given in Figure 2-38 for $U = 280$ ft/sec. The zero angle root locus for the glide slope control system is shown in Figure 2-39. For this root locus the coupler sensitivity is constant at a value of $10°$ of $\theta_{(comm)}$ per degree of ϵ_Γ and the range to the station, R, provides the variable gain. Thus it can be seen that the dynamics of the system change as the aircraft approaches the station. The problem involved in the design of a satisfactory glide slope coupler results from the two poles at the origin, one from the coupler, the other from the geometry. Without the zero at $s = -0.1$, from the proportional plus integral network, the poles at the origin would move immediately into the right-half plane. Even with the zero at $s = -0.1$, if the pole at $s = -0.5$ was not canceled by the zero of the compensator, the system would, at best, be only marginally stable for a limited range of R. Since the range to the station provides the variable gain for the glide slope control system, the root locus can be used to determine the minimum range for stable operation. For this case, the magnitude condition for the point where the root locus crosses the imaginary

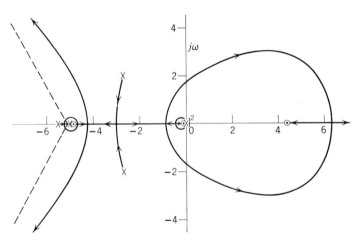

Figure 2-39   Root locus for the automatic glide slope control system (zero angle root locus).

axis yields a K of 27 which must be equal to 3600 S_c/R. Therefore $R = 134S_c$, which for $S_c = 10$ gives a minimum range of 1340 ft. From the computer simulation the minimum range was determined as 1300 ft, thus further verifying the validity of using the short-period approximation for the root locus analysis, if velocity control is assumed.

The results of the computer simulation of the complete longitudinal equations of motion plus the basic autopilot and the velocity control system (see Figure 2-32) are shown in Figures 2-40 and 2-41. The aircraft was placed initially at 1400 ft above the station (100 ft below the glide

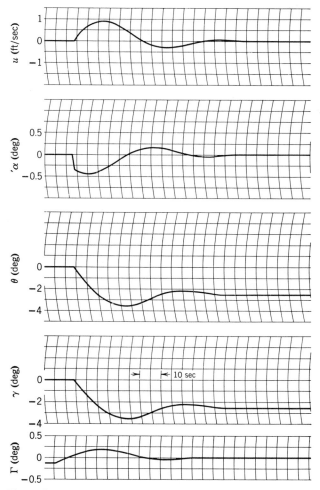

Figure 2-40   Time response of the automatic glide slope control system for $S_c = 10$ deg/deg.

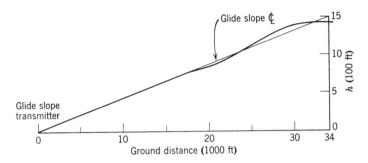

Figure 2-41   Response of the automatic glide slope control system for $S_c = 10$ deg/deg.

slope) in straight and level flight.   The glide slope control system was engaged as the aircraft passed through the glide slope center line.   As can be observed in Figure 2-41, the aircraft has settled onto the glide slope while still about three miles from the station.   From Figure 2-40 it can be seen that the maximum overshoot is only 0.15°, which is much less than the maximum deflection of $\pm 0.5°$ for the pilot's glide slope indicator.

As mentioned earlier, the forward loop gain is dependent upon the range to the station, which leads to instability as the range decreases.   To combat this the coupler sensitivity can be decreased as the range is decreased; however, to do this accurately requires the instantaneous range to the station which is normally not available.   This difficulty can be overcome by using a self-adaptive control system to be discussed in Chapter 6.

The final phase of the landing is the transition from the glide slope to the actual touchdown, generally referred to as "the flare."   The rest of this section deals with the analysis and design of an automatic flare control.

Flight test data has shown that when a pilot performs the flare from the approach glide to the final touchdown he generally decreases his rate of descent in an exponential manner, thus tending to make the aircraft fly an exponential path during the flare.   For the automatic flare control then, the aircraft is commanded to fly an exponential path from the initiation of the flare until touchdown.   Thus

$$h = h_0 e^{-t/\tau} \tag{2-22}$$

Where h is the height above the runway, and h_0 is the height at the start of the flare.   Differentiating Eq. 2-22 yields

$$\dot{h} = -\frac{h_0}{\tau} e^{-t/\tau} = -\frac{h}{\tau} \tag{2-23}$$

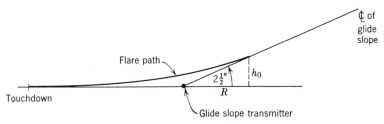

Figure 2-42   Geometry of flare path.

The geometry of the flare path is illustrated in Figure 2-42. Equation 2-23 can be used to determine h_0 if τ is known, that is, for $t = 0$, $h_0 = -\tau \dot{h}_0$. Since $\dot{h}_0$ is the rate of descent at the initiation of the flare, then for a $2\frac{1}{2}°$ glide slope, and a velocity of 280 ft/sec, $\dot{h}_0 = -12.2$ ft/sec. The desired value of τ can be obtained by specifying the distance to the touchdown point from the glide slope transmitter. If this distance is to be 1000 ft and if it is assumed that the aircraft touches down in four time constants, the ground distance traveled during the flare, if the aircraft's velocity is constant, is

$$R + 1000 = (280)(4\tau) \tag{2-24}$$

But $h_0 = 12.2\tau$ and from Figure 2-42, $\tan 2\frac{1}{2}° = h_0/R = 12.2\tau/R$ therefore $R = 280\tau$.

Substituting for R in Eq. 2-24 and solving for τ yields $\tau = 1.18$ and $\dot{h} = -h/1.18 \simeq -0.8h$. Then, $h_0 = 12.2/0.8 = 15.2$ ft. The control equation for the automatic flare control system is then

$$\dot{h}_r = -0.8h \tag{2-25}$$

The block diagram for the automatic flare control system can be seen in Figure 2-43. The outer loop simply supplies the rate-of-descent command, $\dot{h}_r$. The transfer function for the aircraft and autopilot is the same as the

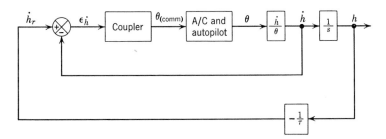

Figure 2-43   Block diagram of automatic flare control system.

one used for the glide slope analysis and as given in Eq. 2-21. The $\dot{h}(s)/\theta(s)$ transfer function can be obtained from the $\gamma(s)/\theta(s)$ transfer function and is

$$\frac{\dot{h}(s)}{\theta(s)} = \frac{U}{57.3}\left[\frac{\gamma(s)}{\theta(s)}\right] = \frac{-7.73(s + 4.85)(s - 4.35)}{57.3(s + 0.585)} \tag{2-26}$$

The coupler transfer function is

$$[TF]_{\text{coupler}} = 10S_c\left(1 + \frac{0.1}{s}\right)\left(\frac{s + 0.5}{s + 5}\right)\left(\frac{s + 5.5}{s + 55}\right) \tag{2-27}$$

which is the same as the coupler for the glide slope control system, plus an additional lead network. The second lead network is added to obtain a higher value of coupler sensitivity, thus preventing the aircraft from flying into the runway too soon. The open loop transfer function from Eqs. 2-21, 2-26, and 2-27 is

$$[TF]_{OL} = \frac{63S_c(s + 0.1)(s + 4.85)(s - 4.35)}{s(s + 5)(s + 55)(s^2 + 5.4s + 11.4)} \tag{2-28}$$

The root locus for the automatic flare control system is shown in Figure 2-44, with the location of the complex poles indicated for a coupler sensitivity of $3°$ of $\theta_{(\text{comm})}$ per ft/sec error in rate of descent. The results of the computer simulation are shown in Figures 2-45 and 2-46. It is found that the command equation must be changed to $\dot{h}_r = 0.6h$ for

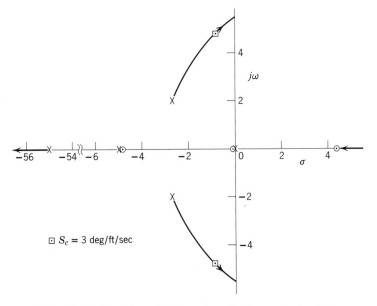

□ $S_c = 3$ deg/ft/sec

Figure 2-44   Root locus for the automatic flare control system.

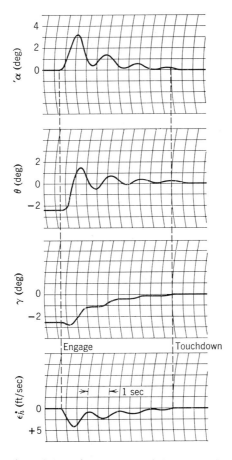

Figure 2-45 Time response of the automatic
flare control system for $S_c = 3$ deg/ft/sec and

$$\dot{h}_r = 0.6h.$$

satisfactory operation. An examination of Figure 2-45 yields a natural
frequency of 4.6 rad/sec as compared with 4.8 rad/sec from the root locus
analysis. The damping ratio is difficult to measure due to the fact that
the oscillation is superimposed on a first-order response, but it is probably
between 0.15 and 0.2. The damping ratio from the root locus is 0.19.
The agreement is considered excellent. In Figure 2-46 the actual path is
compared with the desired flare path and is considered more than acceptable. The rate of descent at touchdown is approximately 30 ft/min,
which does not include the ground effect that would be present for an
actual landing and thus reduce the rate of descent further. This would be
considered an excellent landing by any pilot.

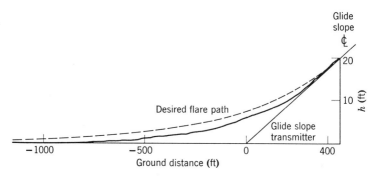

Figure 2-46   Response of the automatic flare control system for $S_c =$ 3 deg/ft/sec and $\dot{h}_r = 0.6h$.

2-5   Flight Path Stabilization

The two modes of operation of a complete longitudinal flight control system investigated here are referred to as "Mach hold" and "altitude hold." These modes are generally used during cruise operation and one or the other selected depending on the mission requirement. In the Mach hold mode the aircraft is made to fly at a constant Mach number by automatically controlling the flight path angle through the elevators. For this mode of operation the aircraft is first trimmed to fly straight and level and the power adjusted to yield the desired Mach number. The Mach hold mode of the flight control system is then engaged. As the aircraft (usually a jet type) cruises, fuel is used, the weight of the aircraft decreases, and the speed tends to increase. The increase in speed is sensed by the control system and corrected for by an up-elevator signal causing the aircraft to climb. The net result of operation in the Mach hold mode is that the aircraft is made to climb slowly as fuel is consumed in order to maintain a constant Mach number. Since the fuel consumption of a jet aircraft decreases as altitude increases, this mode of operation is desirable for long-range operation; however, there are many times when it is necessary to fly at a constant altitude. Under these conditions the altitude hold mode is selected and the aircraft's flight path angle, and thus its altitude, is controlled by the elevators and the airspeed or Mach number is controlled, either manually or automatically, by use of the throttle. For the analysis of these two modes, the transfer functions for the jet transport cruising at 600 ft/sec at 40,000 ft are used.

For the analysis of the Mach hold mode it is assumed that the altitude variations are small so that the variations in Mach number can be represented by variations in velocity. Figure 2-47 is a block diagram of the

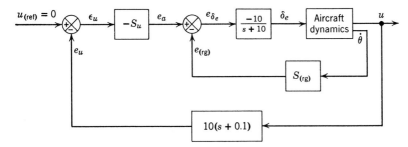

Figure 2-47   Mach hold mode of the flight control system.

Mach hold mode of the flight control system.   As the change in the velocity is needed for the outer loop, the complete three-degree-of-freedom longitudinal equations must be used for the root locus analysis of both the inner and outer loops.   For this aircraft, the required $\dot{\theta}(s)/\delta_e(s)$ transfer function can be obtained from Eq. 1-126, which is repeated here.

$$\frac{\dot{\theta}(s)}{\delta_e(s)} = \frac{-1.31s(s + 0.016)(s + 0.3)}{(s^2 + 0.00466s + 0.0053)(s^2 + 0.806s + 1.311)} \frac{\text{deg/sec}}{\text{deg}}$$

(2-29)

Because the short-period portion of Eq. 2-29 is almost identical to the short-period approximate transfer function for $\dot{\theta}(s)/\delta_e(s)$, Figure 2-8 is still valid for the short-period portion of the inner loop of the Mach hold mode.   If the remaining poles and zeros from Eq. 2-29 were plotted on Figure 2-8, they would all appear as if located at the origin and therefore would not change the root locus in Figure 2-8.   Since the aircraft gain for the three-degree-of-freedom transfer function is 1.31 instead of the 1.39 from the short-period transfer function (see Figure 2-7), the rate gyro sensitivity corresponding to $S_{(rg)} = 1.98$ for Figure 2-8 is 2.1 volt/deg/sec.

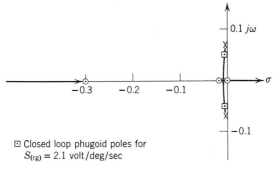

□ Closed loop phugoid poles for
$S_{(rg)} = 2.1$ volt/deg/sec

Figure 2-48   Sketch of expanded portion of the root locus for the inner loop of Figure 2-47.

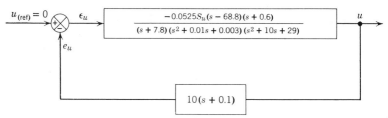

Figure 2-49 Block diagram for the root locus analysis of the outer loop.

It is still necessary to determine the location of the phugoid poles for this value of rate gyro sensitivity. To do this the portion of the root locus near the origin must be expanded as shown in Figure 2-48. To obtain the portion of the root locus in the vicinity of the origin, it is assumed that the angle and magnitude contributions from the poles and zeros not shown in Figure 2-48 are constant. The closed loop location of the phugoid poles is shown in Figure 2-48. Combining the results of Figures 2-8 and 2-48, the closed loop transfer function for the inner loop for $S_{(rg)} = 2.1$ volt/deg/sec can be obtained and is given in Eq. 2-30.

$$\frac{\dot{\theta}(s)}{e_a(s)}\bigg]_{CL} = \frac{13.1s(s + 0.3)(s + 0.016)}{(s + 7.8)(s^2 + 10s + 29)(s^2 + 0.01s + 0.003)} \frac{\text{deg/sec}}{\text{volt}}$$

(2-30)

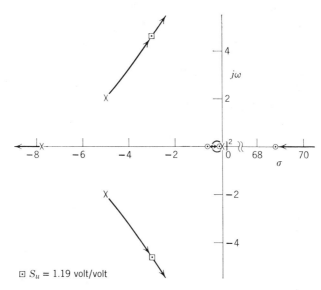

□ $S_u = 1.19$ volt/volt

Figure 2-50 Root locus for the outer loop of the Mach hold mode
(zero angle root locus).

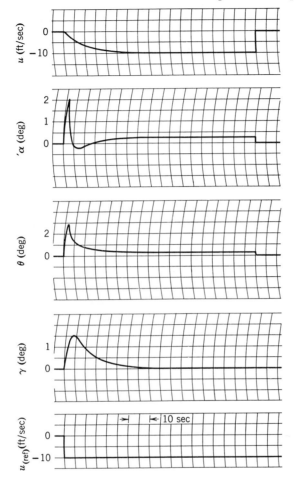

Figure 2-51 Response of the Mach hold control system to a step input of $u_{(ref)}$ for $S_u = 1$.

As in the case of the glide slope, the $\dot{\theta}$ output of the inner loop must be changed to u, the required output for the outer loop. This requires the $u(s)/\dot{\theta}(s)$ transfer function which is U times the ratio of the $'u(s)/\delta_e(s)$ transfer function to the $\dot{\theta}(s)/\delta_e(s)$ transfer function. The transfer function is

$$\frac{u(s)}{\dot{\theta}(s)} = \frac{0.23(s - 68.8)(s + 0.6)}{57.3s(s + 0.016)(s + 0.3)} \frac{\text{ft/sec}}{\text{deg/sec}} \qquad (2\text{-}31)$$

The 57.3 in the denominator of Eq. 2-31 is required to convert the $'u(s)/\delta_e(s)$ from radians to degrees. Using Eqs. 2-30 and 2-31, the simplified block diagram for the outer loop root locus can be drawn and is

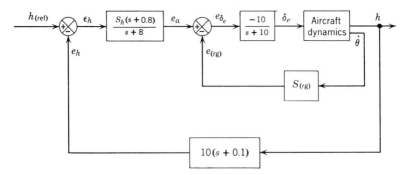

Figure 2-52   Altitude hold mode of flight control system.

shown in Figure 2-49. Due to the zero in the right-half plane of the forward transfer function in Figure 2-49, the steady-state sign of u for a positive input is positive. Therefore the minus sign is required at the summer for the feedback signal, and the zero angle root locus must be plotted. The root locus is seen in Figure 2-50 with the location of the complex roots for $S_u = 1.19$ volt/volt. As these poles move into the right-half plane at some value of S_u, their closed loop location determines the degree of stability of the system. The poles near the origin representing the phugoid move around, break into the real axis, and are the dominant roots. The response of the system to a step command of $u_{(ref)}$ is depicted in Figure 2-51. The dominance of the real pole near the origin is evident, especially in the velocity response. The fact that the velocity response exhibits a pure first-order response characteristic and that the steady state condition is reached in 40 sec seems to be an almost ideal response.

The block diagram for the altitude hold mode is drawn in Figure 2-52. As in the case of the acceleration control system, the short-period approximation can be used for the analysis of the altitude hold mode. Since the jet transport is being used for this analysis, the root locus for the inner loop is the same as the one shown in Figure 2-8, and for $S_{(rg)} = 1.98$ the closed loop transfer function for the inner loop is

$$\frac{\theta(s)}{e_a(s)} = \frac{13.9(s + 0.306)}{(s + 0.8)(s^2 + 10s + 29)} \tag{2-32}$$

The block diagram for the outer loop analysis can now be drawn and is shown in Figure 2-53. The minus sign that appears in the $h(s)/\theta(s)$ transfer function arises from the sign convention used. As γ is taken as positive for an aircraft in a climb, the rate of climb ($\dot{h}$) resulting from a positive γ is taken as positive, and therefore h is considered positive for an aircraft whose altitude has increased. This is also consistent with the

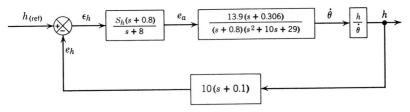

Figure 2-53 Block diagram for the outer loop of the altitude hold mode where

$$\frac{h(s)}{\dot{\theta}(s)} = -\frac{1}{s^3}\frac{a_z(s)}{\theta(s)} = \frac{-0.135(s - 4.89)(s + 4.89)}{s^2(s + 0.306)}.$$

measurement of altitude. Since a_z is taken as positive downward, $h = (-1/s^2)a_z$. The sign at the summer for the feedback signal must be negative for the same reasons as those discussed for the acceleration control system. The zero angle root locus for the altitude hold mode can be seen in Figure 2-54, with the location of the closed loop poles for $S_h = 0.493$.

In Section 2-4 it was observed that it is impossible to control the flight path angle without simultaneous control of the velocity. Therefore, although the short-period equations are used for the analysis, the complete three-degree-of-freedom equations are employed for the simulation combined with the velocity control system that was used for the glide slope control system. The response of the altitude hold mode is shown in Figure 2-55, for a step input of 100 ft for $h_{(ref)}$. This value is equivalent to engaging the altitude hold mode with an error of 100 ft between the

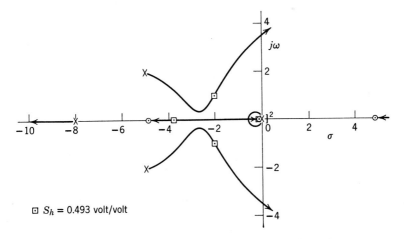

Figure 2-54 Root locus for the outer loop of the altitude hold mode (zero angle root locus).

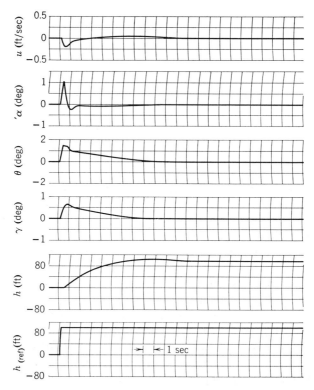

Figure 2-55   Response of the altitude hold mode control system
to a step input of $h_{(ref)}$ for $S_h = 0.5$.

actual and reference altitude.   As can be seen from the computer traces,
the change in altitude is smooth with no violent changes in angle of attack
or pitch attitude.   This discussion completes the analysis of the two modes
of flight path stabilization.

2-6   Vertical Gyro as the Basic Attitude Reference

For some of the autopilot configurations discussed in this chapter it is
necessary to determine the actual pitch attitude of the aircraft.   This can
be accomplished by use of a vertical gyro.   The vertical gyro also provides
an indication of bank angle for use in conjunction with the lateral autopilots to be discussed in Chapter 4.   A vertical gyro is basically a two-
degree-of-freedom gyro constructed so that the angular momentum
vector (direction of the spin axis) is along the local vertical.   The two

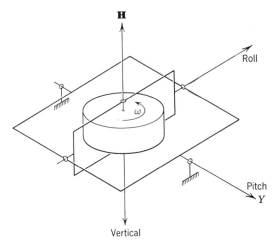

Figure 2-56   Orientation of a vertical gyro.

gimbal axes are then aligned such that the roll axis is horizontal and lies in the *XZ* plane of the aircraft and the pitch axis is parallel to the aircraft's *Y* axis, as illustrated schematically in Figure 2-56.   Due to its gyroscopic action, the spin axis of the gyro tends to remain nonrotating with respect to inertial space or to the so-called "fixed" stars.   Therefore, to maintain the spin axis vertical, a suitable torque must be applied to precess the gyro.   This torque must be applied about the pitch axis to precess the gyro about the roll axis, and about the roll axis for precession about the pitch axis in obedience to the law of the gyro as discussed in Appendix B. To precess the spin axis of the gyro into alignment with the vertical (the direction as indicated by a plum bob whose pivot is stationary with respect to the earth), commonly referred to as "erection," there must be some method for detecting the misalignment and then generating the required torque.   Such a method requires some form of pendulous or acceleration-sensitive device. As the erection schemes rely on pendulous or acceleration-sensitive devices, and as these devices are incapable of distinguishing between accelerations caused by vehicle motion (kinematic accelerations) and the acceleration of gravity, the vertical gyro attempts to align its spin axis in the direction of the total acceleration (vector sum of the kinematic and gravitational accelerations), usually referred to as the "apparent vertical."   If the kinematic accelerations are random, such as atmospheric turbulence, with a zero average, then, due to the slow erection rates (around 2 to 6 deg/min) of the gyro, the average direction of the spin axis will align with the direction of the vertical.   However, during periods of sustained horizontal acceleration (take-off, landing, acceleration from subsonic to supersonic, etc.) or lateral acceleration (sustained turn), the gyro tends to erect to a false vertical.   As the autopilot is not generally engaged during take-off and landing there is no problem encountered here; however, the autopilot may be engaged during accelerations from subsonic to supersonic flight and back again, and the resulting misalignment would cause an error in the pitch attitude.

The behavior of a vertical gyro during a sustained turn is much more complicated than in the case of sustained longitudinal acceleration.   For a coordinated turn the lift vector is normal to the plane of the wings (see Figure 4-13).   The gyro then considers this the direction of the vertical and attempts to align its spin axis with the lift vector.   However, because the aircraft is in a turn, the lift vector is rotating; thus if the spin axis of the gyro is to align itself with the lift vector, the momentum vector (**H**) of the gyro must have an angular velocity which in turn requires a torque to be applied.   This torque can be developed only by a pitch misalignment. For this reason it is desirable to be able to cut out the erecting torque during turns when the aircraft is under control of the autopilot.   One method of accomplishing this is to use mercury switches in the form of levels.   If the gyro is erect, the ball of mercury will lie in the center of the tube and no erection signal will be generated; however if the gyro is misaligned the ball of mercury will move to one end of the tube and close the circuit to an electromagnetic torque generator on the proper axis for erection (this must be done about both axes).   With this type of erection scheme the erection signals can be cut out any time a heading command is fed to the autopilot.   Using this type of erection mechanism the gyro can be made to indicate the vertical to within $\frac{1}{8}°$.

Almost as important as the gyro itself, in determining the ultimate accuracy of a vertical gyro, is the type of pickoff.   There are numerous types of pickoffs, some of the more common being synchros, resolvers, potentiometers, and differential pressure pickoffs.   The ultimate choice depends upon the desired linearity and accuracy of the pickoff signal and the amount of torque, if any, produced by the pickoff or signal generator, as it is sometimes called.

2-7   Gyro Stabilized Platform as the Basic Attitude Reference

Some military aircraft, especially long-range bombers of the B-52 and B-58 type, are equipped with an inertial navigation system, the heart of which is a gyro stabilized platform that is maintained in a level attitude. The ability of either two two-degree-of-freedom or three single-degree-of-freedom gyros to provide the basic gyro stabilization of the platform is

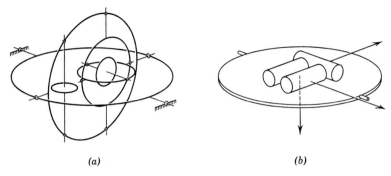

(a) (b)

Figure 2-57 (a) Stable platform utilizing two two-degree-of-freedom gyros; (b) stable platform utilizing three single-degree-of-freedom gyros.

discussed first, followed by a discussion of the technique used to provide the proper signals to the gyros to keep the platform level in the presence of aircraft motion.

The general configuration for the use of the two-degree-of-freedom gyro or the single-degree-of-freedom gyro to stabilize the platform is shown in Figure 2-57. As the single-degree-of-freedom gyro is easier to balance than the two-degree-of-freedom gyro, three single-degree-of-freedom integrating gyros are usually used for platform stabilization. For this application, if any one of the integrating gyros senses an angular velocity about its input axis, an output voltage is generated (the technique used to accomplish this is discussed later in this section) by the gyro which, after being amplified, is used to rotate the gimbals relative to the base so the platform remains fixed, relative to inertial space. One complication that arises is that the platform axis, as indicated by the input axes of the three gyros, may not be aligned with the gimbal axes at all times; thus, the signals from the gyros must be sent through resolvers so that the proper signal is sent to the proper gimbal axis. However, to study the operation and analyze the performance of the stable platform, it is necessary to consider only one axis. Thus, the rest of this discussion deals with an analysis of a single-axis stabilized platform utilizing an integrating gyro with the associated amplifier and servo drive.

A single axis stabilized platform has been referred to by personnel of the Instrumentation Laboratory, MIT, as a "single-axis space integrator."[3] The reason for this becomes more obvious later; however, for brevity, the term space integrator will be used when referring to the single-axis stabilized platform. Figure 2-58 illustrates the components of a single-axis space integrator. The integrating gyro is oriented so that its input axis is parallel to the axis of rotation of the table. If the base starts

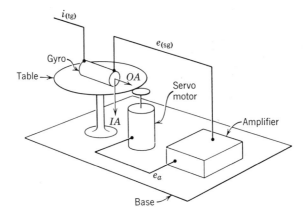

Figure 2-58   Single-axis space integrator.

to rotate, the table is carried along with it.   This causes the gyro to be rotated about its input axis generating a gimbal angle which is sensed by the signal generator.   The voltage from the signal generator $e_{(sg)}$, is then amplified and applied to the servo motor to drive the table in the opposite direction to the base.   The net result is that the table remains non-rotating in inertial space.   The table can be rotated at a desired rate by feeding a current, $i_{(tg)}$, to the torque generator.   The operation of the space integrator is explained in detail in the following paragraphs.

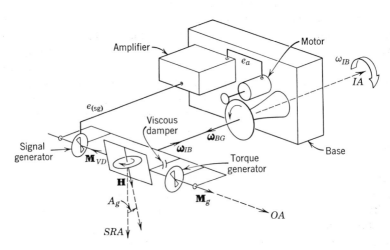

Figure 2-59   Single-axis space integrator, base motion isolation mode.

Figure 2-59 illustrates the sequence of events that results when the base starts to rotate with respect to inertial space, and the platform, represented by the gyro, remains nonrotating with respect to inertial space. This action is referred to as "base motion isolation," that is, the platform is isolated from the motion of the base about the axis indicated. As the base starts to rotate about IA with respect to inertial space, $\boldsymbol{\omega}_{IB}$, the gyro is rotated at the same angular velocity since the motor and gear train allow no motion of the gyro relative to the base as long as the voltage to the motor, e_a, is zero. The rotation of the gyro about its IA results in a gyroscopic output torque, $\mathbf{M}_g$, which is absorbed by the viscous damping torque, $\mathbf{M}_{vd}$. $\mathbf{M}_{vd}$ can result only from a rate of change of the gimbal angle, $\dot{A}_g$; thus, a gimbal angle, A_g, is developed and increases with time. As the gimbal angle increases, the voltage out of the signal generator, $e_{(sg)}$, which is proportional to A_g increases from a null, is amplified, and applied to the servo motor causing the gyro to be rotated relative to the base, $\boldsymbol{\omega}_{BG}$. For proper operation $\boldsymbol{\omega}_{BG}$ must be opposite to $\boldsymbol{\omega}_{IB}$; the resulting angular velocity of the gyro about its input axis is $\boldsymbol{\omega}_{IB} + \boldsymbol{\omega}_{BG}$. As A_g increases, $\boldsymbol{\omega}_{BG}$ increases and $\boldsymbol{\omega}_{IA}$ decreases causing $\dot{A}_g$ to decrease until the steady-state condition is reached, in which $\boldsymbol{\omega}_{IB} = -\boldsymbol{\omega}_{BG}$ and $\boldsymbol{\omega}_{IA}$ goes to zero. With $\boldsymbol{\omega}_{IA} = 0$, $\dot{A}_g = 0$ and A_g remains constant. The steady-state value of A_g is dependent upon the amplifier and servo motor gain. The higher the gain, the smaller the gimbal angle for a given value of $\boldsymbol{\omega}_{IB}$. During the transient period while $\dot{A}_g \neq 0$ there is some net rotation of the gyro about its input axis which results in an angular orientation error in the platform; however, when $\boldsymbol{\omega}_{IB}$ goes to zero, the gimbal angle must return to zero so that $\boldsymbol{\omega}_{BG}$ will be zero. This necessitates an $\boldsymbol{\omega}_{IA}$ in the same direction as $\boldsymbol{\omega}_{BG}$, which eliminates the angular orientation error that was developed during the presence of $\boldsymbol{\omega}_{IB}$. Thus, for an ideal gyro, the platform has no net angular motion with respect to inertial space.

So far it has been assumed that it is desirable to maintain the platform nonrotating with respect to inertial space; however, there are times when it is desirable to command the platform to rotate relative to inertial space. If an earth reference is desired, the platform should be nonrotating relative to the earth or if the platform is to remain normal to the gravity vector while carried in a vehicle (vertical indicator, to be discussed later in this section), the platform must be rotated relative to the earth. Both of these requirements dictate that the platform rotate in some manner relative to inertial space.

Figure 2-60 illustrates how this is accomplished. If it is desired to cause the platform, and thus the gyro, to rotate about its input axis, a command current is fed to the torque generator which develops a torque

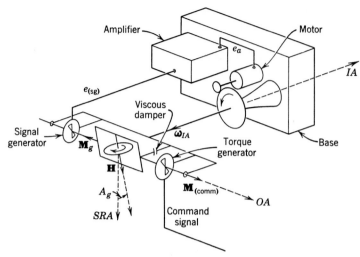

Figure 2-60   Single-axis space integrator, commanded angular velocity input mode.

about the OA, $\mathbf{M}_{(comm)}$.   From here on the sequence of events is the same as those described for the base motion isolation mode of operation. However, in this case the steady-state condition is reached when the gyroscopic output torque $\mathbf{M}_g$ developed by the $\boldsymbol{\omega}_{IA}$ of the gyro is equal and opposite to the $\mathbf{M}_{(comm)}$ developed by the current into the torque generator. Through this technique the platform can be commanded to rotate at any desired angular rate relative to inertial space.   Although the two modes of operation have been discussed individually, in practice they can occur simultaneously.

Without a signal to the torque generator the space integrator remains nonrotating with respect to inertial space except for gyro drift, which causes a small rotation of the platform.   If it is desired that the platform remain nonrotating with respect to the earth and thus remain level (normal to the gravity vector), the platform must be rotated at an angular velocity, with respect to inertial space, equal to the horizontal component of earth's rate (the total average angular velocity of the earth with respect to the sun is usually used for this value, which is $0.07292115 \times 10^{-3}$ rad/sec) for the latitude at which the gyro is located.   In addition, if the platform is then mounted in an aircraft and the aircraft is in motion, for the platform to remain level it must be rotated at an angular velocity, with respect to the earth, equal to the velocity of the aircraft with respect to the earth (ground speed) divided by the radius of the earth.   A signal therefore must be generated and fed as a torquing signal to the integrating gyro to rotate the platform at the required rate.

The portion of the signal required to compensate for earth's rate is computed based on the latitude of the aircraft as indicated by the navigation system. The remainder of the signal is usually supplied from the output of integrating accelerometers corrected for the coriolis acceleration resulting from the earth's angular velocity.

For this operation, two integrating accelerometers are used, with their sensitive axis aligned with the two horizontal platform axes. The platform axes are usually pointed North (X axis) and East or West (Y axis) with the Z axis along the vertical so as to form a right-handed system. The X axis integrating accelerometer provides the torquing signal for the Y axis gyro and vice versa.

By properly adjusting the dynamics of each loop of the vertical indicator (Schuler tuning, which results in an undamped period of approximately 84 min) it can be made to track the vertical very accurately, even under the influence of sustained horizontal accelerations. Such a system is usually at least an order of magnitude more accurate than the vertical gyro for indication of the vertical.

2-8 Effects of Nonlinearities

In all the analyses, thus far presented in this chapter, no mention has been made of the nonlinearities that are usually present in any physical system. These nonlinearities include such things as amplifier saturation, backlash in gear trains or mechanical linkage, control cable stretch, hysteresis in electromagnetic devices, and dead space in actuators and sensors. If any of these nonlinearities are severe enough, the system may be completely unstable. Thus in the physical design of the hardware all nonlinearities should be kept to a minimum. The analytic analysis of a complete autopilot including nonlinearities is usually impractical, if not impossible. Describing function techniques can be used to determine whether or not a limit cycle will exist, but this is about all. However, if there are known or suspected nonlinearities in the system, they can be included in the computer simulation and thus their effects determined during early design phases. Wherever possible it is always advisable to include any physical hardware available in the analog computer simulation. If the simulation indicates that the existing nonlinearities will make the performance of the system unacceptable, the equipment must be redesigned to reduce the nonlinearities to an acceptable level. In some cases a lead compensator can be used to offset some of the effects of backlash.

2-9 Summary

In this chapter a technique for improving the damping of the short-period oscillation was discussed. This pitch damper was used as the inner loop for some typical basic autopilot configurations. This discussion was followed by the analysis of an automatic glide slope and flare control system and the altitude and Mach hold modes of a complete longitudinal flight control system. The chapter concluded with a discussion of the use of a vertical gyro or stabilized platform as a vertical reference.

References

1. J. Bicknell, E. E. Larrabee, R. C. Seamans, Jr., and H. P. Whitaker, "Automatic Control of Aircraft," presented at Journees Internationales de Sciences Aeronautiques, *J.I.S.A.*, 57, Paris, May 27–29, 1957.
2. *Automatic Pilot for High Performance Aircraft*, Project MX–1137, Final Report, TR–54A031, General Electric Co.
3. C. S. Draper, W. Wrigley, and L. R. Grohe, "The Floating Integrating Gyro and Its Application to Geometrical Stabilization Problems on Moving Bases," presented at the 23rd Annual Meeting of the Institute of Aeronautical Sciences, January 1955 (published as *Sherman M. Fairchild Fund Paper*, FF–13).

Lateral Dynamics

3-1 Lateral Equations of Motion

In Chapter 1 the six-degree-of-freedom rigid body equations were separated into three longitudinal and three lateral equations. In this chapter the three lateral equations are linearized and combined with the aerodynamic terms to yield the lateral equations of motion for the rigid aircraft. From these equations, the lateral modes can be determined and the various transfer functions for both rudder and aileron input derived and analyzed. The chapter concludes with a study of the effects of changes in airspeed and altitude on the lateral dynamics in addition to the study of the effects of variations of the stability derivatives.

Before deriving the lateral equations of motion, it is necessary to define the sideslip angle. As in the longitudinal case, stability axes are used. If during the perturbations from equilibrium the aircraft X axis is displaced about the Z axis from the aircraft velocity vector, a sideslip angle will be generated. The positive direction of this angle is indicated in Figure 3-1. The sideslip angle β should not be confused with the yaw angle ψ. An aircraft can have a yaw angle and a yaw rate with zero sideslip. However, if there is an angle of sideslip there must be a yaw angle. In general ψ and β are not equal, but if the flight path of the aircraft is straight and the aircraft is slipping with its wings level, then $\beta = -\psi$.

As mentioned in Section 1-5, there are certain assumptions which, if employed, permit the decoupling of the six equations of motion. The three lateral equations can be obtained from Eq. 1-33 and are

$$\sum \Delta F_y = m(\dot{V} + UR - WP)$$

$$\sum \Delta \mathscr{L} = \dot{P}I_x - \dot{R}J_{xz} + QR(I_z - I_y) - PQJ_{xz}$$

$$\sum \Delta \mathscr{N} = \dot{R}I_z - \dot{P}J_{xz} + PQ(I_y - I_x) + QRJ_{xz} \qquad (3\text{-}1)$$

The pitching moment equation that is neglected is also rewritten here

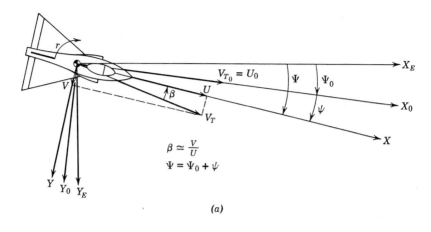

$$\beta \simeq \frac{V}{U}$$

$$\Psi = \Psi_0 + \psi$$

(a)

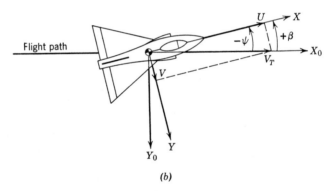

(b)

Figure 3-1   (a) Disturbed aircraft stability axes showing yaw angle ψ and sideslip angle β; (b) aircraft slipping with wings level, $\beta = -\psi$.

from Eq. 1-32: $\sum \Delta \mathcal{M} = \dot{Q}I_y + PR(I_x - I_z) + (P^2 - R^2)J_{xz}$.   This equation contains both P and R but if the perturbations are small, and it is assumed that P and R are so small that their products and squares can be neglected, the equations can be decoupled.   This assumption was made in Section 1-5 and it applies here.   As the equations have been decoupled, Q is assumed zero.   By requiring that the equilibrium direction of the X axis be along the flight path and that there be no sideslip while in equilibrium, then $V_0 = W = 0$ and $U = U_0 + u$.   Then $\dot{V} = \dot{v}$, $\dot{W} = 0$, $\dot{U} = \dot{u}$.   As the aircraft initially is in unaccelerated flight, P_0 and R_0 must be zero. Then $P = p$ and $R = r$. Making these substitutions, Eq. 3-1 becomes

$$\sum \Delta F_y = m(\dot{v} + U_0 r + ur)$$

$$\sum \Delta \mathscr{L} = \dot{p} I_x - \dot{r} J_{xz}$$

$$\sum \Delta \mathscr{N} = \dot{r} I_z - \dot{p} J_{xz} \qquad (3\text{-}2)$$

However, since the perturbations are assumed small, the products of the perturbations can be neglected. Equation 3-2 then reduces to

$$\sum \Delta F_y = m(\dot{v} + U_0 r)$$

$$\sum \Delta \mathscr{L} = \dot{p} I_x - \dot{r} J_{xz}$$

$$\sum \Delta \mathscr{N} = \dot{r} I_z - \dot{p} J_{xz} \qquad (3\text{-}3)$$

Factoring U_0 from the right-hand side of the $\sum \Delta F_y$ equation, the equation becomes $\sum \Delta F_y = m U_0(\dot{v}/U_0 + r)$ but $\dot{v}/U_0 \simeq \beta$, as $U_0 \simeq U$ for small perturbations. After substituting for p and r, using the above relation, Eq. 3-3 becomes

$$\sum \Delta F_y = m U_0(\dot{\beta} + \dot{\psi})$$

$$\sum \Delta \mathscr{L} = \ddot{\phi} I_x - \ddot{\psi} J_{xz}$$

$$\sum \Delta \mathscr{N} = \ddot{\psi} I_z - \ddot{\phi} J_{xz} \qquad (3\text{-}4)$$

It is necessary to expand the forces and moments in terms of the changes in them resulting from the perturbations in the linear and angular velocities. The forces in the Y direction are functions of β, ψ, ϕ, $\dot{\phi}$, and $\dot{\psi}$. Then $\sum F_y$ can be written as the total derivative

$$\sum dF_y = \frac{\partial F_y}{\partial \beta} d\beta + \frac{\partial F_y}{\partial \psi} d\psi + \frac{\partial F_y}{\partial \phi} d\phi + \frac{\partial F_y}{\partial \dot{\phi}} d\dot{\phi} + \frac{\partial F_y}{\partial \dot{\psi}} d\dot{\psi} \qquad (3\text{-}5)$$

If the partial derivatives are assumed linear over the range of the perturbations, the differentials can be replaced by actual increments. Thus Eq. 3-5 becomes

$$\sum \Delta F_y = \frac{\partial F_y}{\partial \beta} \Delta\beta + \frac{\partial F_y}{\partial \psi} \Delta\psi + \frac{\partial F_y}{\partial \phi} \Delta\phi + \frac{\partial F_y}{\partial \dot{\phi}} \Delta\dot{\phi} + \frac{\partial F_y}{\partial \dot{\psi}} \Delta\dot{\psi} \qquad (3\text{-}6)$$

Two of these partial derivatives result from the change in the component of gravity along the Y axis as the aircraft attitude is changed. From

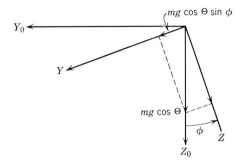

Figure 3-2   Component of gravity along the disturbed Y axis due to ϕ.

Eq. 1-35 the components of gravity along the equilibrium X and Z axes
with $\Phi = 0$ are

$$F_{x_g} = -mg \sin \Theta$$
$$F_{z_g} = \quad mg \cos \Theta \tag{3-7}$$

If the aircraft is disturbed by rolling about the X axis, there is a component
of gravity along the Y axis as shown in Figure 3-2.   From Figure 3-2,

$$F_{y_g} = mg \cos \Theta \sin \phi \tag{3-8}$$

and

$$\frac{\partial F_{y_g}}{\partial \phi} = mg \cos \Theta \cos \phi \tag{3-9}$$

Similarly, if there is a perturbation about the Z axis, there is a component
of gravity along the Y axis, as illustrated in Figure 3-3.   From Figure 3-3,

$$F_{y_g} = mg \sin \Theta \sin \psi \tag{3-10}$$

and

$$\frac{\partial F_{y_g}}{\partial \psi} = mg \sin \Theta \cos \psi \tag{3-11}$$

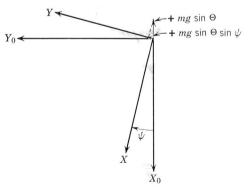

Figure 3-3   Component of gravity along the disturbed Y axis due to ψ.

There will be no other forces in the Y direction due to ψ and ϕ. If either of these angles results in a sideslip angle, it will be included in $\partial F_y/\partial\beta$. Thus Eqs. 3-9 and 3-11 can be rewritten using the small angle assumption for the cos ϕ and cos ψ.

$$\frac{\partial F_y}{\partial\psi} = mg \sin \Theta$$

$$\frac{\partial F_y}{\partial\phi} = mg \cos \Theta \tag{3-12}$$

Since, in Eq. 3-6, the initial values of β, ψ, and ϕ are zero, the Δ's may be dropped. Dropping the Δ's and substituting Eq. 3-6 into the $\sum\Delta F_y$ equation of Eq. 3-4 yields

$$mU_0\dot{\beta} + \frac{-\partial F_y}{\partial\beta}\beta + \left(mU_0 - \frac{\partial F_y}{\partial\dot\psi}\right)\dot\psi - \frac{\partial F_y}{\partial\psi}\psi - \frac{\partial F_y}{\partial\dot\phi}\dot\phi - \frac{\partial F_y}{\partial\phi}\phi = F_{y_a} \tag{3-13}$$

where F_{y_a} is an aerodynamic force of unspecified origin, and is explained in Section 3-2. As in the derivation of the longitudinal equations $U_0 \simeq U$, the sub zero is dropped (see pp. 17 and 21). Dividing Eq. 3-13 by Sq, substituting from Eq. 3-12, and using the definitions of the stability derivatives in Table 3-1, Eq. 3-13 becomes

$$\frac{mU}{Sq}\dot\beta - C_{y_\beta}\beta + \left(\frac{mU}{Sq} - \frac{b}{2U}C_{y_r}\right)\dot\psi - C_{y_\psi}\psi - \frac{b}{2U}C_{y_p}\dot\phi - C_{y_\phi}\phi$$

$$= \frac{F_{y_a}}{Sq} = C_{y_a} \tag{3-14}$$

The $\mathscr{L}$ and $\mathscr{N}$ equations can be expanded in like manner; however, the partial derivatives of $\mathscr{L}$ and $\mathscr{N}$ with respect to ϕ and ψ are zero. Thus expanding $\sum\Delta\mathscr{L}$ and $\sum\Delta\mathscr{N}$ and substituting the results into Eq. 3-4 yields

$$\ddot\phi I_x - \frac{\partial\mathscr{L}}{\partial\dot\phi}\dot\phi - \ddot\psi J_{xz} - \frac{\partial\mathscr{L}}{\partial\dot\psi}\dot\psi - \frac{\partial\mathscr{L}}{\partial\beta}\beta = \mathscr{L}_a$$

$$-\ddot\phi J_{xz} - \frac{\partial\mathscr{N}}{\partial\dot\phi}\dot\phi + \ddot\psi I_z - \frac{\partial\mathscr{N}}{\partial\dot\psi}\dot\psi - \frac{\partial\mathscr{N}}{\partial\beta}\beta = \mathscr{N}_a \tag{3-15}$$

Dividing through by Sqb where b is the wing span and going to coefficient form Eq. 3-15 becomes

$$\frac{I_x}{Sqb}\ddot\phi - \frac{b}{2U}C_{l_p}\dot\phi - \frac{J_{xz}}{Sqb}\ddot\psi - \frac{b}{2U}C_{l_r}\dot\psi - C_{l_\beta}\beta = \frac{\mathscr{L}_a}{Sqb} = C_{l_a}$$

$$-\frac{J_{xz}}{Sqb}\ddot\phi - \frac{b}{2U}C_{n_p}\dot\phi + \frac{I_z}{Sqb}\ddot\psi - \frac{b}{2U}C_{n_r}\dot\psi - C_{n_\beta}\beta = \frac{\mathscr{N}_a}{Sqb} = C_{n_a} \tag{3-16}$$

Table 3-1   Definitions and Equations for Lateral Stability Derivatives

Symbol	Definition	Origin	Equation	Typical Values
C_{l_β}	$\dfrac{1}{Sqb}\dfrac{\partial \mathscr{L}}{\partial \beta}$	Dihedral and vertical tail	Ref. 1, Chapter 9 Ref. 2, Section 3.10	-0.06
C_{l_p}	$\dfrac{1}{Sqb}\left(\dfrac{2U}{b}\right)\dfrac{\partial \mathscr{L}}{\partial p}$	Wing damping	Ref. 1, Chapter 9	-0.4
C_{l_r}	$\dfrac{1}{Sqb}\left(\dfrac{2U}{b}\right)\dfrac{\partial \mathscr{L}}{\partial r}$	Differential wing normal force	$\dfrac{C_L{}^w}{4}$	0.06
C_{n_β}	$\dfrac{1}{Sqb}\dfrac{\partial \mathscr{N}}{\partial \beta}$	Directional stability	Ref. 1, Chapter 8 Ref. 2, Section 3.9	0.11
C_{n_p}	$\dfrac{1}{Sqb}\left(\dfrac{2U}{b}\right)\dfrac{\partial \mathscr{N}}{\partial p}$	Differential wing chord force	$-\dfrac{C_L{}^w}{8}\left(1-\dfrac{d\epsilon}{d\alpha}\right)$	-0.015
C_{n_r}	$\dfrac{1}{Sqb}\left(\dfrac{2U}{b}\right)\dfrac{\partial \mathscr{N}}{\partial r}$	Damping in yaw	$-\dfrac{C_D{}^w}{4}-2\eta_v\dfrac{S_v}{S}\left(\dfrac{l_v}{b}\right)^2\left(\dfrac{dC_L}{d\alpha}\right)^v$	-0.12
C_{y_β}	$\dfrac{1}{Sq}\dfrac{\partial F_y}{\partial \beta}$	Fuselage and vertical tail	No simple equation	-0.6
C_{y_ϕ}	$\dfrac{1}{Sq}\dfrac{\partial F_y}{\partial \phi}$	Gravity	$\dfrac{mg}{Sq}\cos\Theta$	
C_{y_p}	$\dfrac{1}{Sq}\left(\dfrac{2U}{b}\right)\dfrac{\partial F_y}{\partial p}$	Vertical tail	Neglect	
C_{y_ψ}	$\dfrac{1}{Sq}\left(\dfrac{\partial F_y}{\partial \psi}\right)$	Gravity	$\dfrac{mg}{Sq}\sin\Theta$	
C_{y_r}	$\dfrac{1}{Sq}\left(\dfrac{2U}{b}\right)\dfrac{\partial F_y}{\partial r}$	Vertical tail	Neglect	

It should be noted that the same assumptions used in the derivation of the longitudinal equations apply to the derivation of the lateral equations. Equations 3-14 and 3-15 are rewritten here for convenience.

$$-\frac{b}{2U}C_{y_p}\dot{\phi}-C_{y_\phi}\phi+\left(\frac{mU}{Sq}-\frac{b}{2U}C_{y_r}\right)\dot{\psi}-C_{y_\psi}\psi+\frac{mU}{Sq}\beta-C_{y_\beta}\beta=C_{y_a}$$

$$\frac{I_x}{Sqb}\ddot{\phi}-\frac{b}{2U}C_{l_p}\dot{\phi}-\frac{J_{xz}}{Sqb}\ddot{\psi}-\frac{b}{2U}C_{l_r}\dot{\psi}-C_{l_\beta}\beta=C_{l_a}$$

$$-\frac{J_{xz}}{Sqb}\ddot{\phi}-\frac{b}{2U}C_{n_p}\dot{\phi}+\frac{I_z}{Sqb}\ddot{\psi}-\frac{b}{2U}C_{n_r}\dot{\psi}-C_{n_\beta}\beta=C_{n_a}$$

$$(3\text{-}17)$$

These equations are the uncoupled, linearized, lateral equations of motion. As in the case of the longitudinal equations, they are non-dimensional.

3-2   Derivation of Equations for the Lateral Stability Derivatives

The explanation for the indication in Table 3-1 that C_{y_p} and C_{y_r} can be neglected is that C_{y_p} results from the side force on the vertical tail caused by a rolling velocity.   This rolling velocity causes lift to be produced by the vertical tail causing a force in the Y direction; however, this force is small since the rolling velocity must be small in order to decouple the equations.   Similarly, C_{y_r} results from the lift generated by the vertical tail due to a yawing velocity. C_{y_r} is, in general, larger than C_{y_p}, and the magnitudes of both should be checked before neglecting them.   The evaluation of the rest of the stability derivatives follows:

C_{l_β}, the change in the rolling moment due to sideslip, results from the lift produced by the vertical tail due to the sideslip and the effective dihedral of the wing.   There is no simple relation for this stability derivative, mainly because of the contribution of the dihedral effect of the wing. The effect of this dihedral is to produce a rolling moment to the left if the aircraft is sideslipping to the right and the aircraft has positive dihedral. This point is discussed in detail in Chapter 9 of Ref. 1, where it is referred to as C_{l_v}, and in Section 3.10 of Ref. 2.

C_{n_β}, the change in the yawing moment due to sideslip, results from the yawing moment produced by the vertical tail, wing, fuselage, and nacelles in the presence of sideslip.   There is also some effect due to the sidewash created by the wing-fuselage combination.

The wing contribution is very small and is greater for swept wings and approaching zero for zero sweep of the quarter chord points.   The fuselage and nacelles usually have a destabilizing effect which must be overcome by the vertical tail.   The combination of these various effects is covered in detail in Chapter 8 of Ref. 1, and in Section 3-9 of Ref. 2.   In general, C_{n_β} can best be obtained from actual wind tunnel test.   This stability derivative determines the static directional stability of the aircraft, and it must be positive for a stable aircraft.   In Ref. 1, C_{n_β} is referred to as C_{n_ψ} and as $\psi = -\beta$ for pure sideslip, C_{n_β} is positive for an aircraft with directional stability.

C_{y_β}, the change in the force in the Y direction due to sideslip, is the force that, if multiplied by the proper moment arm, generates C_{n_β}, and as in the case of C_{n_β} there is no simple relation.

C_{l_p}, the change in the rolling moment due to a rolling velocity, arises from the change in the angle of attack on the wings caused by a rolling velocity.   The down-going wing experiences an increase in angle of attack while the up-going wing is subjected to a decrease in angle of attack. These changes in angle of attack cause changes in the lift and drag of the up- and down-going wings.   The amounts of the changes in the angle of attack in relation to the rolling velocity are shown in Figure 3-4.

If the new lift and drag vectors are resolved into the equilibrium X and Z axes, it can be seen that the Z components produce a moment opposing

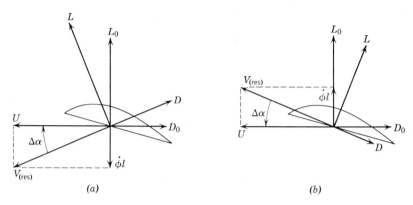

Figure 3-4   Change in lift and drag on a section of the down-going wing (a) and the up-going wing (b) at a distance l from the center of gravity due to $\dot{\phi}$.

the rolling velocity.   Normally the Z component of drag is neglected and only the change in lift considered; however, no simple expression can be obtained for C_{l_p}.   This subject is discussed in Chapter 9 of Ref. 1 and also in NACA TR 635.

C_{n_p} is the change in the yawing moment due to a rolling velocity.   Its cause is the same as that for C_{l_p}.   As can be seen from Figure 3-4, the lift vector is tilted forward on the down-going wing and rearward on the up-going wing.   The components of these lift vectors when resolved into the X axis are greater than the change in the X component of drag, thus producing a negative yawing moment for a positive roll rate.   This is referred to as "adverse yaw," and is the reason rudder is required to keep an aircraft coordinated while rolling into or out of a steady-state turn. For supersonic flight (wing leading edge supersonic) the lift vector remains normal to the chord; therefore the difference in the drag produces a positive C_{n_p}.   As seen from Table 3-1, assuming an elliptical lift distribution,

$$C_{n_p} = -\frac{C_L{}^w}{8}\left(1 - \frac{d\epsilon}{d\alpha}\right) \qquad (3\text{-}18)$$

where $d\epsilon/d\alpha$ is the rate of change of the downwash angle with angle of attack, and can be approximated by Eq. 3-19 (see NACA TR 648).

$$\frac{d\epsilon}{d\alpha} = \frac{2}{\pi e \textit{R}} \frac{dC_L{}^w}{d\alpha} \qquad (3\text{-}19)$$

where e is an efficiency factor varying from 0.8 to 0.9, and $\textit{R}$ is the aspect ratio of the wing (see Appendix D).

C_{l_r}, the change in the rolling moment due to a yawing velocity, arises from the changes in the lift on the wings resulting from a yawing velocity. If the aircraft is subject to a yawing velocity, the relative velocity of the left and right wing panels changes, with respect to the air mass. The forward-going wing experiences an increase in lift while the lift on the rearward-going wing decreases. This factor causes a positive rolling moment for a positive yawing velocity. By strip integration over the wing, and assuming an elliptical lift distribution,

$$C_{l_r} = \frac{C_L^{\,w}}{4} \tag{3-20}$$

Both C_{n_p} and C_{l_r} are extremely difficult to obtain from wind tunnel test and are usually evaluated analytically.

C_{n_r}, the change in the yawing moment due to a yawing velocity, is caused partially by the same phenomenon as that for C_{l_r} and partly by the lift generated by the vertical stabilizer. This moment opposes the yawing velocity, is therefore negative, and is the damping in yaw. The contribution from the vertical tail can be derived as follows.

The change in the angle of attack of the vertical tail due to a yawing velocity r is $\Delta\alpha_v = rl_v/U$, where l_v is the distance from the center of gravity to the aerodynamic center of the vertical tail. The lift generated by this $\Delta\alpha_v$ is

$$L_v = \left(\frac{rl_v}{U}\right) qS_v \left(\frac{dC_L}{d\alpha}\right)^v \eta_v \tag{3-21}$$

where η_v is the vertical stabilizer efficiency factor to compensate for the interference between the vertical tail and the fuselage. The yawing moment caused by this lift is

$$\mathscr{N}_v = -l_v L_v = -\left(\frac{rl_v^{\,2}}{U}\right) qS_v \left(\frac{dC_L}{d\alpha}\right)^v \eta_v \tag{3-22}$$

Differentiating with respect to r, Eq. 3-22 becomes

$$\left(\frac{\partial\mathscr{N}}{\partial r}\right)^v = -\frac{l_v^{\,2}}{U} qS_v \left(\frac{dC_L}{d\alpha}\right)^v \eta_v \tag{3-23}$$

But

$$(C_{n_r})^v = \frac{1}{Sqb}\left(\frac{2U}{b}\right)\left(\frac{\partial\mathscr{N}}{\partial r}\right)^v = -2\frac{S_v}{S}\left(\frac{l_v}{b}\right)^2\left(\frac{dC_L}{d\alpha}\right)^v \eta_v \tag{3-24}$$

The contribution of the wing to C_{n_r} is usually taken as $-C_D^{\,w}/4$. Therefore

$$C_{n_r} = -\frac{C_D^{\,w}}{4} - 2\frac{S_v}{S}\left(\frac{l_v}{b}\right)^2\left(\frac{dC_L}{d\alpha}\right)^v \eta_v \tag{3-25}$$

The external applied aerodynamic forces and moments arise from aileron or rudder deflection.   Thus

$$C_{y_a} = C_{y_{\delta_a}}\delta_a + C_{y_{\delta_r}}\delta_r$$

$$C_{l_a} = C_{l_{\delta_a}}\delta_a + C_{l_{\delta_r}}\delta_r$$

$$C_{n_a} = C_{n_{\delta_a}}\delta_a + C_{n_{\delta_r}}\delta_r \tag{3-26}$$

3-3 Solution of Lateral Equations (Stick Fixed)

As in the case of the longitudinal equations, the homogeneous equation is obtained first.   Taking the Laplace transform of Eq. 3-17 and neglecting C_{y_p} and C_{y_r}, Eq. 3-27 is obtained.

$$\left(\frac{I_x}{Sqb}s^2 - \frac{b}{2U}C_{l_p}s\right)\phi(s) + \left(-\frac{J_{xz}}{Sqb}s^2 - \frac{b}{2U}C_{l_r}s\right)\psi(s) - C_{l_\beta}\beta(s) = 0$$

$$\left(-\frac{J_{xz}}{Sqb}s^2 - \frac{b}{2U}C_{n_p}s\right)\phi(s) + \left(\frac{I_z}{Sqb}s^2 - \frac{b}{2U}C_{n_r}s\right)\psi(s) - C_{n_\beta}\beta(s) = 0$$

$$- C_{y_\phi}\phi(s) + \left(\frac{mU}{Sq}s - C_{y_\psi}\right)\psi(s) + \left(\frac{mU}{Sq}s - C_{y_\beta}\right)\beta(s) = 0$$

$$\tag{3-27}$$

For this example, the stability derivatives for a jet transport flying straight and level at 300 mph at sea level are used.   For this aircraft the values are as follows:

Θ	$= 0$
m	$= 5900$ slugs
U	$= 440$ ft/sec
S	$= 2400$ sq ft
I_x	$= 1.955 \times 10^6$ slug ft^2
I_z	$= 4.2 \times 10^6$ slug ft^2
J_{xz}	$= 0$ by assumption
C_{l_p}	$= -0.38$
C_{l_r}	$= \dfrac{C_L}{4} = \dfrac{0.344}{4} = 0.086$
b	$= 130$ ft
C_{n_p}	$= -0.0228$
C_{n_β}	$= 0.096$
C_{n_r}	$= -0.107$

$C_{y_\beta} \quad \simeq -0.6$

$C_{y_\phi} \quad = \dfrac{mg}{Sq} = C_L = 0.344$

$C_{y_\psi} \quad = 0$

$C_{l_\beta} \quad = -0.057$

$\dfrac{b}{2U} \quad = \dfrac{130}{2(440)} = 0.148 \text{ sec}$

$\dfrac{b}{2U} C_{l_p} = -0.0553 \text{ sec}$

$\dfrac{b}{2U} C_{l_r} = 0.0128 \text{ sec}$

$\dfrac{b}{2U} C_{n_p} = -0.00338 \text{ sec}$

$\dfrac{b}{2U} C_{n_r} = -0.0158 \text{ sec}$

$q \quad = \dfrac{\rho}{2} U^2 = \dfrac{(0.002378)(440)^2}{2} = 230 \text{ lb/sq ft}$

$\dfrac{I_x}{Sqb} \quad = \dfrac{1.955 \times 10^6}{(2400)(230)(130)} = 0.02725 \text{ sec}^2$

$\dfrac{I_z}{Sqb} \quad = \dfrac{4.2 \times 10^6}{(2400)(230)(130)} = 0.0585 \text{ sec}^2$

$\dfrac{mU}{Sq} \quad = \dfrac{(5900)(440)}{(2400)(230)} = 4.71 \text{ sec}$

Substituting these values into Eq. 3-27 and writing the equations in determinant form yields

$$\begin{vmatrix} 0.02725s^2 + 0.0553s & -0.0128s & 0.057 \\ 0.00338s & 0.058s^2 + 0.0158s & -0.096 \\ -0.344 & 4.71s & 4.71s + 0.6 \end{vmatrix} = 0$$

$$(3\text{-}28)$$

Expanding,

$$0.00748s^5 + 0.01827s^4 + 0.01876s^3 + 0.0275s^2 - 0.0001135s = 0$$

$$(3\text{-}29)$$

The fact that one root of the equation is zero indicates that the aircraft is insensitive to heading. This means that once disturbed there is no moment trying to return the aircraft to its original heading. Dividing Eq. 3-29 by 0.00748 yields

$$s^5 + 2.44s^4 + 2.51s^3 + 3.68s^2 - 0.0152s = 0 \tag{3-30}$$

Factoring,

$$s(s^2 + 0.380s + 1.813)(s + 2.09)(s - 0.004) = 0 \tag{3-31}$$

The modes are identified as follows: $s^2 + 0.380s + 1.813$ is the Dutch roll and in this case $\zeta_D = 0.14$, $\omega_{nD} = 1.345$ rad/sec, $s + 2.09$ is the roll subsidence, and $s - 0.004$ is the spiral divergence. These modes are typical of present-day aircraft. The damping and natural frequency of the Dutch roll vary with aircraft and flight conditions and may become objectionable due to light damping. It is also possible for the natural frequency to become fast enough, and combined with light damping to require artificial damping of the Dutch roll mode. The roll subsidence is the rolling response of the aircraft to an aileron input. The spiral divergence is not usually objectionable since the time constant is so large that the pilot has little trouble controlling it. The resulting motion consists of a combination of an increase in yaw angle and roll angle, and the aircraft eventually falls into a high speed spiral dive if uncorrected. The condition for spiral stability and its effect on the Dutch roll mode is discussed in Section 3-8.

3-4   Lateral Transfer Function for Rudder Displacement

Before evaluating the transfer function it is necessary to define positive rudder deflection. Left rudder, which produces a force in the positive Y direction and a negative yawing moment, is defined as "positive rudder." Then let $C_{y_{\delta_r}} = 0.171$, $C_{n_{\delta_r}} = -0.08$, $C_{l_{\delta_r}} = 0.0131$, and $\delta_a = 0$. Taking the Laplace transform of Eq. 3-17 with the initial conditions equal to zero, and substituting the proper values yields

$$\begin{cases} (0.02725s^2 + 0.0553s)\phi(s) - (0.0128s)\psi(s) + 0.057\beta(s) = 0.0131\delta_r(s) \\ (0.00338s)\phi(s) + (0.0585s^2 + 0.0158s)\psi(s) - 0.096\beta(s) = -0.08\delta_r(s) \\ \quad -0.344\phi(s) + (4.71s)\psi(s) + (4.71s + 0.6)\beta(s) = 0.171\delta_r(s) \end{cases} \tag{3-32}$$

Then the transfer function for δ_r input and ϕ output in determinant form can be written

$$\frac{\phi(s)}{\delta_r(s)} = \frac{\begin{vmatrix} 0.0131 & -0.0128s & 0.057 \\ -0.08 & 0.0585s^2 + 0.0158s & -0.096 \\ 0.171 & 4.71s & 4.71s + 0.6 \end{vmatrix}}{\nabla} \tag{3-33}$$

where
$$\nabla = 0.00748s(s^2 + 0.380s + 1.813)(s + 2.09)(s - 0.004) \quad (3\text{-}34)$$

Expanding the numerator, Eq. 3-33 becomes
$$\frac{\phi(s)}{\delta_r(s)} = \frac{0.00355s(s^2 + 0.0535s - 4.18)}{\nabla}$$

Factoring,
$$\frac{\phi(s)}{\delta_r(s)} = \frac{0.475(s + 2.07)(s - 2.02)}{(s^2 + 0.380s + 1.813)(s + 2.09)(s - 0.004)} \quad (3\text{-}35)$$

Going to the alternate form,
$$\frac{\phi(s)}{\delta_r(s)} = \frac{131\left(\dfrac{s}{2.07} + 1\right)\left(\dfrac{s}{2.02} - 1\right)}{\left[\left(\dfrac{s}{1.345}\right)^2 + \dfrac{2(0.14)}{1.345}s + 1\right]\left(\dfrac{s}{2.09} + 1\right)\left(\dfrac{s}{0.004} - 1\right)} \quad (3\text{-}36)$$

The transfer function for ψ output δ_r input in determinant form is

$$\frac{\psi(s)}{\delta_r(s)} = \frac{\begin{vmatrix} 0.02725s^2 + 0.0553s & 0.0131 & 0.057 \\ 0.00338s & -0.08 & -0.096 \\ -0.344 & 0.171 & 4.71s + 0.6 \end{vmatrix}}{\nabla} \quad (3\text{-}37)$$

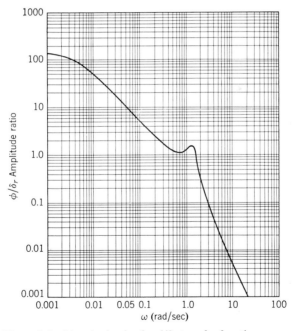

Figure 3-5   Magnitude plot for ϕ/δ_r transfer function versus ω for $s = j\omega$.

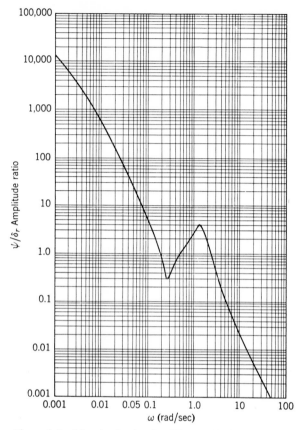

Figure 3-6   Magnitude plot for ψ/δ_r transfer function versus ω for $s = j\omega$.

Expanding,

$$\frac{\psi(s)}{\delta_r(s)} = \frac{-0.0103(s^3 + 2.12s^2 + 0.169s + 0.128)}{\nabla}$$

Factoring,

$$\frac{\psi(s)}{\delta_r(s)} = \frac{-1.38(s + 2.07)(s^2 + 0.05s + 0.066)}{s(s^2 + 0.380s + 1.813)(s + 2.09)(s - 0.004)} \tag{3-38}$$

In the alternate form

$$\frac{\psi(s)}{\delta_r(s)} = \frac{-12.4\left(\dfrac{s}{2.07} + 1\right)\left[\left(\dfrac{s}{0.257}\right)^2 + \dfrac{2(0.097)}{0.257}s + 1\right]}{s\left[\left(\dfrac{s}{1.345}\right)^2 + \dfrac{2(0.14)}{1.345}s + 1\right]\left(\dfrac{s}{2.09} + 1\right)\left(\dfrac{s}{0.004} - 1\right)}$$

$$\tag{3-39}$$

The transfer function for β output δ_r input in determinant form is

$$\frac{\beta(s)}{\delta_r(s)} = \frac{\begin{vmatrix} 0.02725s^2 + 0.0553s & -0.0128s & 0.0131 \\ 0.00338s & 0.0585s^2 + 0.0158s & -0.08 \\ -0.344 & 4.71s & 0.171 \end{vmatrix}}{\nabla} \quad (3\text{-}40)$$

Expanding,

$$\frac{\beta(s)}{\delta_r(s)} = \frac{0.0002725s(s^3 + 39.8s^2 + 77.2s - 1.03)}{\nabla}$$

Factoring,

$$\frac{\beta(s)}{\delta_r(s)} = \frac{0.0364(s - 0.01)(s + 2.06)(s + 37.75)}{(s^2 + 0.380s + 1.813)(s + 2.09)(s - 0.004)} \quad (3\text{-}41)$$

In the alternate form

$$\frac{\beta(s)}{\delta_r(s)} = \frac{1.87\left(\frac{s}{0.01} - 1\right)\left(\frac{s}{2.06} + 1\right)\left(\frac{s}{37.75} + 1\right)}{\left[\left(\frac{s}{1.345}\right)^2 + \frac{2(0.14)}{1.345}s + 1\right]\left(\frac{s}{2.09} + 1\right)\left(\frac{s}{0.004} - 1\right)} \quad (3\text{-}42)$$

Figures 3-5, 3-6, and 3-7 are amplitude ratio plots of Eqs. 3-36, 3-39, and 3-42 against ω for $s = j\omega$.

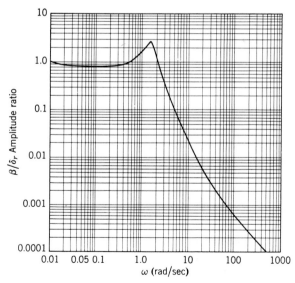

Figure 3-7   Magnitude plot for β/δ_r transfer function versus
ω for $s = j\omega$.

3-5   Lateral Transfer Functions for Aileron Displacement

It is first necessary to define positive aileron deflection.   Right aileron as seen by the pilot (wheel to the right or stick to the right) which produces a positive rolling moment is defined as positive aileron.   Then let $C_{l_{\delta_a}} = 0.6$, $C_{n_{\delta_a}} = -0.01$, $C_{y_{\delta_a}} = 0$, and $\delta_r = 0$.   The Laplace transform of Eq. 3-17 with the initial conditions equal to zero and substitution of the proper values yields

$$(0.02725s^2 + 0.0553s)\phi(s) - (0.0128s)\psi(s) + 0.057\beta(s) = 0.6\delta_a(s)$$
$$(0.00338s)\phi(s) + (0.0585s^2 + 0.0158s)\psi(s) - 0.096\beta(s) = -0.01\delta_a(s)$$
$$-0.344\phi(s) + (4.71s)\psi(s) + (4.71s + 0.6)\beta(s) = 0$$

$$(3\text{-}43)$$

Then the transfer function for δ_a input and ϕ output in determinant form can be written

$$\frac{\phi(s)}{\delta_a(s)} = \frac{\begin{vmatrix} 0.6 & -0.0128s & 0.057 \\ -0.01 & 0.0585s^2 + 0.0158s & -0.096 \\ 0 & 4.71s & 4.71s + 0.6 \end{vmatrix}}{\nabla}$$

$$(3\text{-}44)$$

Expanding,

$$\frac{\phi(s)}{\delta_a(s)} = \frac{0.1625s(s^2 + 1.565s + 2)}{\nabla}$$

Factoring,

$$\frac{\phi(s)}{\delta_a(s)} = \frac{21.7(s^2 + 1.565s + 2)}{(s^2 + 0.380s + 1.813)(s + 2.09)(s - 0.004)}$$

$$(3\text{-}45)$$

Going to the alternate form,

$$\frac{\phi(s)}{\delta_a(s)} = \frac{2865\left[\left(\dfrac{s}{1.414}\right)^2 + \dfrac{2(0.553)}{1.414}s + 1\right]}{\left[\left(\dfrac{s}{1.345}\right)^2 + \dfrac{2(0.14)}{1.345}s + 1\right]\left(\dfrac{s}{2.09} + 1\right)\left(\dfrac{s}{0.004} - 1\right)}$$

$$(3\text{-}46)$$

The transfer function for ψ output δ_a input in determinant form is

$$\frac{\psi(s)}{\delta_a(s)} = \frac{\begin{vmatrix} 0.02725s^2 + 0.0553s & 0.6 & 0.057 \\ 0.00338s & -0.01 & -0.096 \\ -0.344 & 0 & 4.71s + 0.6 \end{vmatrix}}{\nabla}$$

$$(3\text{-}47)$$

Expanding,

$$\frac{\psi(s)}{\delta_a(s)} = \frac{-0.00128(s^3 + 9.6s^2 + 1.2s - 15.3)}{\nabla}$$

Factoring,

$$\frac{\psi(s)}{\delta_a(s)} = \frac{-0.171(s - 1.14)(s + 9.29)(s + 1.45)}{s(s^2 + 0.380s + 1.813)(s + 2.09)(s - 0.004)} \tag{3-48}$$

Going to the alternate form,

$$\frac{\psi(s)}{\delta_a(s)} = \frac{-173\left(\dfrac{s}{1.14} - 1\right)\left(\dfrac{s}{9.29} + 1\right)\left(\dfrac{s}{1.45} + 1\right)}{s\left[\left(\dfrac{s}{1.345}\right)^2 + \dfrac{2(0.14)}{1.345}s + 1\right]\left(\dfrac{s}{2.09} + 1\right)\left(\dfrac{s}{0.004} - 1\right)} \tag{3-49}$$

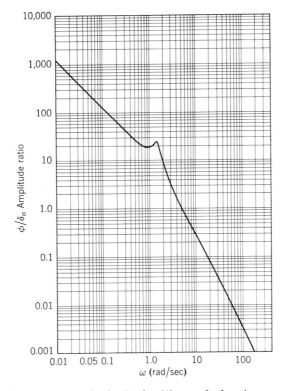

Figure 3-8   Magnitude plot for ϕ/δ_a transfer function versus ω for $s = j\omega$.

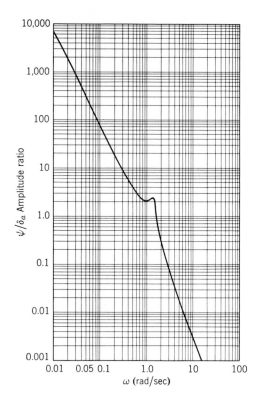

Figure 3-9   Magnitude plot for ψ/δ_a transfer function versus ω for $s = j\omega$.

The transfer function for β output δ_a input in determinant form is

$$\frac{\beta(s)}{\delta_a(s)} = \frac{\begin{vmatrix} 0.02725s^2 + 0.0553s & -0.0128s & 0.6 \\ 0.00338s & 0.0585s^2 + 0.0158s & -0.01 \\ -0.344 & 4.71s & 0 \end{vmatrix}}{\nabla} \tag{3-50}$$

Expanding,

$$\frac{\beta(s)}{\delta_a(s)} = \frac{0.00128s(s^2 + 18.9s + 2.51)}{\nabla}$$

Factoring,

$$\frac{\beta(s)}{\delta_a(s)} = \frac{0.171(s + 18.75)(s + 0.15)}{(s^2 + 0.380s + 1.813)(s + 2.09)(s - 0.004)} \tag{3-51}$$

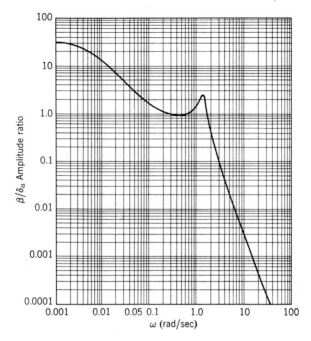

Figure 3-10   Magnitude plot for β/δ_a transfer function versus ω for $s = j\omega$.

Going to the alternate form

$$\frac{\beta(s)}{\delta_a(s)} = \frac{31.7\left(\dfrac{s}{18.75} + 1\right)\left(\dfrac{s}{0.15} + 1\right)}{\left[\left(\dfrac{s}{1.345}\right)^2 + \dfrac{2(0.14)}{1.345}s + 1\right]\left(\dfrac{s}{2.09} + 1\right)\left(\dfrac{s}{0.004} - 1\right)} \qquad (3\text{-}52)$$

Figures 3-8, 3-9, and 3-10 are amplitude ratio plots of Eqs. 3-46, 3-49, and 3-52 against ω for $s = j\omega$.

3-6   Approximate Transfer Functions

As in the case of the longitudinal equations there are some approxima-tions to the lateral modes.   They are as follows:

One-Degree-of-Freedom Dutch Roll Mode.   An examination of the rudder input transfer functions shows that the roll subsidence pole is effectively canceled by the zero in the numerator.   Thus it is considered that the Dutch roll mode consists of only sideslip and yaw.   If pure sideslip exists, $\beta = -\psi$ and ϕ and its derivatives can be assumed zero. Considering only a δ_r input with $\Theta = 0$ and neglecting C_{y_r}, Eq. 3-17 in operator form becomes

$$\frac{mU}{Sq} \dot{\psi}(s) + \left(\frac{mU}{Sq} s - C_{y_\beta}\right)\beta(s) = C_{y_{\delta_r}}\delta_r$$

$$\left(-\frac{J_{xz}}{Sqb} s - \frac{b}{2U} C_{l_r}\right)\dot{\psi}(s) - C_{l_\beta}\beta(s) = C_{l_{\delta_r}}\delta_r(s)$$

$$\left(\frac{I_z}{Sqb} s - \frac{b}{2U} C_{n_r}\right)\dot{\psi}(s) - C_{n_\beta}\beta(s) = C_{n_{\delta_r}}\delta_r(s) \qquad (3\text{-}53)$$

Since ϕ and its derivatives are equal to zero, $\sum \mathscr{L}$ must equal zero, and the rolling moment equation can be eliminated. With $\beta = -\psi$ the Y equation reduces to $-C_{y_\beta}\beta = C_{y_{\delta_r}}\delta_r$ thus the yawing moment equation must be used. The yawing moment equation with $\psi = -\beta$ becomes

$$\left(-\frac{I_z}{Sqb} s^2 + \frac{b}{2U} C_{n_r} s - C_{n_\beta}\right)\beta(s) = C_{n_{\delta_r}}\delta_r(s) \qquad (3\text{-}54)$$

Taking the characteristic equation and dividing through by $-I_z/Sqb$, Eq. 3-54 becomes

$$\left[s^2 - \left(\frac{Sqb}{I_z}\right)\left(\frac{b}{2U}\right)C_{n_r}s + \frac{Sqb}{I_z} C_{n_\beta}\right]\beta(s) = 0$$

then

$$\omega_n = \left(\frac{C_{n_\beta}Sqb}{I_z}\right)^{1/2} = U\left(\frac{C_{n_\beta}S\rho b}{2I_z}\right)^{1/2} \qquad (3\text{-}55)$$

and

$$2\zeta\omega_n = -\left(\frac{Sqb}{I_z}\right)\left(\frac{b}{2U}\right)C_{n_r} = -\frac{S\rho Ub^2}{4I_z} C_{n_r}$$

then

$$\zeta = -\left[\frac{S\rho Ub^2}{2(4I_z)}\left(\frac{2I_z}{C_{n_\beta}S\rho b}\right)^{1/2}\right]\frac{C_{n_r}}{U} = \frac{-C_{n_r}}{8}\left(\frac{2S\rho b^3}{I_zC_{n_\beta}}\right)^{1/2} \qquad (3\text{-}56)$$

From an examination of Eqs. 3-55 and 3-56 it can be seen that for a given altitude the natural frequency is proportional to U while the damping ratio is constant. However, both the natural frequency and the damping ratio are proportional to the $\sqrt{\rho}$ for a given U. This is the same effect as observed for the short-period mode. Equation 3-54 is evaluated to compare the natural frequency and damping ratio with that obtained from the complete equations. Equation 3-54 in transfer function form is

$$\frac{\beta(s)}{\delta_r(s)} = \frac{-C_{n_{\delta_r}}}{\dfrac{I_z}{Sqb} s^2 - \dfrac{b}{2U} C_{n_r}s + C_{n_\beta}} \qquad (3\text{-}57)$$

Evaluating, it becomes

$$\frac{\beta(s)}{\delta_r(s)} = \frac{0.08}{0.0585s^2 + 0.0158s + 0.096}$$

$$\frac{\beta(s)}{\delta_r(s)} = \frac{1.37}{s^2 + 0.27s + 1.64} \tag{3-58}$$

Going to the alternate form,

$$\frac{\beta(s)}{\delta_r(s)} = \frac{0.835}{\left(\dfrac{s}{1.28}\right)^2 + \dfrac{2(0.114)}{1.28}s + 1} \tag{3-59}$$

Comparing this to the values of ζ and ω_n obtained from the complete equations shows good agreement. From the complete equations, $\zeta = 0.14$ and $\omega_n = 1.345$.

The one-degree-of-freedom Dutch roll approximation is then very useful for obtaining the damping ratio and natural frequency of the Dutch roll. It is also generally used for the lateral analysis and simulation of roll stabilized missiles discussed in Chapter 7.

One-Degree-of-Freedom Rolling Mode. Because the rolling mode consists almost entirely of rolling motion, only the rolling moment equation is needed. Since sideslip and yaw angle are being neglected these may be set equal to zero; then the rolling moment equation from Eq. 3-17 in operator form with $\delta_r = 0$ reduces to

$$\left(\frac{I_x}{Sqb}s - \frac{b}{2U}C_{l_p}\right)\dot\phi(s) = C_{l_{\delta_a}}\delta_a(s)$$

Then

$$\frac{\dot\phi(s)}{\delta_a(s)} = \frac{C_{l_{\delta_a}}}{\dfrac{I_x}{Sqb}s - \dfrac{b}{2U}C_{l_p}} \tag{3-60}$$

The time constant is

$$\tau = -\left(\frac{I_x}{Sqb}\right)\left(\frac{2U}{bC_{l_p}}\right) = -\frac{4I_x}{S\rho Ub^2 C_{l_p}} \tag{3-61}$$

Evaluating, $\tau = 0.493$ sec. From the complete equations, $\tau = 1/2.09 = 0.479$ sec, which gives good agreement. The ϕ/δ_a transfer function can be written as

$$\frac{\phi(s)}{\delta_a(s)} = \frac{0.6}{s(0.02725s + 0.0553)}$$

In the alternate form

$$\frac{\phi(s)}{\delta_a(s)} = \frac{10.84}{s\left(\dfrac{s}{2.03} + 1\right)} \tag{3-62}$$

In Chapter 4 it is seen that the one-degree-of-freedom rolling mode is an excellent approximation for the transfer function of the coordinated aircraft defined.

3-7 Transient Response of the Aircraft

In this section the transient response of the aircraft for various airspeeds and altitudes, as obtained by use of the analog computer, is discussed. For the computer simulation, the stability derivatives C_{l_p}, C_{l_β}, C_{n_r}, C_{y_β} are assumed constant, which is a valid assumption for the lower Mach

Table 3-2 *Comparison of the Predicted and Actual Effects of the Variation of Airspeed and Altitude on the Lateral Dynamic Response*

(a) Dutch Roll Mode

Flight Condition		ζ		ω_n(rad/sec)	
Altitude (ft)	U_0 (ft/sec)	Predicted from Eq. 3-56	Actual	Predicted from Eq. 3-55	Actual
Sea Level	236	0.072	0.1	0.98	1.05
Sea Level	600	0.109	0.14	1.7	1.74
40,000	472	0.036	0.035	1.01	1.05
40,000	600	0.045	0.04	0.98	1.03

(b) Roll Subsidence Mode

Flight Condition		τ_r(sec)	
Altitude (ft)	U_0 (ft/sec)	Predicted from Eq. 3-61	Actual
Sea Level	236	0.91	0.8
Sea Level	600	0.353	0.35
40,000	472	1.82	1.6
40,000	600	1.43	1.3

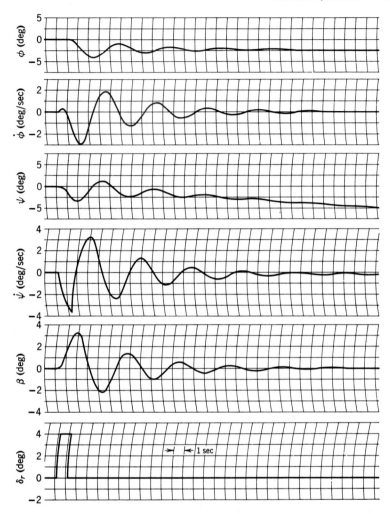

Figure 3-11 Transient response of the aircraft for a pulse rudder deflection
(complete lateral equations for 440 ft/sec at sea level).

numbers. The other stability derivatives are calculated using the equations previously given in Table 3-1. The stability derivatives for the forces and moments caused by rudder and aileron deflection are assumed constant as are the mass and the moments of inertia of the aircraft. The rudder and aileron inputs are applied separately; both consist of pulses 1 sec in duration, with an amplitude of the rudder pulse equal to 4° and that of the aileron equal to 2°. Two different altitudes and four different airspeeds were studied; the results are presented in Figures 3-11 through 3-15.

In the discussion of the one-degree-of-freedom Dutch roll mode it was deduced that the damping ratio was proportional to $\sqrt{\rho}$ and that the

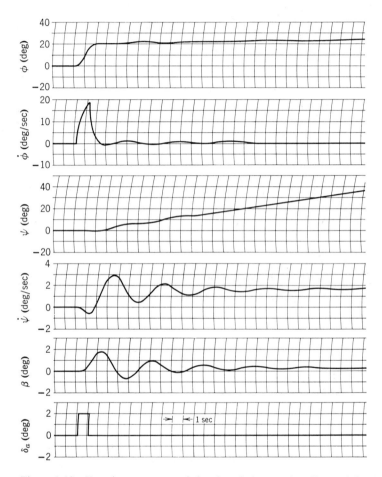

Figure 3-12   Transient response of the aircraft for a pulse aileron deflection (complete lateral equations for 440 ft/sec at sea level).

natural frequency was proportional to $U\sqrt{\rho}$.   The computer study only partially verifies this.   The effects of air speed and altitude on the Dutch roll mode are shown in Figure 3-13 and summarized in Table 3-2(a), where Eqs. 3-56 and 3-55 are used to determine the expected values of the damping ratios and natural frequencies, respectively.   The predicted and actual values compare quite well.   The unpredicted variation of the damping ratio with U is caused by the fact that C_{n_β} is dependent upon airspeed.

In the discussion of the one-degree-of-freedom rolling mode, Eq. 3-61

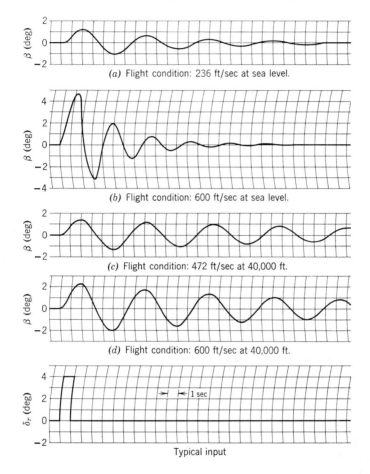

(a) Flight condition: 236 ft/sec at sea level.

(b) Flight condition: 600 ft/sec at sea level.

(c) Flight condition: 472 ft/sec at 40,000 ft.

(d) Flight condition: 600 ft/sec at 40,000 ft.

1 sec

Typical input

Figure 3-13   Computer results showing the effects of changes in airspeed and altitude on the Dutch roll mode (complete lateral equations).

shows that the roll subsidence time constant is inversely proportional to $U\sqrt{\rho}$.   These effects are shown in Figure 3-14 and summarized in Table 3-2(*b*).   The agreement is considered excellent considering the difficulty of measuring the time constant from the computer traces.

The effects of changes in altitude and airspeed on the spiral divergence

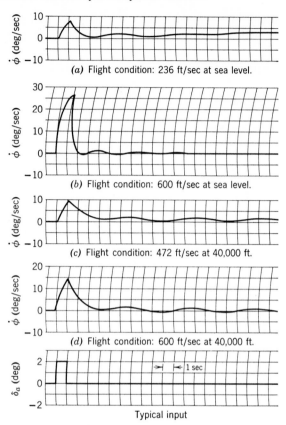

(a) Flight condition: 236 ft/sec at sea level.

(b) Flight condition: 600 ft/sec at sea level.

(c) Flight condition: 472 ft/sec at 40,000 ft.

(d) Flight condition: 600 ft/sec at 40,000 ft.

Typical input

Figure 3-14   Computer results showing the effects of changes
in airspeed and altitude on the roll subsidence time constant
(complete lateral equations).

are graphically illustrated in Figure 3-15.   As the airspeed or air density
and thus the dynamic pressure is decreased, the divergence becomes more
severe; however, at 600 ft/sec at sea level the aircraft is slightly spirally
stable.   (This phenomenon is explained in Section 3-8.)   The adverse yaw
resulting from the aileron deflection and roll rate is clearly shown in the
yaw rate trace of Figure 3-12.

From a study of the computer results, it can be said in general that the
rudder excites the Dutch roll mode primarily, while aileron deflection
produces mostly roll rate and roll angle, with some Dutch roll oscillation
superimposed on this basic motion.   However, considerable Dutch roll
does show up in the yaw rate and sideslip response.   Further use will be
made of the foregoing observations in the chapter on lateral autopilots.

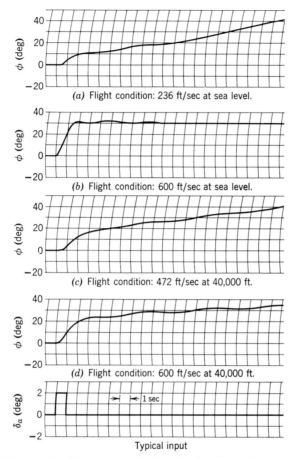

(a) Flight condition: 236 ft/sec at sea level.

(b) Flight condition: 600 ft/sec at sea level.

(c) Flight condition: 472 ft/sec at 40,000 ft.

(d) Flight condition: 600 ft/sec at 40,000 ft.

Typical input

Figure 3-15   Computer results showing the effects of changes
in airspeed and altitude on the spiral mode (complete lateral
equations).

3-8   Effect of Stability Derivative Variation

Before investigating the effect of stability derivative variation, the requirements for spiral stability will be studied.   An expansion of the determinant of the lateral equation yields the following for the constant term

$$C_{y_\phi}\left(C_{l_\beta}\frac{b}{2U}C_{n_r} - C_{n_\beta}\frac{b}{2U}C_{l_r}\right)$$

or

$$\frac{b}{2U}C_{y_\phi}(C_{l_\beta}C_{n_r} - C_{n_\beta}C_{l_r})$$

For an aircraft to be spirally stable, the product of $C_{l_\beta}C_{n_r}$ must be greater than the product of $C_{n_\beta}C_{l_r}$. From the data listed for the jet transport in, Section 3-3,

$$(C_{l_\beta}C_{n_r} - C_{n_\beta}C_{l_r}) = (-0.057)(-0.107) - (0.096)(0.086)$$

or $(C_{l_\beta}C_{n_r} - C_{n_\beta}C_{l_r}) = 0.0061 - 0.00825$

C_{n_r} and C_{n_β} are both affected by the size of the vertical tail so that one cannot be increased without increasing the other. However, C_{l_β} is mainly a function of the effective dihedral of the wing and can be increased

Table 3-3. *Effects of Variation of the Stability Derivatives on the Damping Ratio Natural Frequency, and Time Constants of the Lateral Modes*

Stability Derivative	Basic Value	Change	Dutch Roll ζ	ω_n	Roll Subsidence $1/\tau_r$	Spiral Divergence $1/\tau_s$
C_{l_p}	-0.38	×4	0.13	1.32	-8	*$+0.00105$
		×2			-4	$+0.0021$
		×½			-1.0	$+0.0084$
		×¼			-0.5	$+0.0168$
C_{l_r}	0.086	×4		1.38	-2	$+0.0535$
		×2		1.35		$+0.0206$
		×½		1.32		-0.00406
		×¼				-0.0082
C_{n_r}	-0.107	×4	0.4			-0.0327
		×2	0.22			-0.008
		×½	0.06			$+0.0103$
		×¼	0.04			$+0.0134$
C_{n_β}	0.096	×4	0.05	2.42		$+0.0133$
		×2	0.13	1.87		$+0.0103$
		×½	0.16	0.994		-0.0081
		×¼	0.25	0.738		-0.0328
C_{y_β}	-0.6	×4	0.275	1.36		$+0.0042$
		×2	0.16	1.33		
		×¼	0.094	1.316		
C_{l_β}	-0.057	×4	0.03	1.32	-2.5	-0.034
		×2	0.1		-2.1	-0.008
		×¼	0.16		-1.7	$+0.0134$
C_{n_p}	-0.0228	×4	0.11	1.41	-2.2	$+0.0042$
		×¼	0.13	1.32	-2.0	$+0.0042$

* + sign indicates divergence; − sign indicates convergence.

by increasing the dihedral. Since $C_{l_r} = C_L{}^w/4$, spiral stability can be obtained by decreasing C_L, which can be accomplished by flying faster, thus decreasing the wing angle of attack. This fact indicates that an aircraft can be spirally stable at low angles of attack (high speed) and spirally unstable at low airspeed as shown in Section 3-7. On examining the effects of obtaining spiral stability by changing C_{l_β}, by increasing the dihedral, it can be seen that the damping ratio of the Dutch roll mode is decreased (see Table 3-3). If C_{l_β} is increased sufficiently, the Dutch roll oscillations can go unstable. Thus in most aircraft a slight amount of spiral instability is permitted to obtain other more important flying qualities, such as better damping of the Dutch roll.

The effects of varying the individual stability derivatives on the lateral modes are studied next. The data for Table 3-3 was obtained for an airspeed of 440 ft/sec at sea level through the use of the analog computer by varying each stability derivative individually and then determining the damping ratio and natural frequency of the Dutch roll and the two time constants from the time recordings. In the analysis of the recordings it proves difficult to evaluate the spiral divergence time constant; therefore, this time constant is calculated from the following equation.

$$\tau_s = \frac{mU}{Sq} \left[\frac{C_{l_p} C_{n_\beta}}{C_{y_\phi}(C_{l_\beta} C_{n_r} - C_{n_\beta} C_{l_r})} \right] \tag{3-63}$$

As mentioned in Section 1-11, the varying of the stability derivatives individually is an artificial situation, but it is included to give the reader some feel for the relative importance of the various stability derivatives.

Table 3-4 *List of the Stability Derivatives Having the Largest Effects on the Lateral Modes*

Stability Derivative	Quantity Most Affected	How Affected
C_{n_r}	Damping of the Dutch roll	Increase C_{n_r} to increase the damping
C_{n_β}	Natural frequency of the Dutch roll	Increase C_{n_β} to increase the natural frequency
C_{l_p}	Roll subsidence	Increase C_{l_p} to increase $1/\tau_r$
*C_{l_β}	Spiral divergence	Increase C_{l_β} for spiral stability

* C_{l_β} was selected for the reasons stated at the beginning of Section 3-8.

It must be remembered that the data presented in Table 3-3 were obtained from the analog computer traces, thus extremely small changes in the characteristics of the modes cannot be detected. The basic values of the modes as measured from the computer traces are given here for reference. Dutch roll: $\zeta_D = 0.13$, $\omega_{n_D} = 1.32$; roll subsidence: $1/\tau_r = -2$; spiral divergence: $1/\tau_s = +0.0042$. From Table 3-3 it can be concluded that the stability derivatives in Table 3-4 have the most effect on the lateral modes.

References

1. C. D. Perkins, and R. E. Hage, *Airplane Performance, Stability, and Control*, John Wiley and Sons, New York, 1949.
2. B. Etkin, *Dynamics of Flight: Stability and Control*, John Wiley and Sons, New York, 1959.

4

Lateral Autopilots

4-1   Introduction

As most aircraft are spirally unstable, there is no tendency for the aircraft to return to its initial heading and roll angle after it has been disturbed from equilibrium by either a control surface deflection or a gust.   Thus the pilot must continually make corrections to maintain a given heading. The early lateral autopilots were designed primarily to keep the wings level and hold the aircraft on a desired heading.   A vertical gyro was used for the purpose of keeping the wings level and a directional gyro was used for the heading reference.   Figure 4-1 is a block diagram of such an autopilot.   This autopilot had only very limited maneuvering capabilities. Once the references were aligned and the autopilot engaged only small heading changes could be made.   This was generally accomplished by changing the heading reference and thus yawing the aircraft through the use of the rudders to the new heading.   Obviously such a maneuver was uncoordinated and practical only for small heading changes.   Due to the lack of maneuverability and the light damping characteristic of the Dutch roll oscillation in high performance aircraft, this autopilot is not satisfactory for our present-day aircraft.   Present-day lateral autopilots are

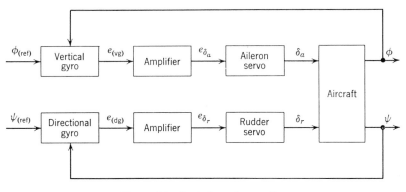

Figure 4-1   Basic lateral autopilot.

much more complex.   In most cases it is necessary to provide artificial
damping of the Dutch roll.   Also to add maneuverability, provisions are
provided for turn control through the deflection of the aileron and coordination is achieved by proper signals to the rudder.   In this discussion
of the lateral autopilots the damping of the Dutch roll is discussed first,
followed by methods of obtaining coordination, and then the complete
autopilot is examined.

4-2   Damping of the Dutch Roll

In the derivation of the lateral equations of motion $\dot{\phi}$ and $\dot{\psi}$ were
substituted for p and r, respectively.   This was done on the assumption
that the perturbations from equilibrium were small.   However, $\dot{\psi}$ is an
angular velocity about the vertical, while r is measured with respect to the
Z axis of the aircraft.   For small roll angles $\dot{\psi}$ and r are approximately
equal, but for larger roll angles this is not true.   In the study of lateral
autopilots, especially in the analysis of methods of obtaining coordination,
the roll angle is not necessarily small; therefore, it is more correct to use r
instead of $\dot{\psi}$ when referring to angular velocities about the aircraft Z axis,
that is, the angular velocity measured by a yaw rate gyro is r not $\dot{\psi}$.   With
reference to p and $\dot{\phi}$, both are angular velocities about the aircraft X axis
and are therefore always identical as long as Θ is zero.   Thus for the
remainder of this chapter, r is used instead of $\dot{\psi}$, but $\dot{\phi}$ and ϕ are used for
the roll rate and roll angle.

In the discussion of the transient response in Section 3-7 it was
stated that the rudder excites primarily the Dutch roll mode; the
Dutch roll that is observed in the yaw rate and sideslip response from
an aileron deflection is caused by the yawing moment resulting from
the aileron deflection.   For these reasons the usual method of damping

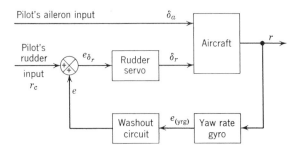

Figure 4-2   Block diagram of Dutch roll damper.

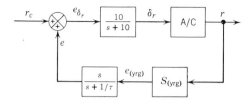

Figure 4-3   Block diagram of the Dutch roll
damper for the root locus study.

$$[TF]_{(A/C)[\delta_r;r]} = \frac{-1.38(s^2 + 0.05s + 0.066)}{(s - 0.004)(s^2 + 0.38s + 1.813)}$$

the Dutch roll is to detect the yaw rate with a rate gyro and use this
signal to deflect the rudder.   Figure 4-2 is a block diagram of the Dutch
roll damper.   The washout circuit produces an output only during the
transient period.   If the yaw rate signal did not go to zero in the steady
state, then for a positive yaw rate, for instance, the output of the yaw
rate gyro would produce a positive rudder deflection.   This would result
in an uncoordinated maneuver and require a larger pilot's rudder input
to achieve coordination.   The transfer function of the washout circuit
is $\tau s/(\tau s + 1)$.

The best way to illustrate the effect of different values of the time constant for the washout circuit is to draw the root locus for two different
values of the time constant.   First, Figure 4-2 is redrawn with the transfer

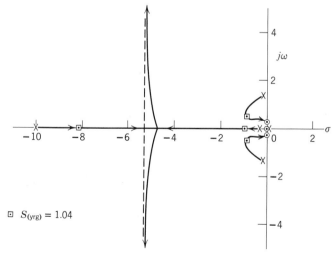

Figure 4-4   Root locus for the Dutch roll damper for τ of the washout
circuit equal to 3 sec.

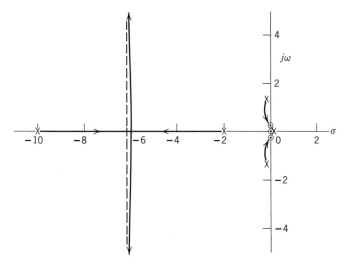

Figure 4-5   Root locus for the Dutch roll damper for τ of the washout circuit equal to 0.5 sec.

functions of the various blocks indicated (see Figure 4-3).   Since the root locus is drawn for a rudder input only, the aileron input is not indicated in Figure 4-3.   The transfer function of the aircraft is obtained from Eq. 3-38 after canceling the rolling mode pole with the appropriate zero. Figures 4-4 and 4-5 are the root loci of the Dutch roll damper as the yaw rate gyro sensitivity is increased from zero for a time constant of the washout circuit of 3 sec and $\frac{1}{2}$ sec, respectively.   As can be seen from the root loci, if the time constant is not large enough, very little increase in the damping ratio can be obtained.   If the time constant is larger, then considerable improvement of the damping ratio can be obtained as the rate gyro sensitivity is increased.   The location of the closed loop poles for a yaw rate gyro sensitivity of 1.04 volt/deg/sec is indicated in Figure 4-4.   These closed loop poles are used in the subsequent lateral autopilot analysis.   Figure 4-6 shows the results of a computer simulation of the aircraft with the Dutch roll damper.   The effectiveness of the Dutch roll damper is evident; however, some sideslip is present, especially during the transient period.   This is due to the adverse yawing moment caused by the roll rate.   It should be noted that the rudder deflection goes to zero shortly after the yaw rate reaches a steady state.   Now that the Dutch roll has been adequately damped, it is necessary to eliminate the sideslip to achieve a coordinated turn.   Section 4-3 discusses four different methods for achieving coordination, and the merits or lack of merits of each system.

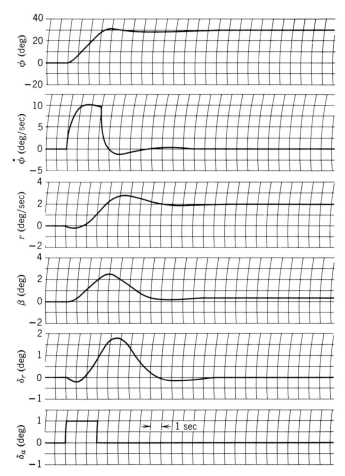

Figure 4-6 Response of the aircraft with Dutch roll damping for $S_{(yrg)} = 1.04$ volt/deg/sec for a pulse aileron deflection (sea level at 440 ft/sec).

4-3 Methods of Obtaining Coordination

1. The most logical method of eliminating sideslip is to feed back a signal proportional to sideslip. Figure 4-7 is a block diagram of this system. The inner loop of the system shown in Figure 4-7 is the Dutch roll damper. Selection of the yaw rate gyro sensitivity determines the closed loop poles of the inner loop, these are in turn the open loop poles of the outer loop. However, there is one additional zero introduced into the closed loop transfer function of the inner loop by the washout circuit.

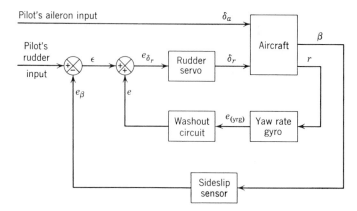

Figure 4-7   Block diagram of system using sideslip to obtain coordination.

Figure 4-7 is redrawn, replacing the inner loop with its appropriate transfer function.   The block containing $\beta(s)/r(s)$ is the ratio of the transfer functions of the aircraft for rudder input to sideslip and yaw rate output, respectively.   The $\beta(s)/r(s)$ transfer function is required to convert the output of the closed loop transfer function of the inner loop from r to β. Because the sideslip and yaw rate transfer functions have the same denominator, their ratio is the ratio of their numerators.   Thus

$$\frac{\beta(s)}{r(s)} = \frac{-0.0364(s - 0.01)(s + 2.06)(s + 37.75)}{1.38(s + 2.07)(s^2 + 0.05s + 0.066)} \text{ sec} \qquad (4\text{-}1)$$

When the forward transfer function shown in Figure 4-8 is multiplied by Eq. 4-1 after canceling the $s + 2.06$ term in the numerator with the $s + 2.07$ term in the denominator and canceling the zero at 0.01 with the pole at 0.004, Figure 4-8 can be redrawn as shown in Figure 4-9.   The poles and zeros that are canceled are so close together that they do not influence the root locus.   The plotting of this root locus will be left as an exercise for the student.

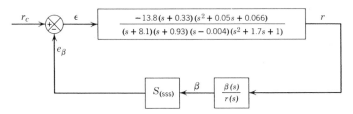

Figure 4-8   Simplified block diagram of coordination system.

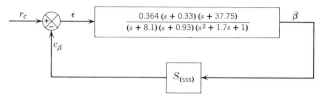

Figure 4-9 Block diagram for root locus study of system using sideslip to obtain coordination.

Before going on to the next method, some problems associated with measurement of the sideslip angle are discussed. In general, it is easier to measure sideslip than to measure angle of attack. In both cases it is not difficult to measure the direction of the local air flow. The problem arises in determining whether the direction of the local air flow measured is the direction of the undisturbed air. Since sideslip is measured with respect to the plane of symmetry, there is less distortion of the local air flow and thus the local air flow is more nearly aligned with the undisturbed air flow ahead of the aircraft. This is not true for angle of attack due to the upwash in front of the wing and the downwash behind the wing. Another problem is the noise produced by the sideslip sensor. This noise can arise either in the signal generator or from disturbances in the air flow. The sensitivity of the sideslip sensor must be kept low or the loop will be too sensitive and the damping too low. The low sensitivity also minimizes the effects of the noise. In spite of the apparent disadvantages, the use of sideslip to achieve coordination is quite extensive.

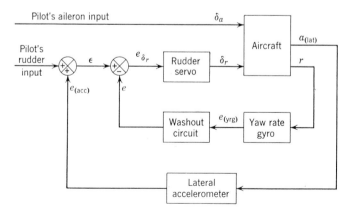

Figure 4-10 Block diagram of system using lateral acceleration to obtain coordination.

2. The second method for obtaining coordination arises from the definition of a "coordinated turn." A coordinated turn is one in which the resulting acceleration lies in the plane of symmetry. This means that there is no lateral acceleration; thus if an accelerometer is used to sense lateral acceleration and its output is fed to the rudder so that the rudder will be deflected in such a direction as to null the output of the accelerometer, coordination can be obtained. Figure 4-10 is a block diagram of the system. To analyze this system, it is first necessary to obtain a transfer function for the aircraft for a rudder input to lateral acceleration output. An expression for the lateral acceleration can be obtained from Eq. 3-14 for $\Theta = 0$, neglecting C_{y_p}. Thus

$$\frac{mU}{Sq}\beta - C_{y_\beta}\beta + \left(\frac{mU}{Sq} - \frac{b}{2U}C_{y_r}\right)r - C_{y_\phi}\phi = C_{y_a} \qquad (4\text{-}2)$$

For the lateral acceleration to be zero the summation of the terms in Eq. 4-2 must be zero. As will be shown in Eq. 4-3, all the terms on the left-hand side of Eq. 4-2, except the term $(b/2U)C_{y_r}r$, add to zero for $\beta = 0$, that is, a steady-state coordinated turn. This means that for the steady-state coordination $C_{y_a} = -(b/2U)C_{y_r}r$. Since $C_{y_a} = C_{y_{\delta_r}}\delta_r$, during a coordinated turn the rudder deflection is such that the side force produced by the turning angular velocity is zero. These terms are then eliminated from the rest of the analysis, and Eq. 4-2 reduces to

$$\frac{Sq}{m}C_{y_\beta}\beta = U(\dot\beta + r) - \frac{Sq}{m}C_{y_\phi}\phi \qquad (4\text{-}3)$$

after multiplying throughout by Sq/m. From Eq. 3-4 it can be seen that $U(\dot\beta + r) = a_y$, and from Table 3-1, $C_{y_\phi} = mg/Sq$ for $\Theta = 0$. Then $(Sq/m)C_{y_\phi}\phi$ is the small angle approximation of $g\sin\phi$, which is the component of gravity along the Y axis as shown in Figure 4-11. Therefore the right-hand side of Eq. 4-3 yields the total lateral acceleration that would be detected by a lateral accelerometer (the kinematic acceleration, a_y, and the gravitational acceleration, $-g\sin\phi$). The lateral acceleration transfer function can be obtained

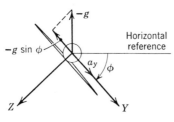

Figure 4-11 Components of acceleration along the Y axis during a turn.

from the sideslip transfer function as shown in Eq. 4-4.

$$\frac{a_{(\text{lat})}(s)}{\delta_r(s)} = \frac{Sq}{m}C_{y_\beta}\frac{\beta(s)\ \text{ft/sec}^2}{\delta_r(s)\ \text{rad}} = \frac{Sq}{(32.2)(57.3)m}C_{y_\beta}\frac{\beta(s)\ \text{g's}}{\delta_r(s)\ \text{deg}} \qquad (4\text{-}4)$$

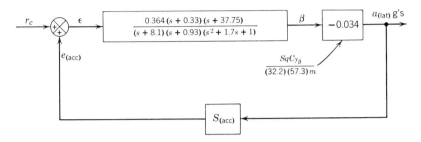

Figure 4-12   Block diagram for root locus study of system using lateral acceleration for coordination.

The simplified block diagram for the system using lateral acceleration to obtain coordination is shown in Figure 4-12.   The reason for the positive sign at the summer for the feedback signal results from the sign of C_{y_β}, which is negative (see p. 112).   This means that positive sideslip results in a negative lateral acceleration.   From this analysis it can be seen that the use of a lateral accelerometer is effectively the same as use of a sideslip sensor; but because the ratio of g's per degree of β is only 0.034, the lateral accelerometer is much less sensitive and is not as effective.   The root locus is the same as for the sideslip sensor and therefore is not drawn here.

Some of the problems associated with the use of the lateral accelerometer are discussed in the following paragraphs.   As in the case of the sideslip sensor, the output of the accelerometer may be noisy thus necessitating a low sensitivity.   In practice, a sensitivity of 10 volt/g would probably be satisfactory, and flight test might show that a higher gain could be used.   Another problem that faces the engineer is determining the minimum acceleration that the accelerometer should be capable of detecting.   If the accelerometer is extremely sensitive, say 10^{-3} g's or lower, the accelerometer, if used in a high performance fighter, could possibly sense the coriolis acceleration resulting from the craft's velocity. Normally an accelerometer capable of sensing 0.1 g's is adequate.   The location of the accelerometer is another factor to be considered.   In general, the accelerometer should be placed as close to the center of gravity of the aircraft as possible.   The farther the accelerometer is removed from the center of gravity of the aircraft the larger will be the tangential acceleration that the accelerometer is subjected to due to yaw angular accelerations of the aircraft.   There is one condition that may dictate a location that is not close to the center of gravity.   This is the case in which the aircraft cannot be assumed rigid.   If there is considerable fuselage bending, the accelerometer should be located closer to the node of the first body bending mode. (This problem is discussed further in Chapter 8.) If body bending is a problem, the final location of the accelerometer will probably be a compromise between locating it near the center of gravity and close to the node of the first body bending mode.

To achieve coordination the use of the lateral accelerometer, like use of sideslip, is also popular. In fact in some cases both have been used together successfully. However, both techniques require that the sensitivities of the closed loop be scheduled as a function of the dynamic pressure, or possibly the Mach number, if the aircraft is to operate over a wide range of air speeds and altitudes.

3. The third method of obtaining coordination is based on the fact that for a certain bank angle and true air speed, there is only one value of yaw rate for which coordination can be achieved. This can be seen by referring to Figure 4-13. For a coordinated turn the lift vector must be perpendicular to the aircraft Y axis. The lift vector is the vector sum of the weight vector (mg) and the centripetal acceleration times the aircraft mass ($m\dot{\Psi}^2 R$), where R is the radius of curvature of the circular flight path of the aircraft.

Now $m\dot{\Psi}^2 R = mV_T\dot{\Psi}$ as $\dot{\Psi} R$ is the tangential velocity of a point moving in a circle. From Figure 4-13 the horizontal component of lift must be equal to $mV_T\dot{\Psi}$, or in equation form,

$$L \sin \phi = mV_T\dot{\Psi} \tag{4-5}$$

The vertical component of the lift must equal mg, thus

$$L \cos \phi = mg \tag{4-6}$$

Dividing Eq. 4-5 by Eq. 4-6, Eq. 4-7 is obtained.

$$\tan \phi = \frac{V_T\dot{\Psi}}{g} \quad \text{or} \quad \dot{\Psi} = \frac{g}{V_T} \tan \phi \tag{4-7}$$

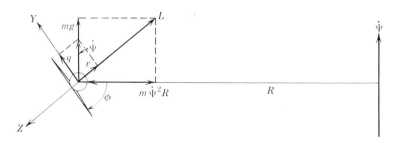

Figure 4-13   Geometry of an aircraft in a coordinated level turn.

From Figure 4-13 for $\Theta = 0$

$$r = \dot{\Psi} \cos \phi \qquad (4\text{-}8)$$

and

$$q = \dot{\Psi} \sin \phi \qquad (4\text{-}9)$$

It should be noted that ϕ is actually a negative angle. Substituting Eq. 4-7 into Eq. 4-8, Eq. 4-8 becomes

$$r = \frac{g}{V_T} \sin \phi \qquad (4\text{-}10)$$

If Eq. 4-7 is substituted into Eq. 4-9, the following equation is obtained.

$$q = \frac{g}{V_T} \tan \phi \sin \phi \qquad (4\text{-}11)$$

Equation 4-10 is of immediate interest, while the usefulness of Eq. 4-11 is discussed in Section 4-7. From Eq. 4-10 it can be seen that if V_T and ϕ are known, the proper value of r can be determined. This can then be the command signal to the rudder. Figure 4-14 is a block diagram of this system. The output of the yaw rate gyro is multiplied by the inverse sensitivity of the yaw rate gyro so that the voltage corresponding to the actual yaw rate, $e_{r(act)}$, has a one to one relation to the r, that is, unity feedback. The variation in gain for this loop is obtained by changing

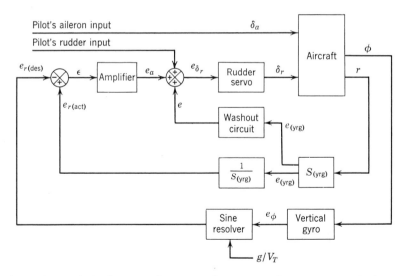

Figure 4-14 Block diagram of system using computed yaw rate to obtain co-ordination.

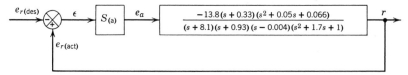

Figure 4-15 Block diagram for root locus study of system using computed yaw rate to obtain coordination.

the gain of the amplifier.   From an observation of Figure 4-14 it might be concluded that two new feedback loops have been added; however, this is not the case.   The apparent feedback through the vertical gyro and the sine resolver is simply to generate the command signal $e_{r(\text{des})}$.   Thus for analysis, Figure 4-14 can be redrawn as shown in Figure 4-15, after replacing the Dutch roll damping loop by its closed loop transfer function. Figure 4-16 is the root locus for the block diagram shown in Figure 4-15. As the gain of the amplifier is increased, the complex poles move toward the complex zeros and the spiral divergent pole moves into the left-half plane toward the real zero.   The closer the poles approach these zeros, the less effect they have on the transient response.   By the time the gain has been increased sufficiently to move the complex poles into the proximity of the zeros, the poles on the real axis have come together and moved out along the 90° asymptotes and off the graph.   These new complex poles are then the dominant roots and determine the character-istics of the transient response.   In general, the gain should be as high as possible consistent with good response.   In this case a gain of 10 volt/volt

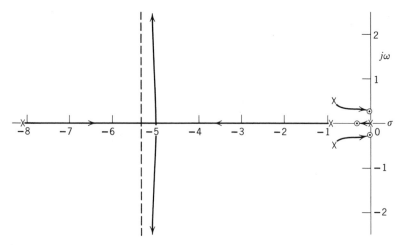

Figure 4-16 Root locus of system using computed yaw rate to obtain coordination.

yields a damped natural frequency of about 11 radians /sec with a damping ratio of about 0.5. This system is obviously more complicated than the first two systems for achieving coordination, and for this reason it is not as popular. As with the other systems, the gains must be changed as the flight conditions vary. Besides this the velocity being fed into the sine resolver (see Figure 4-14) must be the true air speed if the value of the desired yaw rate is to be correct. This requires a voltage proportional to g/V_T to be continuously generated and fed to the sine resolver along with the signal from the vertical gyro. If the quantity g/V_T is not correct, or if the sine resolver is not positioned to an angle equal to the roll angle, the value of $e_{r(\text{des})}$ will be incorrect, thus calling for the incorrect amount of rudder and will cause the turn to be uncoordinated. If the value of $e_{r(\text{des})}$ is too large, too much rudder will be called for resulting in "bottom" rudder being held in the turn. If $e_{r(\text{des})}$ is too low "top" rudder will be held in a steady-state turn. These situations are demonstrated by the computer traces in Section 4-4, along with the effects of holding "top" and "bottom" rudder. The lack of coordination resulting from such a situation is very objectionable to the pilot. This is probably the main reason for the lack of popularity of this system.

4. The last method to be discussed utilizes a component referred to as the "rudder coordination computer," which computes the amount of rudder required for a given amount of aileron. Figure 4-17 is a block diagram of this system. The transfer function of the rudder coordination computer is now derived. The total sideslip experienced by the aircraft is equal to the sideslip resulting from the rudder and the aileron, or in equation form,

$$\beta_{\text{total}} = [TF]_{(A/C)[\delta_r;\beta]}\delta_r + [TF]_{(A/C)[\delta_a;\beta]}\delta_a \qquad (4\text{-}12)$$

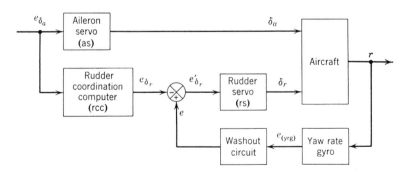

Figure 4-17 Block diagram of system using a rudder coordination computer to achieve coordination.

But if the coordination is perfect, the total sideslip is zero; then from Eq. 4-12

$$\frac{\delta_r}{\delta_a} = \frac{-[TF]_{(A/C)[\delta_a;\beta]}}{[TF]_{(A/C)[\delta_r;\beta]}} \tag{4-13}$$

But from Figure 4-17

$$\delta_r = -e_{\delta_a}[TF]_{(rcc)}[TF]_{(rs)} \tag{4-14}$$

and

$$\delta_a = e_{\delta_a}[TF]_{(as)} \tag{4-15}$$

where the minus sign in Eq. 4-14 comes from the summer and arises from the fact that for positive aileron negative rudder is required. Dividing Eq. 4-14 by Eq. 4-15, Eq. 4-16 is obtained,

$$\frac{\delta_r}{\delta_a} = -[TF]_{(rcc)} \tag{4-16}$$

as the transfer functions for the aileron and rudder servos are the same.

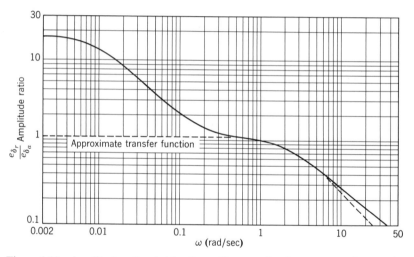

Figure 4-18   Amplitude ratio plot for the rudder coordination computer for $s = j\omega$.

The transfer function for the rudder coordination computer can be obtained by equating Eq. 4-16 and 4-13.   Therefore

$$[TF]_{(rcc)} = \frac{[TF]_{(A/C)[\delta_a;\beta]}}{[TF]_{(A/C)[\delta_r;\beta]}} \tag{4-17}$$

The transfer function for the rudder coordination computer can be obtained by dividing Eq. 3-51 by Eq. 3-41 yielding

$$[TF]_{(rcc)} = \frac{0.171(s + 18.75)(s + 0.15)}{0.0364(s - 0.01)(s + 2.06)(s + 37.75)} \tag{4-18}$$

or in the alternate form,

$$[TF]_{(rcc)} = \frac{17\left(\frac{s}{18.75} + 1\right)\left(\frac{s}{0.15} + 1\right)}{\left(\frac{s}{0.01} - 1\right)\left(\frac{s}{2.06} + 1\right)\left(\frac{s}{37.75} + 1\right)} \qquad (4\text{-}19)$$

An amplitude ratio plot of Eq. 4-19 is shown in Figure 4-18, and the phase angle plot in Figure 4-19. The transfer function given in Eq. 4-18 is comparatively complex; however, it is not usually necessary to design a circuit that gives the complete transfer function. Normally, the desired transfer function can be approximated by a first-order time lag. For this case the approximate transfer function is

$$[TF]_{(rcc)} \simeq \frac{1.1}{\frac{s}{2.06} + 1} = \frac{2.26}{s + 2.06} \qquad (4\text{-}20)$$

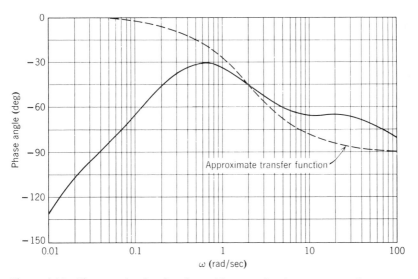

Figure 4-19   Phase angle plot for the rudder coordination computer for $s = j\omega$.

The amplitude ratio and phase angle of the approximate transfer function are also plotted in Figures 4-18 and 4-19 (see dotted line). As long as the frequencies of the input lie between about 0.5 and 10 rad/sec the approximation is quite good. However, at low frequencies, that is, below 0.5 rad/sec, the first-order time lag does not adequately represent the complete transfer function; thus it might be reasoned that the steady-state coordination would be poor. This is not necessarily true, for present-day aircraft, especially jet aircraft, require very little aileron to maintain a desired bank angle, and the amount of rudder required for coordination is often negligible.

There is, however, one serious drawback if this system is to be used on high performance aircraft. It is well known that the various transfer functions for high performance aircraft vary considerably with altitude and airspeed, thus the transfer function of the rudder coordination computer has to be programmed as a function of airspeed, altitude, dynamic pressure, or Mach number. This situation is even more complicated than gain scheduling, because the time constant of the first-order time lag has to be varied also. One practical use for this system is a conventional aircraft that flies at one normal cruise airspeed, within limited ranges of altitudes. For this case a transfer function for the normal cruise condition can be calculated and used for all cruise conditions. The coordination system can be turned off for take-off and landing and other slow speed flying. This system is therefore the least practical of the four systems discussed.

In this section four methods of obtaining coordination have been discussed. Of these, the systems utilizing sideslip and lateral acceleration are the most straightforward and simple. Of these two, the system probably most used is the one utilizing the lateral accelerometer. The main disadvantage of the sideslip sensor is the difficulty often encountered in properly installing the instrument, as mentioned early in this section.

4-4 Discussion of Coordination Techniques

Figure 4-20 shows some computer results of the simulation of the system using sideslip feedback to achieve coordination. The sensitivity of the sideslip sensor is 1 volt/deg. A higher gain is preferred; however, a further increase in the gain leads to a lightly damped oscillatory response which is objectional. This response is confirmed if the root locus of the system shown in the block diagram of Figure 4-9 is drawn. From the root locus it can be seen that 90° asymptotes lie in the right-half plane and the complex poles move outward from the real axis toward the imaginary axis leading to a lightly damped oscillatory response at the higher values of gain. A lead compensator can be added to cause the complex poles to move farther into the left-half plane as they move out from the real axis. This procedure allows the use of a higher gain and maintains sufficient damping. If a gain as high as 10 volt/deg could be used, the maximum sideslip would be only 0.3° instead of 1.4°, as shown in Figure 4-20.

Another form of compensation that could be used is a lag compensator

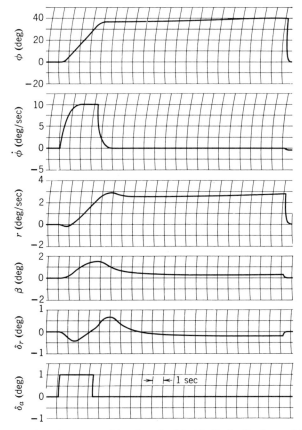

Figure 4-20   Response of the aircraft with sideslip feedback to achieve coordination for a pulse aileron deflection. $S_{(sss)} = 1$ volt/deg.

such as $(s + 0.1)/(s + 0.01)$, which would have little effect on the shape of the root locus, but would reduce the error coefficient by a factor of 10, approximately.   This factor would reduce the steady-state sideslip by the same amount.   However, a lag compensator would not decrease the maximum sideslip experienced while rolling into or out of a turn.   Therefore, the best results are achieved by the use of both a lead and a lag compensator.   A suggested lead compensator is $(s + 0.93)/(s + 9.3)$.

The computer results illustrated in Figure 4-20 also apply to the system using lateral acceleration feedback to achieve coordination for an accelerometer sensitivity of 30.6 volt/g.   Thus it can be seen that the use of sideslip feedback leads to a more sensitive system for the control of sideslip than the use of lateral acceleration.

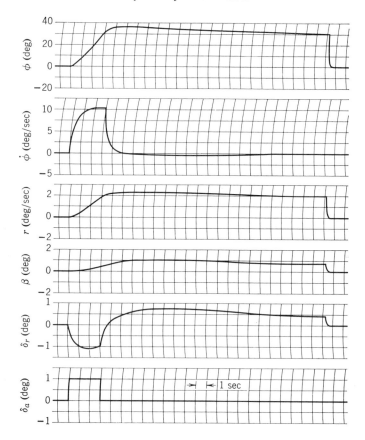

Figure 4-21   Response of the aircraft using computed yaw rate to achieve co-ordination for a pulse aileron deflection.   Top rudder held in the steady state.

The results of the computer simulation of the system using computed yaw rate are shown in Figures 4-21 and 4-22.   The gain used for both cases is 10 volt/volt.   In Figure 4-21 the actual yaw rate is greater than the computed desired yaw rate resulting in top rudder being held, which causes the aircraft to slowly decrease its angle of bank.   In Figure 4-22 the actual yaw rate is less than the computed desired yaw rate resulting in bottom rudder being held causing the aircraft to increase its angle of bank.   The value of the computed desired yaw rate is dependent upon the value of g/V_T that is fed into the sine resolver.   Thus the performance of this system depends on the proper value of g/V_T being available, as mentioned in Section 4-3.   However, the system is very effective for controlling sideslip during a turn.

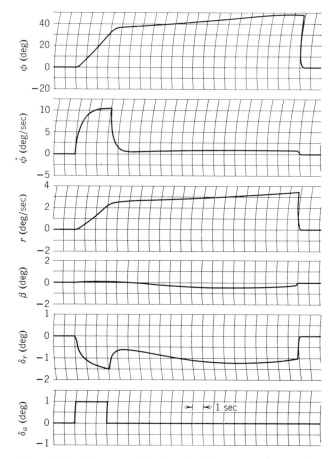

Figure 4-22   Response of the aircraft using computed yaw rate to achieve coordination for a pulse aileron deflection.   Bottom rudder held in the steady state.

Figures 4-23 and 4-24 are the computer results of the simulation of the system using the rudder coordination computer to achieve coordination. Two different transfer functions used are for Figure 4-23 $[TF]_{(rcc)} = 2.26/(s + 2.06)$; for Figure 4-24 $[TF]_{(rcc)} = 2.26/(s + 0.8)$.   The transfer function used in Figure 4-23 has the same time constant as that derived in Section 4-3; however, there is not much reduction in the maximum value of sideslip from the case of no sideslip control (see Figure 4-6). The best control is obtained by using the transfer function with $1/\tau = 0.8$, which leads to some negative sideslip but lowers the maximum sideslip to $1.2°$.   For both transfer functions the steady-state value of sideslip is about $0.5°$.

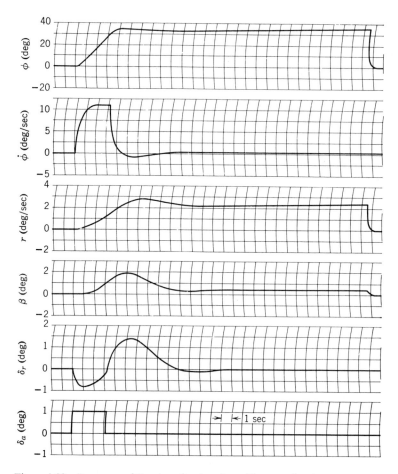

Figure 4-23   Response of the aircraft using the rudder coordination computer to achieve coordination for a pulse aileron deflection. $[TF]_{(\text{rcc})} = 2.26/(s + 2.06)$.

As can be seen from the computer analysis, the use of sideslip to achieve coordination is the most effective method, and even with a compensator to allow for a higher gain this system is simple and straightforward.   All the methods studied do decrease the maximum amount of sideslip developed during a turn and can be used, but the use of sideslip or lateral acceleration yields the simpler systems and thus are probably preferred. Since the use of sideslip is the most effective method, it is used to achieve coordination in the study of the yaw orientational control system in Section 4-5.

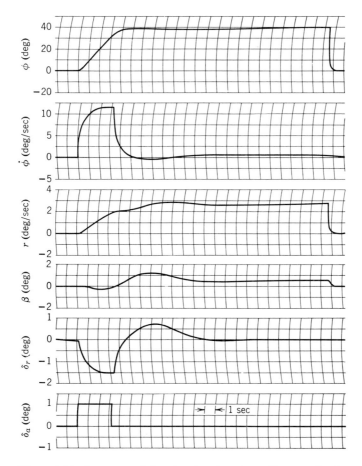

Figure 4-24   Response of the aircraft using the rudder coordination computer to achieve coordination for a pulse aileron deflection. $[TF]_{(\text{rcc})} = 2.26/(s + 0.8)$.

4-5   Yaw Orientational Control System

Figure 4-25 is the block diagram for the yaw orientational control system including the Dutch roll damper and sideslip feedback for coordination. In addition, a yaw rate integrating gyro is used so that heading changes can be obtained by commanding a desired yaw rate. In this case an additional loop is used for stabilization. The need for this is shown in Figure 4-28. For this stabilization a roll rate signal is fed back and summed with the yaw rate command signal. The result is fed as the torquing current to the yaw rate integrating gyro. In the steady

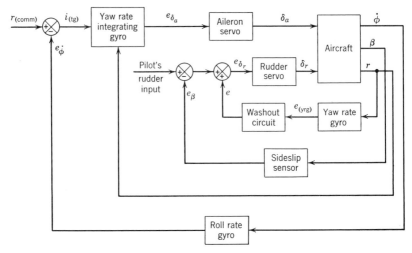

Figure 4-25   Block diagram of yaw orientational control system.

turn this signal goes to zero so that the signal to the yaw rate integrating gyro is the commanded yaw rate.   In Section 3-4 the aircraft with sideslip feedback was analyzed.   With the results of this analysis the root locus study could be continued to analyze the two new loops that have been added; however, the closed loop transfer function of the two inner loops can be greatly simplified.   An examination of the roll rate trace of Figure 4-20 shows that this response is similar to the step response of a first-order system.   This is indeed the case.   Thus the two inner loops (Dutch roll damper and sideslip feedback) of the aircraft autopilot combination can be replaced with one block referred to as the coordinated aircraft.   From an analysis of the roll rate trace of Figure 4-20 the

$$[TF]_{(coord\ A/C)[\delta_a:\dot{\phi}]} = \frac{23}{s + 2.3} \tag{4-21}$$

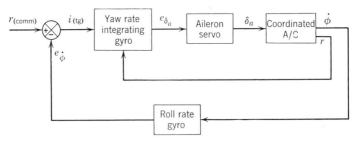

Figure 4-26   Simplified block diagram of yaw orientational control system.

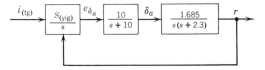

Figure 4-27 Block diagram of inner loop of
Figure 4-26 for root locus study.

The validity of representing the coordinated aircraft for aileron input to roll rate output by a first-order time lag has also been verified by flight test. Now Figure 4-25 can be redrawn to include the coordinated aircraft block as shown in Figure 4-26. A root locus analysis is performed on this simplified system. The inner loop is analyzed first. To do this, the transfer function for the coordinated aircraft for aileron input to yaw rate output is required. Equation 4-21 represented the transfer function for the coordinated aircraft for aileron input to roll rate output. Then

$$[TF]_{\text{(coord A/C)}[\delta_a:\phi]} = \frac{23}{s(s + 2.3)} \tag{4-22}$$

and $$[TF]_{\text{(coord A/C)}[\delta_a:r]} = \frac{g}{V_T} [TF]_{\text{(coord A/C)}[\delta_a:\phi]} \tag{4-23}$$

because at a given airspeed there is only one rate of turn corresponding to a particular bank angle for a coordinated turn. Therefore $[TF]_{\text{(coord A/C)}[\delta_a:r]}$ = 1.685/s(s + 2.3) for V_T = 440 ft/sec. The block diagram of the inner loop used for the root locus analysis is shown in Figure 4-27. Figure 4-28 is the root locus for the inner loop of the yaw orientational control

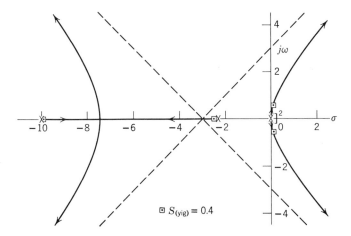

Figure 4-28 Root locus for block diagram shown in Figure 4-27.

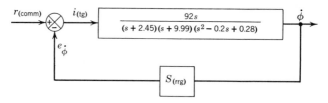

Figure 4-29   Block diagram of outer loop of the yaw orientational control system for root locus analysis.

system, with the location of the closed loop poles shown for a yaw rate integrating gyro sensitivity of 0.4 volt/deg/sec. The reason the roll rate feedback loop is required for stabilization is now evident. From the root locus the

$$[TF]_{(CL)[i_{(tg)};r]} = \frac{6.74}{(s + 2.45)(s + 9.99)(s^2 - 0.2s + 0.28)} \qquad (4\text{-}24)$$

To obtain the transfer function required for the analysis of the outer loop it is necessary to reverse the steps followed when proceeding from Eq. 4-21 to Eq. 4-23. Then

$$[TF]_{(OL)[i_{(tg)};\dot\phi]} = s\,\frac{V_T}{g}\,[TF]_{(CL)[i_{(tg)};r]} \qquad (4\text{-}25)$$

Therefore

$$[TF]_{(OL)[i_{(tg)};\dot\phi]} = \frac{92s}{(s + 2.45)(s + 9.99)(s^2 - 0.2s + 0.28)} \qquad (4\text{-}26)$$

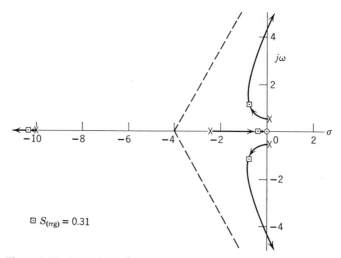

Figure 4-30   Root locus for the block diagram shown in Figure 4-29.

and the block diagram for the outer loop to be used for the final root locus analysis is shown in Figure 4-29. The root locus for the outer loop is shown in Figure 4-30. The locations of the closed loop poles for a roll rate gyro sensitivity of 0.31 volt/deg/sec are shown. Since the inner loop is unstable, its gain on the computer cannot be adjusted without having the outer loop closed. Therefore, after selecting initial values for the sensitivity of the yaw rate integrating gyro and the roll rate gyro from the root locus, the system is optimized by adjusting the sensitivities of the two gyros on the computer simulation. The criterion used is the fastest response of yaw rate for a commanded step input without oscillations.

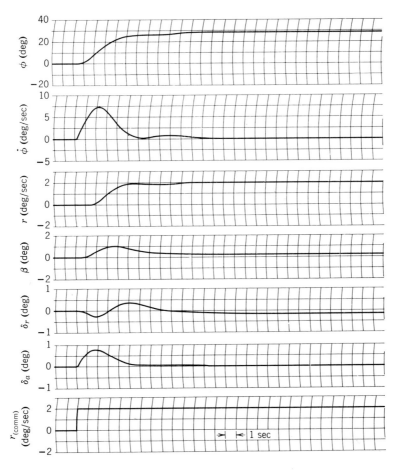

Figure 4-31 Response of the yaw orientational control system for a step input of commanded yaw rate.

No attempt has been made here to completely optimize the system. By adding compensation the response could probably be improved further. The main purpose here is to show the validity of assuming that the two inner loops of the yaw orientational control system can be approximated by a simple first-order system, and that using this simplified model in the subsequent root locus analysis gives sufficiently accurate predictions of the behavior of the complete system. Figure 4-31 shows the result of the computer run of the yaw orientational control system for a commanded yaw rate. For the computer simulation, the linearized lateral three-degree-of-freedom equations are used, with no further simplifying assumptions. The yaw rate integrating gyro sensitivity is 0.4 volt/deg/sec and the roll rate gyro sensitivity is 0.31 volt/deg/sec. Although Figure 4-31 might not convince the reader that the root loci shown in Figures 4-28 and 4-30 do accurately predict the final response, checks made at other values of the gyro sensitivities indicated excellent agreement between the computer results and the root locus analysis.

Until now no mention has been made of the origin of the yaw rate command signal. One source of input could be a control stick steering autopilot in which the signal results from a lateral deflection of the stick in the same manner as that described in Chapter 2. There is one big difference between aileron input for a turn and an elevator input for a pitch rate. Under manual control, elevator stick force must be continuously applied to maintain a desired pitch rate; however, if lateral (aileron) stick force is continuously applied, the result is a sustained roll rate. Thus stick pressure need only be applied during the entry into a turn and the roll out. However, the yaw orientational control system needs a constant yaw rate command to maintain a sustained rate of turn, which can be accomplished by feeding the force stick output to an integrator. Thus the constant voltage required for the commanded input yaw rate signal is the integral of the output of the force stick sensor. In this discussion the control stick was referred to; however, the same technique is applicable to the control column using a wheel for the aileron input.

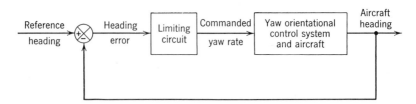

Figure 4-32   Block diagram of lateral autopilot with yaw rate limiting.

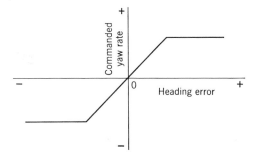

Figure 4-33   Input-output characteristics of limiting circuit.

Another source of input could be a heading reference, which would result in complete automatic flight. For this source another loop would be closed through a directional gyro. Thus any deviation from the desired heading would result in a commanded yaw rate to bring the aircraft back to the desired heading. Turns to new headings could be accomplished by use of a turn control or by resetting the heading reference. Operation of the turn control would disengage the heading reference and command a yaw rate. Upon recentering the turn control the commanded yaw rate returns to zero and the heading reference automatically re-engages at the new heading.

Still another source of input could be the instrument landing and approach system (ILAS) course coupler, which, together with the glide slope coupled into the longitudinal autopilot, would result in a completely automatic approach.

The use of the heading error to generate the yaw rate command can lead to a commanded yaw rate that overstresses the aircraft. For this reason a limiting circuit must be placed in the system as shown in Figure 4-32. The input-output characteristics of the limiting circuit are shown in Figure 4-33. The maximum value of the commanded yaw rate can be dictated by the maximum airspeed at which the autopilot is used and the maximum allowable "g's" for which the aircraft is stressed. This concludes the discussion of the yaw orientational control system.

4-6   Other Lateral Autopilot Configurations

In Section 4-5 a yaw orientational control system was analyzed. In this section some other possible lateral autopilot configurations are discussed. In the case of the yaw orientational control system a turn was obtained by commanding a desired yaw rate; however, it may be desirable to command a desired bank angle, which necessitates the use of a vertical

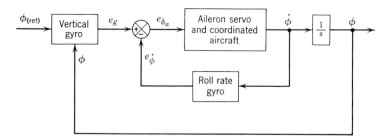

Figure 4-34   Block diagram of roll angle control system.

gyro.   Figure 4-34 is a block diagram of a roll angle control system.   The reader probably wonders why the roll rate feedback is necessary (this could have been added to the yaw orientational control system).   A look at a sketch of the root locus for the system without roll rate feedback shows why it is needed.   As can be seen from Figure 4-35, even a low gain drives the system unstable, and unity feedback results in a too lightly damped response.   Figure 4-36 shows the effect of roll rate gyro feedback. By adding the inner loop with the roll rate feedback, a higher outer loop sensitivity can be used without producing instability.   To provide a heading reference a third loop can be added to Figure 4-34 by feeding heading back through a directional gyro.   With this configuration, a heading error provides a commanded roll angle to the roll angle control system.   With heading feedback providing the command signal for the roll angle control system, it is possible to command an excessive roll angle when the heading error is large.   Thus, as in the case of the yaw orientational control system, it is necessary to limit the maximum commanded roll angle to prevent exceeding the "g" limits of the aircraft.   With respect to the limiting

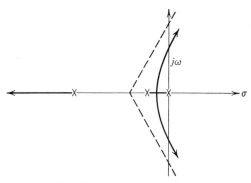

Figure 4-35   Sketch of root locus of roll angle
control system without roll rate feedback.

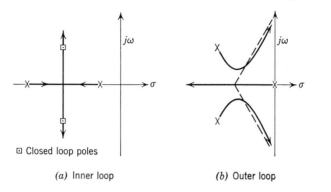

□ Closed loop poles

(a) Inner loop   (b) Outer loop

Figure 4-36   Sketch of root locus of roll angle control system
with roll rate feedback.

required, the roll angle control system is superior to the yaw orientational control system. This can be seen by looking at the equations for the horizontal accelerations resulting from a turn. From Figure 4-13, the horizontal acceleration during a coordinated turn is given by

$$a_{(hor)} = V_T \dot{\Psi} \tag{4-27}$$

From Eq. 4-8 $r = \dot{\Psi} \cos \phi$. Thus

$$a_{(hor)} = \frac{V_T r}{\cos \phi} \tag{4-28}$$

but from Eq. 4-7 $\dot{\Psi} = g/V_T \tan \phi$; then substituting for $\dot{\Psi}$ in Eq. 4-27 yields

$$a_{(hor)} = g \tan \phi \tag{4-29}$$

The total acceleration experienced by the aircraft is the vector sum of the acceleration of gravity and $a_{(hor)}$ (see Figure 4-13). Thus, it can be seen from Eq. 4-29 that the maximum "g"s will be experienced at the same bank angle regardless of velocity. However, for the yaw orientational control

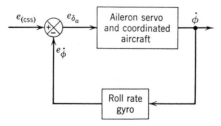

Figure 4-37   Control stick steering mode for
roll angle control system.

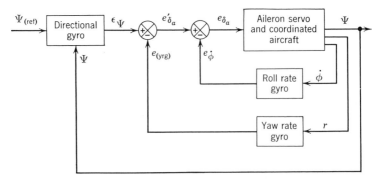

Figure 4-38   Block diagram of a rate stabilized lateral autopilot.

system, a limit on the value of r would provide "g" limiting at the maximum value for only one value of V_T (see Eq. 4-28). Therefore, at lower airspeeds the aircraft would experience less "g"s than the maximum allowable. Another source of actuating signal to the roll angle control system can be the pilot in the form of an aileron control stick steering signal. For this source the block diagram shown in Figure 4-34 would be altered by removing the vertical gyro loop and feeding in the control stick steering signal in place of e_g (see Figure 4-37).

A third lateral autopilot configuration is shown in Figure 4-38. This autopilot is similar to the yaw orientational control system in that a heading error commands a yaw rate. However, the integration resulting from the yaw rate integrating gyro used in the yaw orientational control system is lacking.

This concludes the discussion of some of the different types of lateral autopilots, which was not intended to be exhaustive but only to illustrate some of the more common configurations. The analysis methods discussed in this chapter are applicable to these and any other configurations that might be encountered. There is, however, one problem that is common to all autopilots, and that is the generation of the proper signal to the elevator during turns. This problem is discussed in Section 4-7.

4-7   Turn Compensation

When an aircraft enters a steady-state turn, the vertical component of lift must still support the weight of the aircraft (see Figure 4-13). Thus the lift must be increased by increasing the angle of attack. This requires an elevator displacement, the amount of which depends on the bank angle and the elevator effectiveness. The problem, therefore, for completely automatic flight, is to apply the correct amount of elevator to maintain altitude in a turn.   One possible technique arises from Eq. 4-11, which was derived in Section 4-3 and is repeated here for convenience.

$$q = \frac{g}{V_T} \tan \phi \sin \phi \qquad (4\text{-}30)$$

Also in the same section, from Eq. 4-10,

$$r = \frac{g}{V_T} \sin \phi \qquad (4\text{-}31)$$

If Eq. 4-31 is substituted into Eq. 4-30, an expression for the pitch rate required during a turn is obtained, which is

$$q = r \tan \phi \qquad (4\text{-}32)$$

Thus, if ϕ the roll angle, and r, the yaw rate, are known, the pitch rate can be calculated.   If the longitudinal autopilot is a pitch orientational control system, the required pitch rate can be fed to the longitudinal autopilot. This technique requires a vertical gyro to measure the roll angle and a yaw rate gyro to measure r and a computer to solve Eq. 4-32 for q.   Two of the three lateral autopilot configurations studied did not use a vertical gyro, and all longitudinal autopilots cannot accept a commanded pitch rate.   Thus, this simple scheme does not work in too many cases, and the equation must be approximated.   If the signal is applied directly to the elevator servo, the amount of pitch rate per volt of signal must be known; this, of course, varies with airspeed.   If the longitudinal autopilot has an altitude hold mode, then this feature can be used to provide the necessary elevator signal.

4-8   Automatic Lateral Beam Guidance

It is often desirable to be able to couple the output of an Omni or localizer receiver to the lateral autopilot in order to obtain automatic following of either of these beams.   A localizer receiver can provide an automatic landing system.   The problems of coupling either the Omni or localizer signals to the lateral autopilot are similar, the only difference being in the width of the beams.   For Omni tracking, full-scale deflection of the course indicator represents an angle of $10°$ with respect to the center line while for the localizer full-scale indication represents $2\frac{1}{2}°$.   Thus a different gain may be required for the two cases.   In both cases the systems become more sensitive as the aircraft nears the station, because a given lateral distance off course corresponds to a larger angle off course as the station is approached.   This situation normally requires some gain

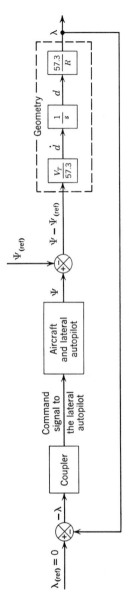

Figure 4-39 Generalized block diagram of lateral beam guidance system.

scheduling as a function of range to the station. The major question to be answered then is the type of command signal that is to be sent to the lateral autopilot as a function of the error signals from the Omni or localizer receivers. If the lateral autopilot is already constructed or designed, this may dictate the type of signal required. If not, the choice can be based on the type of command that provides the best response with the simplest coupler. The basic choices are a heading command, a yaw rate command, or a bank angle command. The last two are similar in that they both result in a yaw rate; however, the mechanization would be different and one choice might yield a simpler coupler. Another factor to be considered is the requirement for an integration in the forward loop so the aircraft remains on course in the presence of a steady crosswind. With these points in mind, different configurations are analyzed here to determine the type of coupler that is required. Figure 4-39 is a generalized block diagram that fits all systems and is used to show how the geometry of the problem is introduced for purposes of analysis and simulation. Figure 4-40 shows this geometry and defines the variables introduced. From Figure 4-40 $\tan \lambda = d/R$ then $57.3d/R \simeq \lambda$ deg (using small angle assumptions). Also $d = V_T \sin (\Psi - \Psi_{(ref)}) \simeq V_T(\Psi - \Psi_{(ref)})$ for small angles. If Ψ and $\Psi_{(ref)}$ are to be measured in degrees, then $\dot{d} \simeq V_T/57.3(\Psi - \Psi_{(ref)})$ and d is negative for the case shown in Figure 4-40. In the generalized block diagram shown in Figure 4-39, $\Psi_{(ref)}$, which represents the heading of the beam to be followed, would be set in by the pilot. To simplify the following explanations, $\Psi_{(ref)}$ is taken as zero.

It is now necessary to select a specific lateral autopilot in order to analyze the system. The first system studied is the roll angle control system

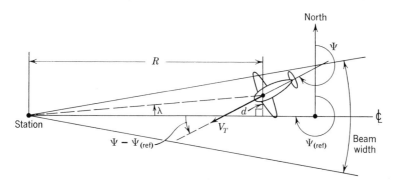

Figure 4-40 Geometry of lateral beam guidance. R = distance from the station to the aircraft; d = lateral distance of the aircraft off course; λ = angular error represented by d (λ and d are positive as shown); $\Psi - \Psi_{(ref)}$ = interception angle is negative when λ and d are positive.

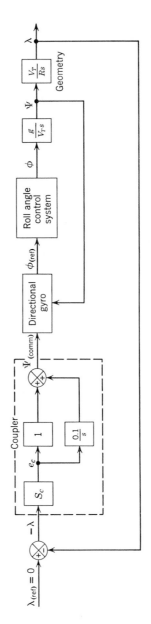

Figure 4-41 More detailed block diagram of lateral beam guidance system. *Note*: S_c = coupler gain, deg of Ψ/deg of λ.

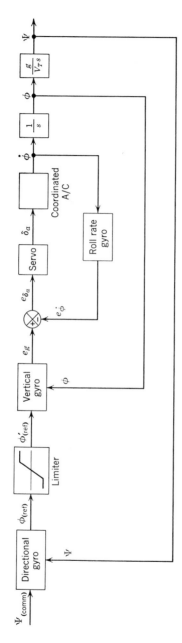

Figure 4-42 Block diagram of lateral autopilot used for the lateral beam guidance study.

$$[TF]_{(servo)[e\delta_a:\delta_a]} = \frac{4}{s + 10} \text{ deg/volt}$$

$$[TF]_{(coord)A/C[\delta_a:\dot{\phi}]} = \frac{23}{s + 2.3} \text{ sec}$$

Limits on limiter = $\pm 30°$ of $\phi'_{(ref)}$

171

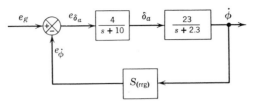

Figure 4-43   Block diagram of inner loop of Figure
4-42 for root locus analysis.

shown in Figure 4-34, with a heading feedback loop added.   Figure 4-39
is redrawn to include more of the details of the system to be studied (see
Figure 4-41).   The gain of the coupler (S_c) is set at 10° of commanded Ψ
per deg of λ; this may sound high, but it proves very satisfactory as will be
seen in Figure 4-51.   The 0.1/s path provides the slow integration necessary
to cope with a steady crosswind.   The value of 0.1 is selected somewhat
arbitrarily, but proves to be a good selection.   A detailed analysis of the
system, starting with the roll angle control system follows.

Figure 4-42 is a detailed block diagram of the roll angle control system
used, complete with heading feedback.   The very inner loop of Figure
4-42 is redrawn for the root locus analysis to determine the sensitivity of
the roll rate gyro (see Figure 4-43).   Figure 4-45 is the root locus for the
inner loop.   It should be noted that the steady-state gain of the aileron
servo is reduced from that previously used.   This is found necessary
because the response of the lateral autopilot to a step heading command
results in excessive aileron displacements and roll rates.   This is dis-
covered from the simulation, not the root locus; however, the root locus
does predict the rapid response times, which would be an indication to the
designer that the loop gains may be too high.   The locations of the closed
loop poles for the inner loop are taken to provide a damping ratio of 0.7.
This results in a roll rate gyro sensitivity of 0.564 volt/deg/sec.   Using the
closed loop poles from Figure 4-45, the block diagram for the root locus
analysis of the vertical gyro loop can now be drawn (see Figure 4-44).
The root locus for Figure 4-44 is shown in Figure 4-46.   The reason for
the choice of the vertical gyro sensitivity is not too obvious from Figure

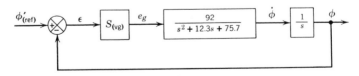

Figure 4-44   Block diagram of vertical gyro loop for root locus analysis.

4-46. This results from the fact that the initial root locus analysis is performed for a

$$[TF]_{(\text{servo})[e_{\delta_a}:\delta_a]} = \frac{10}{s + 10}$$

For both the high and low gain servos the same inner loop closed loop poles are used. Thus, Figures 4-45 and 4-46 are valid for both cases (the rate gyro sensitivity is different for both cases). The vertical gyro sensitivity is then selected using the higher gain servo that yielded a damping ratio of 0.5 for the complex roots in Figure 4-46. This gives a value of 1 volt/deg for the vertical gyro sensitivity. When the lower gain servo is introduced, the vertical gyro gain is unchanged, since an increase in this gain will offset the decrease in the servo gain. Using the closed loop poles from Figure 4-46, the block diagram for the heading feedback loop for root locus analysis can be drawn (see Figure 4-47). In Figure 4-47 the limiter is eliminated and Ψ and r are considered equal. This restricts the analysis to small roll angles, but is necessary as the root locus

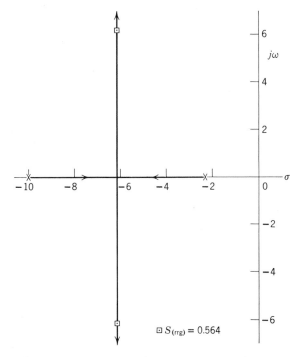

Figure 4-45   Root locus for Figure 4-43, the inner loop
of the roll angle control system.

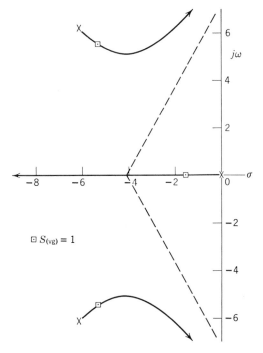

Figure 4-46   Root locus for Figure 4-44, the vertical
gyro loop of the roll angle control system.

is valid only for linear systems.   This restriction need be imposed only
during the analytic analysis and may be removed later in the computer
simulation.   Figure 4-48 is the root locus of Figure 4-47.   The next
problem is to decide on a sensitivity for the directional gyro.   As the
system will be nonlinear for large input signals, the root locus is used only
as a guide and a value is selected for the directional gyro sensitivity from
observation of the computer response.   This procedure results in a
directional gyro sensitivity of 6 volt/deg.   Any higher value results in an
oscillatory response and, finally, in instability.   The optimization is

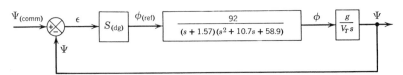

Figure 4-47   Block diagram of heading feedback loop for root locus analysis.
$V_T = 440$ ft/sec.

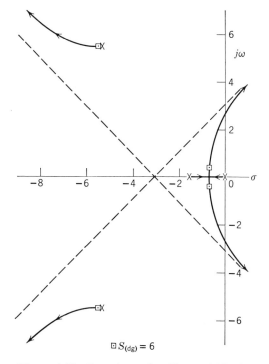

$$\Box\, S_{(dg)} = 6$$

Figure 4-48 Root locus for Figure 4-47, the heading feedback loop of the lateral autopilot.

accomplished using a 10° heading step command. The locations of the closed loop poles for the selected directional gyro sensitivity are indicated in Figure 4-48.

Another factor that should be considered in selecting the directional gyro sensitivity is shown in the root locus analysis of the final loop. Using the closed loop poles from Figure 4-48, the final block diagram can be drawn (see Figure 4-49). The root locus for Figure 4-49 is shown in Figure 4-50. As can be seen from Figure 4-49, the geometry loop adds two poles at the origin: one from the geometry and one from the integration introduced by the coupler. The motion of these two poles is very

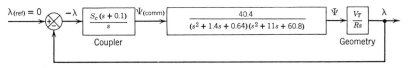

Figure 4-49 Block diagram for geometry loop.

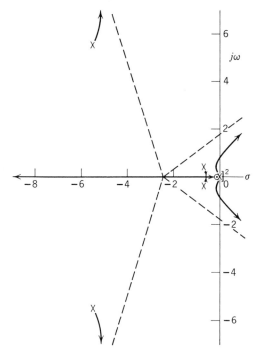

Figure 4-50 Root locus for Figure 4-49, the
geometry loop for the lateral beam guidance
system.

critical; they must move into the left-half plane for at least some value of
closed loop gain. The two complex poles near the origin have a large
effect on the movement of the two poles at the origin. If the complex
poles are too close to the origin, they may cause the two poles at the origin
to move rapidly into the right-half plane, thus resulting in a completely
unstable system. Therefore in closing the directional gyro loop the
sensitivity of the directional gyro should be such that the complex poles
are as far from the origin as possible, but not too close to the imaginary
axis. After obtaining the best value for the sensitivity of the directional
gyro, the root locus of the outer loop enables the designer to determine
whether a compensating network should be added to the coupler. A lead
network of the form of $(1 + \alpha \tau s)/(1 + \tau s)$ could be used with a zero
close to the origin and an α of 10 or larger. In order not to attenuate the
error signal, an active network must be used with a gain of α. Such a
network usually causes the two poles at the origin to circle around and
break into the real axis between the two zeros. The complex poles near

the origin are thus the ones to move into the right-half plane, which may provide a better response for the system or a smaller minimum range to the station for instability; however, it may just add to the complexity of the system with no real significant improvement in performance. Figure 4-49 illustrates that the loop gain is a function of V_T and R as well as of S_c. Normally, a value of S_c that will not result in excessive interception angles is selected. The coupler sensitivity used in the simulation is $10°$ of Ψ for each degree of λ off course. Having selected a value of S_c and V_T for a particular simulation, the root locus can be used to determine at what range the system will go unstable. It is usually necessary to decrease the coupler sensitivity as the range decreases; however, as the range goes to zero, the loop gain goes to infinity for any finite coupler sensitivity, thus there is always a minimum range for satisfactory operation. In actual operation if it is desirable to fly across an Omni station, the system must be disengaged at some minimum range and then reengaged after passing the station by the same minimum range. If the system is to be used in conjunction with a localizer beam for an automatic approach system, the range need not go to zero since normally the localizer transmitter is located at the far end of the landing runway. Thus, if the minimum range for stability is less than the length of the runway, the system should perform satisfactorily. From Figure 4-50 the calculated value of the minimum range for $V_T = 440$ ft/sec and $S_c = 10$ is 3720 ft. This figure is very accurately verified by the computer simulation.

Figure 4-51 shows the results of the computer simulation. Figure 4-51(a) shows the response of the system for an initial displacement of the aircraft from the centerline of the beam but on the beam heading. The values of d_0 were 100 and 500 ft. The first overshoot is largely caused by the integration introduced by the coupler. Figure 4-51(b) shows the response of the system to a step crosswind, with the aircraft initially on course and on the beam heading. The values of the crosswinds were 25 and 100 ft/sec, $90°$ to the course. In all cases the aircraft was back on course by the time the range had decreased to 10,000 ft. For all cases the aircraft was well within the $\pm 2\frac{1}{2}°$ beam width of the localizer beam. The initial conditions imposed in the simulation were probably more severe than would be encountered in practice, as normally the aircraft would be lined up with the centerline of the beam before engaging the coupler. This discussion concludes the analysis of the lateral beam coupler for this particular autopilot. The same analysis can be applied to any of the other lateral autopilots discussed here or encountered in the field. The final root locus is always the most critical and indicates whether some compensating networks are required to insure system stability for the overall system.

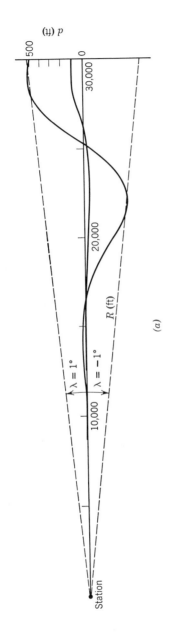

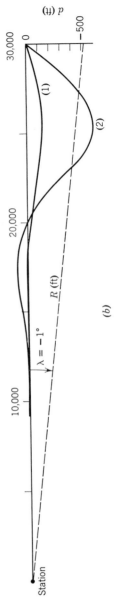

Figure 4-51 (a) Response of the lateral beam guidance system for $d_0 = 100$ and 500 ft and $\Psi_0 = \Psi_{(ref)}$. (b) Response of the lateral beam guidance system with $d_0 = 0$ and $\Psi_0 = \Psi_{(ref)}$, for a $90°$ crosswind of (1) 25 ft/sec; (2) 100 ft/sec.

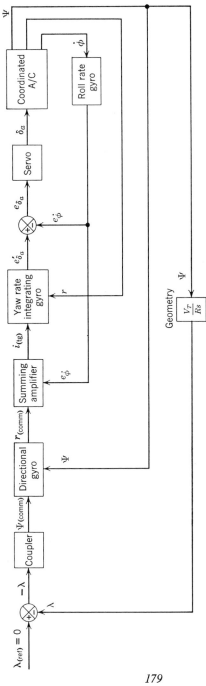

Figure 4-52 Block diagram of lateral beam guidance system using a yaw orientational control system.

$$[TF]_{[\text{servo}][e_{\delta_a} : \delta_a]} = \frac{10}{s + 10}$$

$$[TF]_{[\text{coord A/C}][\delta_a : \dot{\phi}]} = \frac{23}{s + 2.3}$$

An examination of the rate stabilized lateral autopilot (see Figure 4-38) shows that it yields the same pole configuration as the roll angle control system and the same coupler as that developed for the roll angle control system should be satisfactory for the rate stabilized system.

This technique can be used for the design and analysis of the coupler required for the yaw orientational control system studied in Section 4-5; however, roll rate feedback should be added around the coordinated aircraft as the inner loop.

Figure 4-52 is a block diagram of the proposed system. A detailed analysis is not presented here but is left as a design problem for the student; however, some suggestions are offered. In selecting the roll rate gyro sensitivity for the innermost loop, it is recommended that the complex poles have a damping ratio of 0.7. The next loop through the yaw rate integrating gyro is very critical. As shown in Section 4-5, the yaw integrating gyro loop adds two poles at the origin (see Figure 4-28) that move into the right-half plane as the integrating gyro sensitivity is increased. The roll rate feedback into the summing amplifier then brings these poles back into the left-half plane (see Figure 4-30). If the yaw rate integrating gyro sensitivity is selected properly, when these poles move into the left-half plane, they will break into the real axis. Above some critical value the poles will behave as shown in Figure 4-30. The selected value for the sensitivity of the yaw rate integrating gyro should be equal to or less than this critical value. The summing amplifier gain should then be selected to give a damping ratio for these complex poles of between 0.7 and 0.8. The heading feedback loop adds another pole at the origin. The directional gyro sensitivity should then be selected using the same criterion as discussed earlier in this section for the roll angle control system. A limiter will probably be required, after the directional gyro to limit the commanded yaw rate. The final root locus can then be used to determine if a compensator is required as part of the coupler.

This concludes the analysis of the automatic lateral beam guidance system. It should be mentioned, however, that although the first-order time lag was used to approximate the coordinated aircraft, excellent correlation was obtained between the computer results and the root locus.

4-9 Nonlinear Effects

So far, in all the analyses on the lateral autopilot, except for the effect of the limiter, it has been assumed that the system was linear. This is seldom the case for the actual system. Phenomena such as saturation, backlash, deadspot, and cable stretching introduce nonlinearities. These nonlinearities may result in a limit cycle roll oscillation or, if severe enough, in system instability. However, if they are known or suspected, these effects can be simulated during the computer simulation. By adjusting the severity of these nonlinearities during the computer simulation, the allowable limits on the nonlinearities can be determined. This information can then be used in specifying the tolerances on the actual hardware. Normally, very little backlash and deadspot can be tolerated, and every effort should be made to eliminate or minimize them. This can best be effected by installing close tolerance bolts in all linkages, using close tolerance gears or antibacklash gears, maintaining maximum allowable tension in all cables, and minimum deadspot in all hydraulic control valves.

4-10 Summary

In this chapter some of the techniques for artificially improving the lateral transient response of the aircraft have been presented and analyzed. The Dutch roll oscillation was first damped, then coordination improved, and then the complete lateral autopilot studied. The problems involved in providing automatic lateral beam guidance were demonstrated by means of the design and analysis of the lateral beam guidance coupler. After coordination was obtained, the transfer function of the lateral aircraft was reduced to a first-order time lag for aileron input to roll rate output, and the validity of this simplification was evidenced. All the paper analyses were substantiated by an analog computer simulation to show the actual improvements provided by each of the systems added.

Section 3-7 showed that the transient response of the aircraft varies considerably with changes in airspeed and altitude. This indicates changes in the roots of the characteristic equation and changes in the location of the aircraft poles and zeros when plotted in the complex plane. Thus, the root locus for each flight condition will vary as well as the optimum loop gain for any particular loop. Therefore, for the best operation the gains must be varied as the flight conditions change. This situation is discussed in more detail in Chapter 6.

There is one more aspect of autopilot design that may still be bothering the reader. How well does the computer simulation predict the actual behavior of the aircraft? It has been found that, in general, the computer simulation enables the designer to determine quite accurately all the system gains before the aircraft flies. Normally, some small changes in gains are required for optimum performance, but on the first model that flies there are provisions made for making final adjustments on the system gains. If the system does not perform as expected, the cause is probably due to either nonlinearities in the system that were overlooked or to inaccuracies in the values of the stability derivatives used.   Simulation of the basic equations should yield excellent results if the values of the stability derivatives themselves are accurate.   If the assumption of quasisteady flow is considered invalid, its effects may be included by introducing the proper stability derivatives if they are available.   The other assumptions may also be eliminated or modified by introducing the proper terms into the equations.   This procedure may result in nonlinear equations which cannot be studied analytically, but the equations can still be solved and the system analyzed on the computer.   This subject is discussed further in Chapter 6.

5

Inertial Cross-Coupling

5-1 Introduction

The problem of inertial cross-coupling did not manifest itself until the introduction of the century series aircraft. As the fighter plane evolved from the conventional fighter, such as the F-51 and F-47, through the first jet fighter, the F-80, and then to the F-100 and the other century series aircraft, there was a slow but steady change in the weight distribution. During this evolution more and more weight became concentrated in the fuselage as the aircraft's wings became thinner and shorter. This shift of weight caused relations between the moments of inertia to change. As more weight is concentrated along the longitudinal axis, the moment of inertia about the X axis decreases while the moments of inertia about the Y and Z axes increase. This phenomena increases the coupling between the lateral and longitudinal equations, and can best be seen by examining the basic moment equations which, from Eq. 1-32, are

$$\sum \Delta \mathscr{L} = \dot{P}I_x - \dot{R}J_{xz} + QR(I_z - I_y) - PQJ_{xz}$$

$$\sum \Delta \mathscr{M} = \dot{Q}I_y + PR(I_x - I_z) + (P^2 - R^2)J_{xz}$$

$$\sum \Delta \mathscr{N} = \dot{R}I_z - \dot{P}J_{xz} + PQ(I_y - I_x) + QRJ_{xz} \qquad (5\text{-}1)$$

As I_x becomes much smaller than I_y and I_z, the moment of inertia difference terms $(I_x - I_z)$ and $(I_y - I_x)$ become large. If a rolling moment is introduced this results in some yawing moment, and the term $PR(I_x - I_z)$ may become large enough to cause an uncontrollable pitching moment.

Another factor that must be considered is the aerodynamic effects which result if the aircraft is rolled about an axis other than the instantaneous stability axis, or the aircraft's velocity vector. If an aircraft rolls about its longitudinal axis at some angle of attack, the angle of attack after 90° of roll is now a sideslip angle; after 180° of roll it is a negative angle of attack, and after 270° it is again a sideslip angle. This interchange of angle of attack and sideslip angle causes the aerodynamic moments to alternate between pitching and yawing moments, which, coupled with the moment of inertia effects, further aggravates the inertial cross coupling.

5-2   Effects of High-Roll Rates

The design changes that have increased the coupling between the lateral and longitudinal equations were discussed in the Introduction. As a result of these design changes, some aircraft are completely unstable at high-roll rates. A mathematical analysis of this phenomenon would generally be considered impossible since the coupled equations involve products of the dependent variables. However, by making certain simplifying assumptions and with the proper combining of the equations, a linear analysis is possible. This linear analysis predicts the susceptibility of the aircraft to inertial cross-coupling, which is verified to some extent in this chapter, and has been verified in the past by actual flight test. For the purpose of this discussion, an aircraft is considered as being susceptible to inertial cross-coupling if the aircraft goes unstable at high-roll rates. After using the simplified linear analysis to predict the susceptibility of the aircraft to inertial cross-coupling, the analysis is continued to determine the required loop gains for a control system to stabilize the inertially cross-coupled aircraft. The predicted results are compared with the results of an analog computer simulation of the inertially cross-coupled aircraft complete with the control system. The reader may find it difficult to accept some of the assumptions made; however, they are necessary in order to obtain the desired linear equations. The justification stems mainly from the fact that the predicted results are valid, not that the assumptions themselves are normally considered valid.

To show mathematically that some aircraft are completely unstable at high-roll rates, the Y and Z force equations are combined so as to yield two oscillatory modes, which at zero roll rate represent the short-period and Dutch roll modes. It is shown mathematically that as the roll rate is increased, the natural frequency and damping ratio of the two oscillations vary and the lateral oscillations may eventually become divergent.

For this analysis, U is considered constant so that the X force equation can be eliminated. This leaves only the Y and Z force equations which, from Eq. 1-33, are

$$\sum \Delta F_y = m(\dot{V} + UR - WP)$$

$$\sum \Delta F_z = m(\dot{W} + VP - UQ) \tag{5-2}$$

As it is desired to study the effects of high-roll rates, P is assumed to be equal to P_0; and since only the steady state condition is to be evaluated, P_0 is assumed to exist instantaneously. How it is developed or what happens during the transient period is of no interest at this time. As the roll rate has been assumed constant, the changes in the rolling moments must be zero, and the rolling moment equation may be eliminated. Also, if $V_0 = W_0 = Q_0 = R_0 = 0$, then Eqs. 5-1 and 5-2 become

$$\sum \Delta \mathcal{M} = \dot{q} I_y + P_0 r (I_x - I_z) + (P_0{}^2 - r^2) J_{xz}$$

$$\sum \Delta \mathcal{N} = \dot{r} I_z + P_0 q (I_y - I_x) + q r J_{xz}$$

$$\sum \Delta F_y = m(\dot{v} + U_0 r - w P_0)$$

$$\sum \Delta F_z = m(\dot{w} + v P_0 - U_0 q) \qquad (5\text{-}3)$$

If the stability axes are also considered as principle axes, and if $\sum \Delta \mathcal{M} = \sum \Delta \mathcal{N} = \sum \Delta F_y = \sum \Delta F_z = 0$, then Eq. 5-3 reduces to

$$\dot{q} I_y + P_0 r (I_x - I_z) = 0$$

$$\dot{r} I_z + P_0 q (I_y - I_x) = 0$$

$$\dot{\beta} + r - P_0{}'\alpha = 0$$

$${}'\dot{\alpha} + P_0 \beta - q = 0 \qquad (5\text{-}4)$$

because $\beta \simeq v/U_0$ and ${}'\alpha \simeq w/U_0$.

Combining the moment equations from Eq. 5-4 with the aerodynamic moments including all the stability derivatives retained in Chapters 1 and 3 yields

$$\frac{I_z}{Sqb} \dot{r} - \frac{b}{2U} C_{n_r} r - C_{n_\beta} \beta + \frac{I_y - I_x}{Sqb} P_0 q = \frac{b}{2U} C_{n_p} P_0$$

$$\frac{I_y}{Sqc} \dot{q} - \frac{c}{2U} C_{m_q} q - \frac{c}{2U} C_{m_{\dot{\alpha}}}{}'\dot{\alpha} - C_{m_\alpha}{}'\alpha + \frac{I_x - I_z}{Sqc} P_0 r = 0 \qquad (5\text{-}5)$$

Solving the force equations of Eq. 5-4 for r and q, yields

$$r = P_0{}'\alpha - \dot{\beta} \quad \text{and} \quad q = P_0 \beta + {}'\dot{\alpha} \qquad (5\text{-}6)$$

Differentiating,

$$\dot{r} = P_0{}'\dot{\alpha} - \ddot{\beta} \quad \text{and} \quad \dot{q} = P_0 \dot{\beta} + {}'\ddot{\alpha} \qquad (5\text{-}7)$$

Substituting Eqs. 5-6 and 5-7 into Eq. 5-5 yields

$$\frac{I_z}{Sqb}(P_0'\dot{\alpha} - \ddot{\beta}) - \frac{b}{2U}C_{n_r}(P_0'\alpha - \dot{\beta}) - C_{n_\beta}\beta + \frac{I_y - I_x}{Sqb}P_0(P_0\beta + '\dot{\alpha})$$

$$= \frac{b}{2U}C_{n_p}P_0$$

$$\frac{I_y}{Sqc}(P_0\dot{\beta} + '\ddot{\alpha}) - \frac{c}{2U}C_{m_q}(P_0\beta + '\dot{\alpha}) - \frac{c}{2U}C_{m_{\dot{\alpha}}}'\dot{\alpha} - C_{m_\alpha}'\alpha$$

$$+ \frac{I_x - I_z}{Sqc}P_0(P_0'\alpha - \dot{\beta}) = 0 \quad (5\text{-}8)$$

The roots of the characteristic equation determine the stability of the system; thus only the characteristic equation will be considered in the subsequent analysis. Regrouping and taking the Laplace transformation of the characteristic equation, Eq. 5-8 becomes

$$\left[\frac{I_z}{Sqb}s^2 - \frac{b}{2U}C_{n_r}s + C_{n_\beta} - \left(\frac{I_y - I_x}{Sqb}\right)P_0^2\right]\beta(s)$$

$$+ \left[-\left(\frac{I_y - I_x + I_z}{Sqb}\right)P_0s + \frac{b}{2U}C_{n_r}P_0\right]'\alpha(s) = 0$$

$$\left[\left(\frac{I_y - I_x + I_z}{Sqc}\right)P_0s - \frac{c}{2U}C_{m_q}P_0\right]\beta(s)$$

$$+ \left[\frac{I_y}{Sqc}s^2 - \frac{c}{2U}(C_{m_q} + C_{m_{\dot{\alpha}}})s - C_{m_\alpha} + \left(\frac{I_x - I_z}{Sqc}\right)P_0^2\right]'\alpha(s) = 0 \quad (5\text{-}9)$$

An examination of Eq. 5-9 shows that for $P_0 = 0$, the coupled equations reduced to two uncoupled equations which are the same as Eq. 3-54 (the one-degree-of-freedom Dutch roll mode) and Eq. 1-130, after eliminating the C_{z_α} terms (the short-period approximation). For $P_0 \neq 0$, Eq. 5-9 has the form of

$$(a_2s^2 + a_1s + a_0)\beta(s) + (c_1s + c_0)'\alpha(s) = 0$$

$$(d_1s + d_0)\beta(s) + (b_2s^2 + b_1s + b_0)'\alpha(s) = 0 \quad (5\text{-}10)$$

which yields a fourth-order characteristic equation, the roots of which are a function of P_0. These roots can then be plotted in the complex plane as a function of P_0. The values used in evaluating Eq. 5-9 are listed in Table 5-1. After substitution of these values into Eq. 5-9 the characteristic equations for the two aircraft are

Aircraft A: $s^4 + 0.585s^3 + (14.7 + 1.92P_0^2)s^2 + (4.2 + 0.3P_0^2)s$

$$+ (54 - 13.24P_0^2 + 0.817P_0^4) = 0 \quad (5\text{-}11)$$

Table 5-1   Values Used to Evaluate Eq. 5-9

Parameter	Aircraft A	Aircraft B
C_{m_α}	-1	-0.6
C_{n_β}	0.4	0.1
$\dfrac{I_x - I_z}{Sqc}$	-0.121	-0.21
$\dfrac{I_y - I_x}{Sqb}$	0.05	0.056
$\dfrac{I_y + I_z - I_x}{Sqb}$	0.11	0.126
$\dfrac{I_y + I_z - I_x}{Sqc}$	0.25	0.44
$\dfrac{I_y}{Sqc}$	0.13	0.23
$\dfrac{I_z}{Sqb}$	0.057	0.07
$\dfrac{c}{2U} C_{m_q}$	-0.04	-0.02
$\dfrac{c}{2U}(C_{m_q} + C_{m_{\dot\alpha}})$	-0.06	-0.026
$\dfrac{b}{2U} C_{n_r}$	-0.007	-0.0034

Aircraft B: $s^4 + 0.16s^3 + (4 + 1.7P_0{}^2)s^2 + (0.3 + 0.115P_0{}^2)s$
$$+ (3.7 - 3.4P_0{}^2 + 0.73P_0{}^4) = 0 \quad (5\text{-}12)$$

The locus of the roots for Eq. 5-11 is plotted in Figure 5-1.   As P_0 is increased from zero, the locus of the roots is similar to Figure 5-2 with the exception that the real root does not move into the right-half plane and, thus the aircraft remains stable.   The locus of the roots of Eq. 5-12, as P_0 is varied, is plotted in Figure 5-2.*   As P_0 is increased from zero, the natural frequency of one set of roots increases and the damping ratio decreases, while the opposite is true for the other roots.   The latter set of roots then breaks into the real axis forming two real roots, one of which moves into the right-half plane and becomes divergent.   The neutral

* As the curve is symmetrical about the real axis, only the upper half of the plot is shown.

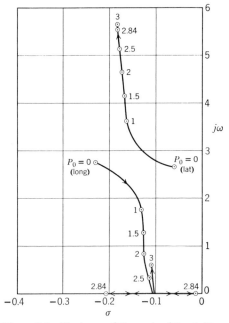

Figure 5-1   The locus of the roots of Eq. 5–11,
Aircraft A, as a function of the steady-state
roll rate, P_0.

stability point occurs at $P_0 = 1.33$ rad/sec (97.4 deg/sec). At the higher
value of P_0 the divergent root moves into the left-half plane, and the
aircraft is again stable. A further increase in P_0 results in the two real
roots becoming oscillatory again. An examination of the coefficient of
the s^0 term shows whether the aircraft will go unstable. These terms, for
both aircraft, are presented here for comparison.

Aircraft A: $54 - 13.24P_0^2 + 0.817P_0^4$

Aircraft B: $3.7 - 3.4P_0^2 + 0.73P_0^4$ (5-13)

Since the algebraic signs of the rest of the terms in the characteristic
equation are positive, the sign of the term represented by Eq. 5-13
determines the stability or instability of the system. A characteristic
equation can have roots in the right-half plane even though all the
coefficients are positive (see Routh's stability criterion p. 78–83, Ref. 1).
If this does occur, the value of P_0 would be so large that it would be of
academic interest only. Because the coefficient of the P_0^2 terms in Eq.
5-13 are negative, it is possible for the whole term to become negative.

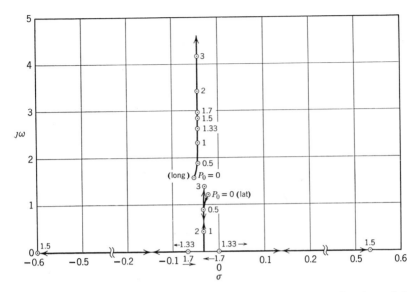

Figure 5-2   The locus of the roots of Eq. 5-12, Aircraft B, as a function of the steady-state roll rate, P_0.

The points of neutral stability can be determined by equating the Eq. 5-13 to zero and solving for P_0.   This operation yields the values of 1.33 and 1.7 rad/sec for aircraft B and imaginary roots for aircraft A.   The imaginary roots indicate there is no real value of roll rate that makes the coefficient of the s^0 term go to zero; therefore the term cannot change sign, and aircraft A is stable for all roll rates.

5-3   Determination of the Aircraft Parameters that Affect Stability

In Section 5-2 it was shown that the sign of the coefficient of the s^0 term of the characteristic equation for steady rolling determined the stability of the aircraft at high-roll rates.   The s^0 term is examined here in its general form to determine which of the aircraft parameters have the largest effect on its sign.   This term can be obtained from the expansion of the determinant formed from Eq. 5-9; the s^0 term is then

$$-(C_{n_\beta} C_{m_\alpha}) + \left\{ \left[C_{m_\alpha}\left(\frac{I_y - I_x}{Sqb}\right) + C_{n_\beta}\left(\frac{I_x - I_z}{Sqc}\right) \right] + \left(\frac{bc}{4U^2} C_{n_r} C_{m_q}\right) \right\} P_0^2$$

$$- \left(\frac{I_y - I_x}{Sqb}\right)\left(\frac{I_x - I_z}{Sqc}\right) P_0^4 \quad (5\text{-}14)$$

From Eq. 5-13 it can be seen that it is the coefficient of the $P_0{}^2$ term that is negative; thus, this term of Eq. 5-14 is examined first. Referring to Table 5-1, it can be seen that the portion of the coefficient of the $P_0{}^2$ term of Eq. 5-14 that is in brackets is negative, while the portion in parentheses is positive. Since the negative portion is larger than the positive term, the net sign of the term is negative. If the moments of inertia of the aircraft about each of the three axes were equal, the coefficient of the $P_0{}^2$ term would be positive, and there would be no inertial cross-coupling. As I_y and I_z increase and I_x decreases (which has been the trend in modern jet fighters) the negative portion of the $P_0{}^2$ term increases; eventually the sign of the $P_0{}^2$ term goes negative as I_y and I_z become increasingly larger than I_x, and the possibility of inertial cross-coupling increases. There is little that can be done about the relative magnitude of the moments of inertia of the present-day high performance fighters; thus, to offset the effects of the increase in the magnitude of the negative portion of the $P_0{}^2$ term, the term in parentheses which contains the product of C_{n_r} and C_{m_q} should be increased. This can be done by increasing the size and lift curve slope of the vertical stabilizer for C_{n_r} and the size and lift curve slope of the horizontal stabilizer for C_{m_q}. However, this operation would require extensive modification of the aircraft and would be expensive. The terms C_{n_r} and C_{m_q} can be increased artificially by adding yaw and pitch rate feedback. It is first necessary to determine how much the product of C_{n_r} and C_{m_q} must be increased so that Eq. 5-14 will be positive for all values of P_0. By examining Eq. 5-13 for aircraft B, which is repeated here for convenience,

$$3.7 - 3.4P_0{}^2 + 0.73P_0{}^4 \qquad (5\text{-}15)$$

it can be seen that if the coefficient of the $P_0{}^2$ term of Eq. 5-15 is decreased to 3.27, the equation will be positive for all values of P_0. This means that

$$\frac{\left[C_{m_\alpha}\left(\dfrac{I_y - I_x}{Sqb}\right) + C_{n_\beta}\left(\dfrac{I_x - I_z}{Sqc}\right)\right] + \dfrac{bc}{4U^2}C_{n_r}C_{m_q}}{\left(\dfrac{I_z}{Sqb}\right)\left(\dfrac{I_y}{Sqc}\right)} = -3.27 \quad (5\text{-}16)$$

Equation 5-16 is divided by $(I_z/Sqb)(I_y/Sqc)$ just as Eq. 5-12 was normalized by dividing by the coefficient of the s^4 term. Substituting the values from Table 5-1 into Eq. 5-16 it is found that $(bc/4U^2)C_{n_r}C_{m_q}$ must be equal to 0.002. The original value of $(bc/4U^2)C_{n_r}C_{m_q}$ was 0.000068; therefore, $C_{n_r}C_{m_q}$ must be increased by a factor of 29.4. This means that if C_{n_r} and C_{m_q} are each increased by a factor of 5.42, the desired results will be obtained. From Eq. 3-56 it can be seen that the damping ratio of the Dutch roll is proportional to C_{n_r}, and Eq. 1-136 shows that the damping ratio of the short-period is proportional to the $\sqrt{C_{m_q}}$. Thus, to obtain the desired increase in C_{n_r} and C_{m_q}, the damping of the Dutch roll should be increased by a factor of 5.42 and the damping ratio of the short-period by a factor of 2.33. In Section 5-4 a control system to achieve the desired increase in the damping ratios of the Dutch roll and short-period oscillations is presented and analyzed to determine its effectiveness in stabilizing the aircraft at high-roll rates.

5-4 System for Controlling an Aircraft Subject to Inertial Cross-Coupling

Figure 5-3 is a block diagram of the system for controlling an aircraft that is subject to inertial cross-coupling. To analyze the system, each of the two loops in Figure 5-3 are studied separately to determine the sensitivities of the two rate gyros required to achieve the required increase in the damping ratios of the short period and Dutch roll modes. Figure 5-4 is the block diagram for the pitch rate feedback loop with the transfer functions of each block indicated. From the aircraft transfer function it can be seen that the damping ratio of the short period is 0.176. As developed in Section 5-3, this amount should be increased by a factor of 2.33, which means a minimum value of 0.41. Figure 5-5 is a root locus for the pitch rate feedback loop. To provide for a margin of safety, a minimum damping ratio of 0.5 is selected. This results in a $S_{(prg)}$ of 1.28 volt/deg/sec.

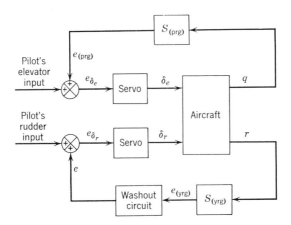

Figure 5-3 System for controlling an aircraft subject to inertial cross-coupling.

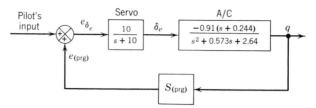

Figure 5-4   Pitch rate feedback loop.

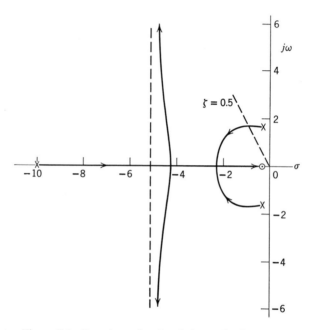

Figure 5-5   Root locus for the pitch rate feedback loop.

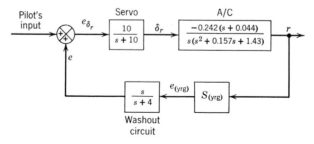

Figure 5-6   Yaw rate feedback loop.

Figure 5-6 is the block diagram for the yaw rate feedback loop, again with the transfer functions indicated.   From the aircraft transfer function it can be seen that the damping ratio of the Dutch roll is 0.066.   To increase this amount by a factor of 5.42 (as indicated in Section 5-4) means that the minimum value of the damping ratio must be 0.358 or, to provide a margin of safety, use 0.4.   Figure 5-7 is the root locus for the yaw rate feedback loop.   The value of the yaw rate gyro sensitivity for maximum damping is 26 volt/deg/sec, which yields a damping ratio of 0.45.   These values are used in the computer simulation, the results of which are shown in Figure 5-8 and 5-9.   Figure 5-8 shows the response of the basic aircraft for a commanded roll rate of 1.5 rad/sec.   The effects of inertial

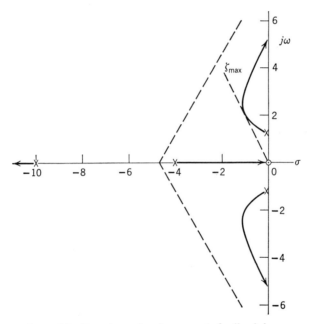

Figure 5-7   Root locus for the yaw rate feedback loop.

cross-coupling are obvious.   Figure 5-9 shows the response of the basic aircraft and the control system.   The improvement in the response is apparent; however, the response is still not optimum as large values of sideslip and angle of attack still exist, which are undesirable.   The sideslip could be eliminated by one of the coordination techniques discussed in Chapter 4.   The steady-state angle of attack is required to balance the Z components of acceleration resulting from the pitch rate and the product of βP.

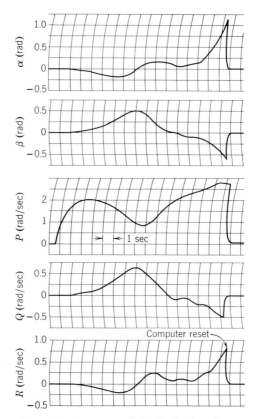

Figure 5-8   Response of the basic aircraft to a
commanded roll rate of 1.5 rad/sec.

5-5 Improved System for Controlling an Aircraft Subject to Inertial Cross-Coupling

The pitch orientational control system is used to reduce the pitch rate to zero so that the steady-state angle of attack is reduced; if necessary a sideslip sensor is used to reduce sideslip. The block diagram of the proposed system is shown in Figure 5-10. Since the pole at the origin introduced by the integrating gyro causes the closed loop complex poles, resulting from the inner loop to move toward the imaginary axis, the gain of the inner loop is increased from 1.28 to 2.37 volt/deg/sec. This increase in gain increases the damping ratio of the complex poles to 0.8 and results in the following closed loop transfer function for the inner loop.

$$[TF]_{(CL)[e_{(ig)};q]} = \frac{-9.1(s + 0.244)}{(s + 6.9)(s^2 + 3.6s + 5.3)}$$

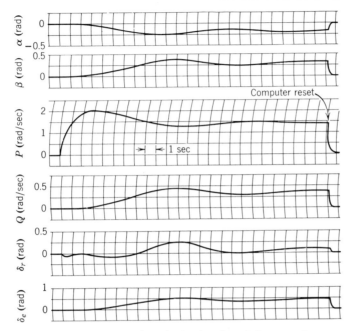

Figure 5-9   Response of the basic aircraft and the control system
to a commanded roll rate of 1.5 rad/sec.

The block diagram for the outer loop is shown in Figure 5-11.   The root
locus for the outer loop is shown in Figure 5-12 with the location of the
closed loop complex poles for an integrating gyro gain of 10 volt/deg/sec.
This gain was determined from the simulation because the validity of the
root locus for the outer loop was questionable.   Several other gains were
used to determine the best integrating gyro sensitivity.   An integrating
gyro sensitivity of 0.62 volt/deg/sec, that should have yielded a damping
ratio of 0.7 for the complex roots, resulted in a rapid divergence after

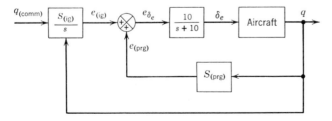

Figure 5-10   Pitch orientational control system for controlling an
aircraft subject to inertial cross-coupling.

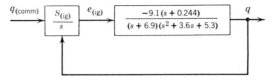

Figure 5-11 Block diagram of outer loop for the
inertial cross-coupling control system.

about 19 sec.   This is really not too surprising considering the assumptions
made for the linear analysis.   However, the root locus did predict the
integrating gyro gain for instability very accurately, as can be seen in
Table 5-2.   The results of the computer simulation are shown in Figure
5-13.   The effects of the integrating gyro are apparent; not only has the
steady-state pitch rate been reduced to zero but the sideslip also has been
greatly reduced.   The maximum sideslip is 1.25° with a steady-state value
of 0.75°.   It is doubtful that a sideslip sensor would be very effective in
reducing the remaining sideslip.   An examination of the roll-rate response
shows that it has the characteristics of a first-order response.   The one

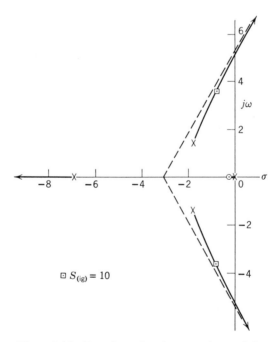

Figure 5-12 Root locus for the outer loop of the
inertial cross-coupling control system.

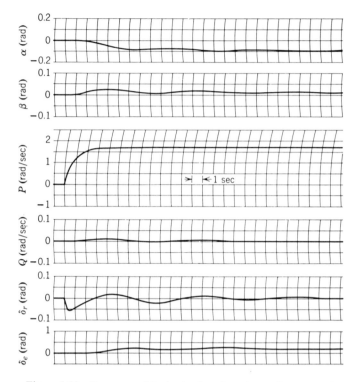

Figure 5-13   Response of the basic aircraft and complete control
system to an aileron step command of 0.2 rad.

Table 5-2   *Comparison of Actual and Predicted Values of $S_{(lg)}$ for Instability*

	Predicted	Actual
$S_{(lg)}$ for $\zeta = 0$ (volt/deg/sec)	27.7	26.2
ω_n for $\zeta = 0$ (rad/sec)	5.2	5.23

degree-of-freedom rolling mode transfer function can be obtained by
using Eq. 3-60 and is

$$\frac{P(s)}{\delta_a(s)} = \frac{10}{s + 1.2} \tag{5-17}$$

For the computer simulation an aileron step input of 0.2 rad is used.
Then from Eq. 5-17 the steady-state roll rate should be 1.67 rad/sec,
which checks perfectly with the computer results.   From Eq. 5-17 the

roll subsidence time constant is 0.833 sec. The measured value of the time constant from the computer trace is 0.8 sec, which provides excellent agreement. Thus, now that the effects of inertial cross-coupling have been overcome, the coordinated aircraft transfer functions can be used for any further lateral autopilot analysis for this aircraft, as was done in Chapter 4 for the jet transport.

This concludes the analysis of an inertially cross-coupled aircraft and a control system to stabilize such an aircraft. Although the simplified linear analysis was used to study the effects of inertial cross-coupling, and to determine the minimum values of the pitch and yaw rate gyro sensitivities, the complete coupled five-degree-of-freedom equations were used in the computer simulation (the airspeed was assumed constant).

References

1. J. J. D'Azzo, and C. H. Houpis, *Feedback Control System Analysis and Synthesis*, McGraw-Hill Book Co., New York, 1960.

6

Self-Adaptive Autopilots

6-1 Introduction

As discussed in Chapters 1 and 3, the transient response of the aircraft varies considerably with changes in airspeed and altitude. This means that the roots of the characteristic equation, and thus the location in the complex plane, of the poles and zeros of the aircraft transfer functions change widely. Thus, the root locus for each flight condition and the optimum gain for each loop vary. For this reason, until recently, the gains of all autopilots were scheduled as a function of Mach number or dynamic pressure. This required extensive system analysis on the ground to determine the gains for the various flight conditions. To perform this analysis the stability derivatives for the entire flight regime had to be known accurately. These figures were usually obtained from wind tunnel tests of scale models of the aircraft. After the aircraft was built, provisions had to be made for changing the autopilot gains in flight. This procedure was then followed by extensive flight testing to determine the final optimum gain settings. All this analysis was time consuming and expensive. Also, any changes in airplane configuration probably would require more autopilot testing and adjustment.

For these reasons, the Air Research and Development Command initiated a program to determine methods of adjusting the loop gains of the flight control systems without the necessity of sensing any air data. This program lead to a closed loop system operation of automatic gain adjustment by constantly monitoring the performance of the system. In January 1949 a symposium was held at Wright-Patterson AFB to present to industry the results of this program and to encourage applications of the techniques developed. The proceedings of this symposium were published in the form of a WADC Technical Report.[1] This report defines a self-adaptive system as one "which has the capability of changing its parameters through an internal process of measurement, evaluation, and adjustment to adapt to a changing environment, either external or internal, to the vehicle under control."

With this introduction as background, the general philosophy of the operation of self-adaptive control systems is discussed, followed by a description of three of the systems presented at the Wright-Patterson symposium.

6-2   General Philosophy of the Self-Adaptive Control System

To better understand the action of the self-adaptive control system, the effects of very high forward loop gain are discussed. Figure 6-1 is the block diagram of a generalized control system. The controller contains the dynamics of the systems necessary to drive the controlled element including compensators, if required, and a variable gain. The controlled element would consist of the dynamics of the unit being controlled, in this case, the aircraft. The closed loop transfer function relating the output to the input is

$$\frac{\text{Output}}{\text{Input}} = \frac{G_1(s)G_A(s)}{1 + G_1(s)G_A(s)} \simeq 1 \quad \text{for} \quad G_1(s)G_A(s) \gg 1 \qquad (6\text{-}1)$$

The transfer function of the system for a disturbance input is

$$\frac{\text{Output}}{\text{Disturbance}} = \frac{G_A(s)}{1 + G_1(s)G_A(s)} \simeq \frac{1}{G_1(s)} \quad \text{for} \quad G_1(s)G_A(s) \gg 1 \quad (6\text{-}2)$$

Thus, Eq. 6-1 shows that if the gain is very high, the ratio of the output to the input is approximately 1 or that the output always follows the input. Equation 6-2 shows that the response of the system to an external disturbance is approximately equal to $1/G_1(s)$, which, if $G_1(s)$ is very large, means that the response of the system to a disturbance is negligible. The basic system can be improved by adding an input filter ahead of the summer. This input filter, when used with self-adaptive control systems, is usually referred to as a "model." The revised block diagram is shown in Figure 6-2. Now the input to the summer is the output of the model, and as the closed loop transfer function for the basic

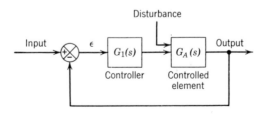

Figure 6-1   Block diagram of a generalized control system.

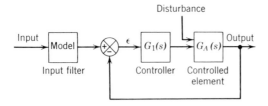

Figure 6-2   Block diagram of basic system with input filter added.

system is approximately 1, the output approximates the output of the model.   The dynamics of the model can therefore be adjusted to yield an optimum response; thus the system always behaves in an optimum fashion.   All this theory would be fine if the extremely high gain required could be realized at all times; however, as seen in Chapters 2 and 4, there is usually a practical limitation on the gain if the system is to remain stable. In spite of this problem, the idea of the high forward loop gain is used in most of the self-adaptive systems.   For systems of this type various schemes are used to obtain the maximum gain and still maintain stability. There is one precaution that must be taken when utilizing this principle: the poles that move toward the imaginary axis are the poles of the controller and not those of the controlled element.   This often requires the addition of a compensator to assure that the controlled element poles break into and remain on the negative real axis.   With this introduction complete, Sections 6-3, 6-4, and 6-5 describe three different self-adaptive control systems.

6-3   Sperry Self-Adaptive Control System[1,2]

The Sperry self-adaptive control system relies on maintaining the maximum gain possible with light damping of the servo poles.   Figure 6-3 is a block diagram of the basic system without the automatic gain changer.

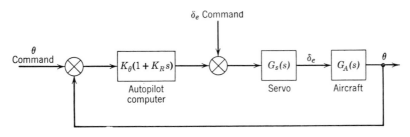

Figure 6-3   Basic pitch attitude system.

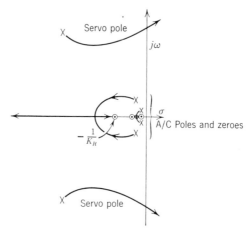

Figure 6-4   Typical root locus for the basic attitude system. *Note:* This figure is not to scale.

The basic autopilot employs both pitch rate and pitch attitude feedback with K_R the ratio of the rate to displacement gain ($K_{\dot{\theta}}/K_\theta$), with $K_{\dot{\theta}}$ the rate gain and K_θ the displacement gain.   Figure 6-4 shows the form of the root locus as K_θ is increased with K_R held constant.   If the gain is high enough, the aircraft poles will be driven into the real axis, one moving out along the real axis to infinity and the other three moving into close proximity of the three zeros.   Thus the servo poles become the dominant ones.   For a given damping ratio, the higher the natural frequency of the servo poles, the higher the gain possible before instability.   Therefore, it would be desirable to have the natural frequency of the servo as high as possible with a damping ratio of about 0.7.   In Section 6-2 it was shown that the higher the forward loop gain, the more nearly the closed loop transfer function approximated unity.   Thus, it is desirable to increase the gain to move the servo poles as close to the imaginary axis as possible without decreasing the damping ratio too much.   The Sperry system was designed to keep the damping ratio of the servo poles between about 0.11 and 0.23.

Figure 6-5 is a block diagram of the Sperry self-adaptive control system. To determine the location of the servo poles, small amplitude narrow pulses are fed to the servo actuator and the response of the servo measured.   The amplitude and duration of the pulses is such that they mainly excite the oscillatory mode associated with the lightly damped high frequency servo poles.   The output of the pulse shaper is a series of constant amplitude pulses, with the number of pulses corresponding to the number of reversals

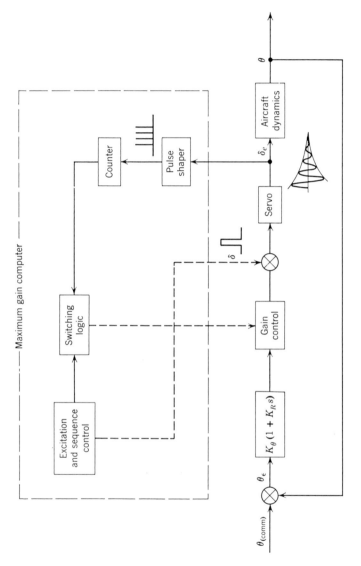

Figure 6-5   Block diagram of the Sperry self-adaptive control system.

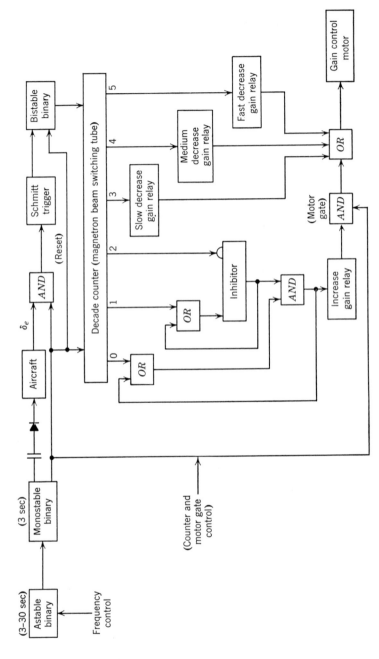

Figure 6-6   Block diagram of the logic circuit of the maximum gain computer.

of δ_e. The pulses are counted by the counter which controls the switching logic that adjusts the system gain. Since the number of pulses is a measure of the damping, the switching logic is able, by sensing the output of the counter, to control the gain and maintain the damping of the servo poles between the desired bounds (sometimes referred to as process identification). The sampling period and the amplitude and shape of the excitation pulse are established by the excitation sequence control. With the aircraft flying in still air, the pilot will probably be able to detect the tremors of the aircraft resulting from the pulsing. The repetition rate of the pulse is made a function of the changing environment, with a higher repetition rate occurring when the environment is changing rapidly. The period between pulses was varied from 3 to 30 sec. This was accomplished by sensing vertical rate and forward acceleration.

Figure 6-6 is a block diagram of the logic circuit of the maximum gain computer. The output of the Schmitt Trigger is a series of pulses, the number of which is equal to the number of control surface reversals which exceed the threshold sensitivity of the Schmitt Trigger. The number of pulses from the Schmitt Trigger controls the gain control motor through the decade counter. If the count of the counter is three, four, or five, the gain is decreased at a slow, medium, or fast rate, respectively. A slow increase in gain is commanded if a count of zero or one is obtained from two successive sampling periods. If the count is two, no change in gain occurs. The three-second pulse from the monostable binary stage or circuit allows the counter and motor gate to remain open for three seconds. Thus, after the three-second pulse from the monostable binary circuit, no change in gain is effected until the next pulse is generated by the astable binary circuit. This prevents gain changes due to random disturbances during the period between pulses.

The system was simulated by Sperry on an analog computer using a hypothetical aircraft, the physical equipment for the maximum gain control, and the servo actuator. For this simulation K_θ was held constant and also varied as a function of the error (θ_ϵ). The latter method, referred to as "nonlinear error control," resulted in tighter control of the aircraft.

A self-adaptive system is ideally suited to determine the switchover point from aerodynamic control to reaction controls for vehicle exit and vice versa on reentry. As the vehicle reaches the less dense atmosphere, the system gains continue to increase to compensate for the reduction in elevator effectiveness. By designing the system to switch to reaction controls when the gain reaches a certain value, successful control of the vehicle during exit and reentry is possible. This was also demonstrated by analog simulation. Thus, the Sperry system demonstrated the practicality of the self-adaptive control system utilizing a maximum forward gain controlled by a self-contained process of measurement, evaluation, and adjustment.

6-4   Minneapolis-Honeywell Self-Adaptive Control System[1,3]

The desirability of using a model as an input filter was discussed in Section 6-2. This concept was utilized by Minneapolis-Honeywell for their self-adaptive control system, the block diagram of which is shown in Figure 6-7. In order for the response of the aircraft to be equal to that of the model, the forward loop gain must be very high. As the gain required is usually larger than can be achieved by a linear servo system, a nonlinear system of the bang-bang type was used. In order to eliminate the limit cycle oscillations characteristic of a bang-bang system, an A-C dither signal was introduced. The effect of the dither signal on the apparent output of the relay is explained later in this section.

Figure 6-7 indicates that the system can operate as a pitch-rate command system or a pitch-attitude command system. The operation as a pitch-rate command system is discussed first.

The commanded pitch rate is fed to the model, the dynamics of which were selected from values established by NACA and Cornell Aeronautical Laboratories as acceptable for this type of aircraft. The output of the model is summed with the actual pitch rate of the aircraft, which is detected by a rate gyro. The output of the summer is the error between the aircraft's response and that of the model's. This error signal (ϵ) is fed to the switching logic circuit and the gain changer. The denominator time constant of the switching logic circuit is ideally zero, thus yielding a proportional plus derivative circuit. Also included in the switching logic block, but not shown, is an isolation amplifier. The output of the switching logic circuit is summed with the A-C dither signal, which is a 2000 cps sine wave, and fed to the electronic relay. Obviously the servo and the aircraft cannot follow the 2000 cps dither signal; thus, the effective result is an averaging of the actual output of the relay so that the system responds as if the output of the relay were the arc sine characteristics shown in Figure 6-8.

The output of the relay is a commanded elevator rate. The limiter is designed to control the output of the relay as a function of the absolute magnitude of the error (see Figure 6-7). As long as the magnitude of the error is greater than some fixed quantity B, the output of the limiter is at its maximum value of 9.2 deg/sec of commanded elevator rate. When the magnitude of the error decreases to B, the gain changer starts decreasing the limits of the limiter at an exponential rate, with a time constant of

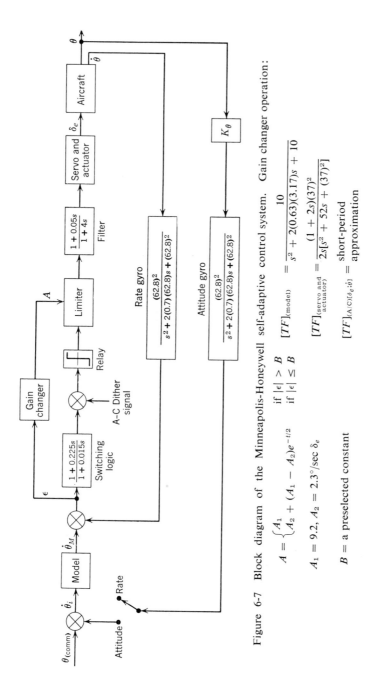

Figure 6-7 Block diagram of the Minneapolis-Honeywell self-adaptive control system. Gain changer operation:

$$A = \begin{cases} A_1 & \text{if } |\epsilon| > B \\ A_2 + (A_1 - A_2)e^{-t/2} & \text{if } |\epsilon| \le B \end{cases}$$

$$A_1 = 9.2, A_2 = 2.3°/\text{sec } \delta_e$$

B = a preselected constant

$$[TF]_{(model)} = \frac{10}{s^2 + 2(0.63)(3.17)s + 10}$$

$$[TF]_{\substack{(servo\ and \\ actuator)}} = \frac{(1 + 2s)(37)^2}{2s[s^2 + 52s + (37)^2]}$$

$$[TF]_{[A/C][\delta_e;\dot\theta]} = \text{short-period approximation}$$

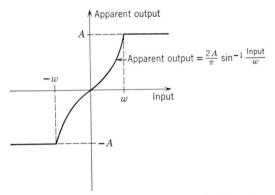

Figure 6-8   Apparent output of an ideal relay with sinusoidal dither.

2 sec to the minimum value of 2.3 deg/sec of commanded elevator rate. The output of the limiter is fed through a filter before being applied to the servo. The pole of the filter was selected to cancel approximately the zero of the servo and actuator. The zero or lead of the filter is introduced to compensate for the backlash of the control gearing. This operation is an example of compensation required to counteract the nonlinear effects and would not be predicted by any linear analysis. The servo is represented by a second-order system with a natural frequency of 37 rad/sec and a damping ratio of 0.7. The $(1 + 2s)/s$ in the servo transfer function results from a proportional plus integral network in the forward loop ahead of the servo and actuator. The aircraft is represented by the short-period approximation. By maintaining a very high forward loop gain, the actual pitch rate is made to follow the output of the model with an acceptable error; thus, the aircraft's response is effectively that of the model at all operating conditions.

The system was flight tested in a F-94C. The first set of tests were conducted with the limiter set at 4.6 deg/sec of commanded elevator rate. For this condition the performance was excellent except for the extreme ranges of the flight envelope. At low dynamic pressure the damping was lower than that of the model, while at high dynamic pressure the system followed the model, but there was a low amplitude limit cycle that was only marginally acceptable. Introduction of the variable gain improved the performance at both extremes of the flight envelope by increasing the gain at conditions of low dynamic pressure and decreasing it at conditions of high dynamic pressure. The limits of the gain of 9.2 and 2.3 deg/sec of elevator rate were found to be completely adequate to provide acceptable performance throughout the flight envelope.

For the attitude control mode the switch is moved to the attitude position. Since the closed loop transfer function for $\dot{\theta}_M$ in to $\dot{\theta}$ out is effectively unity, the closed loop transfer function for $\theta_{(comm)}$ in to θ out is simply a third-order system, the two complex poles from the model and the $1/s$ necessary to go from $\dot{\theta}$ out to θ out. The closed loop poles for the attitude control system for $K_\theta = 1$ are a real pole at -1.63 and two complex poles with a natural frequency of 2.48 rad/sec and a damping ratio of 0.48. The flight test of this configuration substantiated the assumption that the inner loop transfer function can be assumed to be unity. Thus, this system is also self-adaptive and adjusts the forward loop gain as a function of the error between the model response and the system response.

6-5 MIT Model-Reference Adaptive Control System for Aircraft[1,4]

As the name implies, the MIT system also uses a model but not as an input filter as in the case of the Minneapolis-Honeywell system. The output of the model is compared to the output of the system, and the gains of the system are adjusted as a function of the system error. However, the gains are not kept at the highest possible level consistent with a certain stability level, but they are adjusted so that certain error criteria are satisfied. These error criteria and the operation of the system are discussed in detail in the following paragraphs.

Figure 6-9 is the block diagram of the system. An examination of Figure 6-9 shows that the basic control system is the same as the yaw orientational control system discussed in Chapter 4, with the addition of the roll damping loop. The three loop gains are varied by the output of the error criteria logic system through the three rate servos that drive the three potentiometers. To better understand the error criteria used to control each loop gain and the necessity for the roll damping loop, the effects of the loop gains are illustrated by examining the root locus of each loop.

The root loci of the system are first drawn using the optimum gains as selected by the adaptive system for the F-94A at 22,000 ft, at Mach 0.6 and then drawn for the system without the roll damping loop. Figure 6-10 is the block diagram of the roll damping loop. Figure 6-11 is the root locus for the roll damping loop with the location of the closed loop poles for a $S_{(rd)}$ of 0.352. Using the closed loop poles from Figure 6-11, the block diagram for the yaw orientational control loop is shown in Figure 6-12. The root locus for the yaw orientational control loop is shown in Figure 6-13 with the location of the closed loop poles for

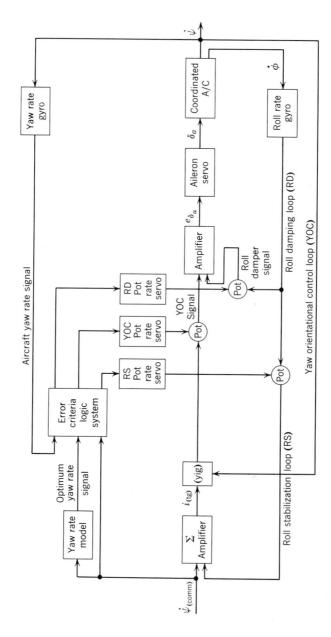

Figure 6-9 Block diagram of the MIT Model-Reference adaptive control system.

$$[TF]_{m_{[c \cdot c]}} = \frac{1}{(1 + 0.7s)\left[1 + \dfrac{2(0.8)}{1.65}\,s + \left(\dfrac{s}{1.65}\right)^2\right]}$$

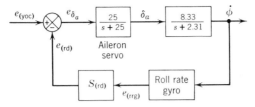

Figure 6-10 Block diagram of the roll damping loop.

$$[TF]_{(rrg)(\dot{\phi};e)} = 1$$

$S_{(yoc)} = 11.75$. Figure 6-14 is the block diagram of the roll stabilizing loop using the closed loop poles from Figure 6-13. Figure 6-15 is the root locus for the roll stabilization loop with the location of the closed loop poles for $S_{(rs)} = 0.086$. Figures 6-16 and 6-17 are the root loci for the yaw orientational control and roll stabilization loops for $S_{(rd)} = 0$ and $S_{(yoc)} = 11.75$.

A comparison of Figures 6-15 and 6-17 shows the effect of the roll damping loop on the overall system response. From Figure 6-15 it can be seen that a roll damping loop gain of only 0.352 volt/volt makes it possible to achieve any damping ratio desired in the final roll stabilization loop; whereas Figure 6-17 indicates that when the gain of the roll damping loop is zero, the maximum damping ratio obtainable by the roll stabilization loop is about 0.22. This illustration graphically demonstrates the effect of the roll damping loop.

Having discussed the root loci of the three loops and demonstrated the necessity for the roll damping loop, the error criteria used to set the three loop gains can now be discussed. The yaw orientational control loop gain is in the forward loop; therefore, it affects the magnitude of the signal being fed to the aileron servo and thence to the aileron. This gain has a large effect on the initial response of the aircraft. For this reason the error criterion used for the yaw orientational control loop is to make

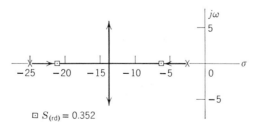

$\square\ S_{(rd)} = 0.352$

Figure 6-11 Root locus for the roll damping loop.

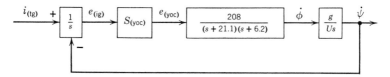

Figure 6-12   Block diagram of the yaw orientational control loop where $g/U = 0.0524 \text{ sec}^{-1}$.

the integral of the yaw rate error signal zero, over a period starting with the initiation of the input signal and ending at the time at which the model output reached 70 per cent of the input.   Thus, if the gain is too low, the aircraft lags the model, and the signal from the error logic criteria system commands an increase in the gain.   The opposite would be true if the gain were too high.   As the aircraft enters flight conditions resulting in low dynamic pressures, the response of the aircraft in roll will be slower (see Eq. 3-61); thus, the system tries to compensate for this by increasing the yaw orientational loop gain.   If this gain is driven too high, the other two loop gains may be driven to excessively large values in trying to optimize the system, with detrimental results on the system response.   For this reason an upper limit must be placed on the yaw orientational control loop gain, but this alone is not the solution.   If the yaw orientational control loop gain is limited, the error criterion used to control that gain will not be satisfied.   This fact plus the concept that for the flight conditions yielding low dynamic pressures it might be desirable to slow down the response of the model, led to the idea of adjusting the time constant of the model under these conditions.   Thus, the gain of the yaw orientational control loop was varied linearly from its minimum value of about 3.5 to a

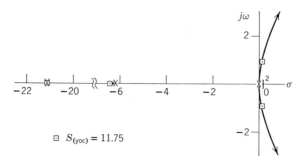

Figure 6-13   Root locus for the yaw orientational control loop for $S_{(rd)} = 0.352$.

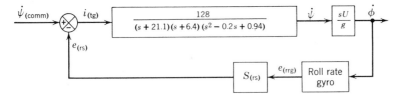

Figure 6-14 Block diagram of the roll stabilization loop.
$$[TF]_{(rrg)(\dot{\phi};e)} = 1.$$

maximum value of 7.5. At this point the time constant of the first-order term of the model was increased linearly from 0.7 until the error criterion was satisfied. The maximum value of the time constant used was 1.2 sec. In this manner both the system and the model were made self-adaptive.

As seen from the root loci, the roll stabilization loop is responsible for stabilizing the overall system and determining the final closed loop poles of the system. Since the location of the closed loop poles determines the response time of the system, the integral of the error sampled over the response time was used to vary the roll stabilization loop gain. The response time was taken as the time between the initiation of an input signal and the time for the output of the model to reach 95 per cent of the input. However, for this loop the integral of the error was nulled rather than made equal to zero, as for the case of the yaw orientational control loop.

As was shown in Section 4-6, the roll damping loop determines the amount of damping possible in the overall system. The system may be underdamped with its response oscillating about the desired response, yet the other two error criteria may be satisfied; thus the integral of the absolute value of the error was chosen as this criterion. The integral of the

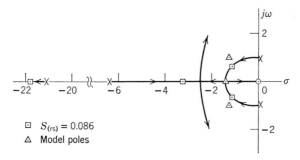

$\square$ $S_{(rs)} = 0.086$
$\triangle$ Model poles

Figure 6-15 Root locus for the roll stabilization loop
for $S_{(rd)} = 0.352$.

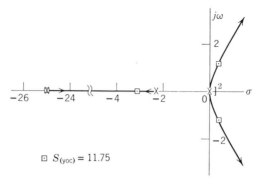

Figure 6-16   Root locus for the yaw orientational
control loop for $S_{(rd)} = 0$.

absolute value of the error is large if the system is underdamped or so overdamped that the system response is sluggish. However, using the absolute value of the error introduces the problem of which direction to adjust the gain of the roll damping loop. In order to determine in which direction to adjust the roll damping loop gain, each successive value of the integral of the absolute value of the error is compared with the previous value. As long as each successive value of the error is less, the gain is adjusted in the same direction. If the successive value of the error is greater than the previous value, the adjustment of the gain is reversed. In this manner the gain of the roll damping loop is adjusted to minimize the integral of the absolute value of the error.

Having discussed the various error criteria the effectiveness of the optimum values of the various gains can now be discussed. As mentioned in the preceding paragraphs, Figures 6-11, 6-13, and 6-15 are the root loci

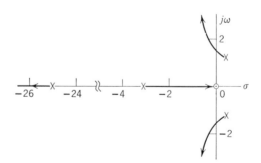

Figure 6-17   Root locus for the roll stabilization
loop for $S_{(rd)} = 0$.

for the three autopilot loops, and the closed loop poles indicated are for the optimum values of each loop gain as determined by the system. For this case there was no limiting of the yaw orientational control loop gain. Figure 6-15 shows the location of the outer loop closed loop poles as well as the model poles. From an examination of Figure 6-15 it can be seen that the closed loop complex system poles are very close to the model's complex poles, while the real closed loop system pole is much farther to the left of the model pole. From a comparison of the system and model poles it is not evident that the error criteria have been satisfied. To check this, the inverse Laplace transforms of the system and the model for a step input of $\psi_{(comm)}$ were taken. A comparison of the two responses showed that the error between the aircraft and the model varied from 0.08 to 2.6 per cent with the system response leading the model slightly. This was caused by the real system pole; however, if the complex system poles were slightly closer to the imaginary axis the results would have been better. The difference between the two responses is probably due to the inaccuracies in the root loci plots. However, it is evident from Figure 6-15 that, for this flight condition, the real pole at -6.4 will never move to the location of the model real pole, if the system's complex poles are to remain in the vicinity of the model's complex poles. This situation will probably be the case for most of the flight conditions; thus this system is in general incapable of forcing the dominant system poles to become identical with the model poles. This means that complete optimization is impossible for all flight conditions; however, the system is apparently capable of positioning the closed loop poles of the system in such a manner that the error criteria are satisfied. The capabilities of this system were demonstrated by actual flight test using a F-94C, the results of which are shown in Ref. 4.

6-6 MH-90 Adaptive Control System

Since the writing of the references listed at the end of the chapter, new and improved systems have been developed; however, the same basic idea of the highest forward loop gain consistent with some set stability criterion is still prevalent. A good example of this is the MH-90 flight control system developed by Minneapolis-Honeywell for the F-101. This system is an outgrowth of the Minneapolis-Honeywell system discussed in Section 6-4. The maximum possible forward loop gain is still used so that the closed loop transfer function of the system will be approximately one, thus the output of the system theoretically will follow the output of the model. However, the technique for controlling the maximum allowable gain is different.   The MH-90 system maintains the forward loop gain at a sufficient level so as to keep the complex servo poles on the imaginary axis.   This results in a limit cycle, the frequency of which is the natural frequency of the servo poles where they cross the imaginary axis.   To control the forward loop gain, the limit cycle is fed through a narrow band filter which is tuned to 25/rad/sec, the ideal limit-cycle frequency.   The output of the filter is rectified and compared to a limit-cycle amplitude "set point."   The difference between the set point and the limit cycle is used to adjust the variable forward loop gain.   As the operation of the gain changer is not pertinent to the discussion to follow, it will not be explained further.   The basic system is shown in Figure 6-18.   Several Master of Science theses have been written under the author's supervision by students of the Graduate Guidance and Control course of the Air Force Institute of Technology, under the sponsorship of the Flight Control Laboratory, Aeronautical Systems Division at Wright-Patterson AFB, Ohio (Refs. 5, 6, 7).   These theses have disclosed some very interesting operating characteristics of the MH-90 adaptive control system.   The main results of these theses are discussed in the following paragraphs.

An investigation performed by Lt. Blum[5] shows the effect of noise on the operation of the MH-90 adaptive control system.   In the investigation, white noise, in the form of an electrical signal, was fed to the input of the servo which was simulated along with the aircraft and the rest of the system on an analog computer.   With the forward loop gain at the proper level to assure a limit cycle with no noise present, the input to the filter in the presence of noise would be the limit cycle, plus the noise response of the servo.   The output of the filter and rectifier of the gain control would then be larger than the set point.   This situation would result in a decrease in the forward loop gain, and with sufficient noise present the limit cycle would disappear completely, being replaced by the noise signal passed by the filter, and the system operation would deteriorate rapidly.

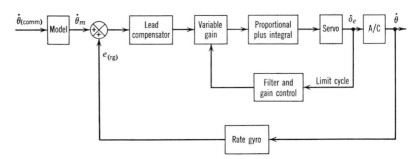

Figure 6-18   General block diagram of the MH-90 adaptive control system.

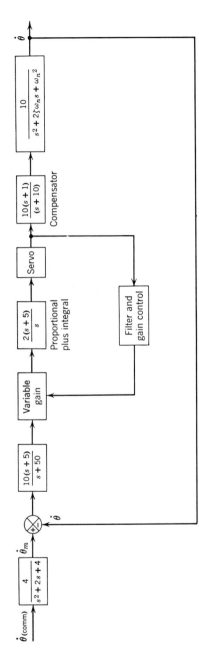

Figure 6-19 Block diagram for analysis of MH-90 adaptive control system (Ref. 6).

For a limit cycle with a peak-to-peak amplitude of 0.1 volts, the maximum allowable RMS noise response of the servo was found to be 0.34°.

Lt. Johannes in his thesis[6] investigated the effects of using a low-frequency servo on the performance of the MH-90 adaptive control system. Figure 6-19 is the block diagram of the system as analyzed by Lt. Johannes. At the request of the Flight Control Laboratory, he used a simple second-order system as the controlled element and added a compensator between the servo and the controlled element. The results of this study would still be valid if the controlled element were the short-period dynamics of an aircraft. The natural frequency of the servo was varied from 40 to 15 rad/sec with the damping ratio held constant at 0.8. Figure 6-20 shows the form of the root locus for the system with the natural frequency of the servo equal to 40 rad/sec. From Figure 6-20 it can be seen that for high natural frequencies of the servo, the branch that crosses the imaginary axis originated from the two real poles of the two compensators. The frequency of the limit cycle was 33 rad/sec. If the natural frequency of the servo is reduced to 35 rad/sec, the branches from the servo poles cross the imaginary axis. Finally, for a servo natural frequency of 15 rad/sec the servo poles break into the two real zeros, and the branches from the controlled element poles cross the imaginary axis. For this condition the limit-cycle frequency has decreased to about 19 rad/sec. The fact that the controlled element poles cross the imaginary axis makes the operation of the MH-90 system, with a servo having a natural frequency of 15 rad/sec, unsatisfactory. This is caused by the fact that, as the dynamics of the controlled element change, the natural frequency at which they cross the imaginary axis changes excessively and degrades the operation of the system.

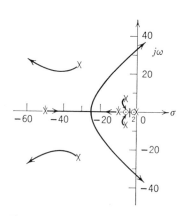

Figure 6-20   Sketch of root locus for system with the natural frequency of the servo equal to 40 rad/sec.

Captains Christiansen and Fleming in their thesis[7] investigated, among other things, the improvement gained by introducing a variable compensator when using a low-frequency servo (ω_n of 12 rad/sec). In this investigation the damping ratio of the controlled element was varied from 0.2 to 0.9, while the natural frequency was varied from 4 to 10 rad/sec. A total of 9 combinations was used. With the compensator as shown in Figure 6-19, the imaginary axis crossing frequency varied from 9.7 to 18.4 rad/sec. The two extreme cases were caused by a ζ of 0.2 and ω_n of 4 for the lower value of imaginary axis crossing frequency and a ζ of 0.9 and ω_n of 10 for the higher value. Since the band pass filter at the input to the gain control has a very narrow band pass, centered on 25 rad/sec, the operation of the adaptive loop is unsatisfactory unless the limit cycle remains near 25 rad/sec. It was found that by moving the pole of the compensator to the left, the imaginary axis crossing frequency was increased. A system was then devised by these two students to control the location of the compensator pole as a function of the limit-cycle frequency. The location of the pole was varied from -10 to -95 automatically to assure a limit-cycle frequency of 25 rad/sec. Through

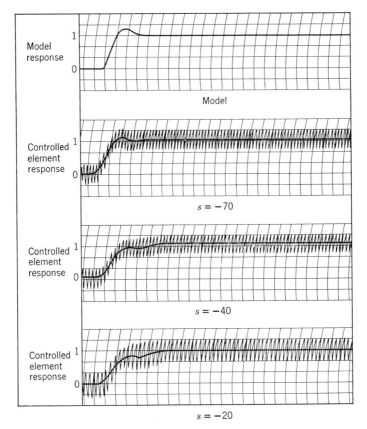

Figure 6-21 System response with $\zeta_{ce} = 0.4$, $\omega_{nce} = 10$ rad/sec for 4 compensator pole locations, for a unit step input (Ref. 7).

the use of the dynamic compensator the MH-90 system performed adequately with the low natural frequency servo.

As mentioned in Section 6-2, proper operation of an adaptive control system similar to the MH-90 depends on a very high forward loop gain so that the closed loop transfer function of the system, exclusive of the model, can be considered 1. The high gain required cannot in all cases be obtained without driving the system completely unstable. This is especially true for the MH-90 system when used with a low-frequency servo. Referring to Figure 6-20, it can be seen that there is a pole at the origin that moves toward the zero at -1. The closed loop location of this pole relative to the zero determines the effect of this large time constant on the overall system response. The transient resulting from this pole persists the longest and is of the form of $Ae^{-t/\tau}$. If A is large, the slow decay of this term prevents the system from following the model. The value of A is dependent upon the distance from this closed loop pole to the zero. The closer the pole is to the zero, the smaller A becomes and the less effect this large time constant has on the behavior of the system. With the compensator pole at -10, the gain of the system cannot be raised sufficiently to drive the pole at the origin close enough to the zero to prevent this adverse performance; however, the addition of the dynamic compensator permits a larger gain and decreases the effect of this troublesome pole. This effect is illustrated in Figure 6-21 from Ref. 7.

In addition to the problems introduced by noise, structural flexibility can result in unsatisfactory operation if one of the body bending modes is near the limit-cycle frequency. If the amplitude of the body bending mode oscillation that appears at the input to the filter of the adaptive control is of sufficient amplitude, then the gain changer reduces the loop gain and system operation deteriorates. This effect has been investigated by Cornell Aeronautical Laboratory, and is reported in Refs. 8 and 9. This subject is discussed in more detail in Chapter 8.

6-7 Summary

In this chapter the basic concept of the self-adaptive autopilot was discussed, and several examples were examined. In all cases, except the MIT system, adaptation was accomplished by maintaining the maximum forward loop gain consistent with some set margin of stability. The MIT system, however, adjusted the loop gains by satisfying certain error criteria, and in this respect may be considered more truly self-adaptive. One factor that should be remembered is that no system will be able to maintain optimum system performance for all flight conditions. In most cases some form of compensation will be required to obtain the best adaptation possible, and as shown in Section 6-6, the compensator may be dynamic.

The introduction of the self-adaptive control system has been a great step forward in the field of automatic control systems. Self-adaptive autopilots will play a large role in the conquest of space. The X-15 is now equipped with a later model of the MH-90 self-adaptive control system and the Dyna Soar vehicle will be equipped with such a system. For reentry vehicles, the adaptive control system automatically switches control from the reactive controls to the aerodynamic controls during reentry, and from aerodynamic to reactive control on exit (as in the case of the X-15). For this type of operation there usually will be a period when both aerodynamic as well as reactive controls are used.

References

1. *Proceedings of the Self-Adaptive Flight Control Systems Symposium*, WADC-TR-59-49, March 1959.
2. *Final Technical Report Feasibility Study Automatic Optimizing Stabilization System*, Part I, Sperry Gyroscope Company, WADC-TR-58–243, June 1958.
3. *A Study to Determine an Automatic Flight Control Configuration to Provide a Stability Augmentation Capability for a High-Performance Supersonic Aircraft*, Minneapolis-Honeywell Regulator Company, Aeronautical Division, WADC-TR-57-349 (Final), May 1958.
4. H. P. Whitaker, J. Yarmon, and A. Kezer, *Design of Model-Reference Adaptive Control Systems for Aircraft*, Instrumentation Laboratory, Massachusetts Institute of Technology, Report R-164, September 1958.
5. J. J. Blum, *A Study to Determine the Effects of Random Disturbances in an Adaptive Flight Control System*, M.S. thesis, Air Force Institute of Technology, Wright-Patterson AFB, Ohio, March 1961.
6. R. P. Johannes, *Study of a Self-Adaptive Control System Including Simulation and Effects of Variation of the System Dynamics*, M.S. thesis, Air Force Institute of Technology, Wright-Patterson AFB, Ohio, March 1961.
7. R. H. Christiansen, and R. J. Fleming, *Variable Compensator Loop Application to a Self-Adaptive Flight Control System*, M.S. thesis, Air Force Institute of Technology, Wright-Patterson AFB, Ohio, March 1962.
8. *Application of Self-Adaptive Control Techniques to the Flexible Transport*, First Quarterly Technical Report, CAL Report No. IH-1696-F-1, October 1, 1962, AF33(657)8540.
9. Clark, Notess, Pritchard, Reynolds, Schuler, *Application of Self-Adaptive Control Techniques to the Flexible Supersonic Transport*, Cornell Aeronautical Laboratory, CAL Report No. IH-1696-F-5, Vol. I, July 1, 1963.

7

Missile Control Systems

7-1 Introduction

Having discussed the control systems for aircraft in the preceding chapters, this chapter deals with the discussion of control systems for guided missiles. The name "guided missile" implies that the missile is controlled either by an internal guidance system or by commands transmitted to the missile by radio from the ground or launching vehicle.

Before discussing some typical guidance systems, the terms "navigation system," "guidance system," and "control system" are defined. A navigation system is one that automatically determines the position of the vehicle with respect to some reference frame, for example, the earth, and displays this to an operator. If the vehicle is off course, it is up to the operator to make the necessary correction. A guidance system, on the other hand, automatically makes the necessary correction to keep the vehicle on course by sending the proper signal to the control system or autopilot. The guidance system then performs all the functions of a navigation system plus generating the required correction signal to be sent to the control system. The control system controls the direction of the motion of the vehicle or simply the orientation of the velocity vector.

The type of guidance systems used depends upon the type and mission of the missile being controlled, and they can vary in complexity from an inertial guidance system for long-range surface-to-surface or air-to-surface winged missiles to a simple system, where the operator visually observes the missile and sends guidance commands via a radio link. In any case, the guidance command serves as the input to the missile control system. The command may be in the form of a heading or attitude command or a pitching or turning rate depending upon the type of guidance scheme used. This chapter deals with the control system in the missile that receives the signal from the guidance system and not with the guidance system itself. For details of various guidance systems the reader is referred to the references at the end of this chapter.

Of the various types of guided missiles, those that are flown in the same manner as manned aircraft (that is, missiles that are banked to turn such as the Bomarc, Mace, etc.) will not be discussed. Control systems and methods of analysis previously discussed can be used for these types of missiles. Of interest in this chapter are other "aerodynamic missiles" (one which uses aerodynamic lift to control the direction of flight), such as the Falcon and Nike and "ballistic missiles" (one that is guided during powered flight by deflecting the thrust vector and becomes a free-falling body after engine cut off). One feature of these missiles is that they are roll stabilized; thus there is no coupling between the longitudinal and the lateral modes, which simplifies the analysis. The basic assumptions made in Chapter 1 are still valid for studying the control systems; however, for analyzing the guidance system for ballistic missiles, the rotating earth cannot be assumed as an inertial reference. Additional assumptions will be made as required. Since the missiles to be studied in this chapter are roll stabilized, a system for accomplishing this is discussed first.

7-2 Roll Stabilization

Roll stabilization can be accomplished by different means depending on the type of missile. For aerodynamic missiles, the required rolling moment is achieved by differential movement of the control surfaces. For ballistic missiles, the rolling moment can be obtained by differential swivelling of small rockets mounted on the side of the missile, as is done on the Atlas, or by differential swivelling on the two main rocket engines if more than one engine is used. The next problem is the detection of the rolling motion so that it can be controlled, and the reduction of roll rate to zero or maintenance of the roll angle equal to some specified reference. The use of a roll rate gyro would not be satisfactory unless it was desirable only to reduce the roll rate considerably. The use of the roll rate gyro would result in a Type 0 system, which would further result in a steady-state error in roll rate in the presence of a constant disturbing rolling moment. To maintain a desired roll angle, some form of an attitude reference must be used. This can be a vertical gyro for air-to-air missiles or a stable platform for surface-to-air missiles. In either case the feedback would be a signal proportional to the roll angle about the longitudinal axis of the missile. Still another possible method is the use of an integrating gyro with its input axis along the longitudinal axis of the missile. Figure 7-1 is the block diagram of a possible roll stabilization control system. The servo might be represented by a first-order time lag or a second-order system. The transfer function of the missile for δ_a input to roll angle

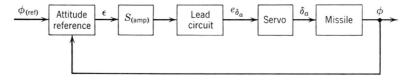

Figure 7-1   General block diagram of a roll stabilization system.

$$[TF]_{\text{(lead circuit)}} = \frac{1}{\alpha}\left[\frac{1 + \alpha\tau s}{1 + \tau s}\right]$$

output would be the same as the one-degree-of-freedom rolling mode derived in Section 3-6 for aircraft, which is repeated here

$$\frac{\phi(s)}{\delta_a(s)} = \frac{C_{l_{\delta_a}}}{s\left(\dfrac{I_x}{Sqd}s - \dfrac{d}{2U}C_{l_p}\right)} \tag{7-1}$$

where $C_{l_{\delta_a}}$ is the rolling moment generated by the aerodynamic controls or the reaction controls divided by Sqd, and d is the diameter of the missile. The reference area, S, is usually taken as the cross-sectional area of the missile. C_{l_p} for the aerodynamic missile results from the same cause as discussed in Section 3-2; for the ballistic missile C_{l_p} would be due mostly to aerodynamic friction and thus is negligible. The requirement for the lead network is demonstrated by drawing the root locus of the system for the worst condition, that is, $C_{l_p} = 0$ (roll rate feedback could also be used for stabilization). Figure 7-2 shows the block diagram used for the root locus. The root locus of the system is shown in Figure 7-3. Without the lead circuit the two poles at the origin would move directly into the right-half plane; the effect of the lead circuit is evident. If C_{l_p} is not zero, the transfer function of the missile consists of a pole at the origin and a pole at $s = -1/\tau = -(Sqd^2/2UI_x)C_{l_p}$. The larger C_{l_p}, the larger $1/\tau$, and the less is the requirement for the lead circuit.

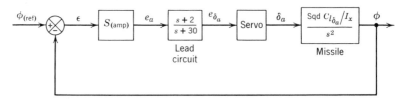

Figure 7-2   Block diagram of roll stabilization system for root locus analysis.

$$[TF]_{\text{(servo)}[e_{\delta_a}:\delta_a]} = \frac{2750}{s^2 + 84s + (52.5)^2}$$

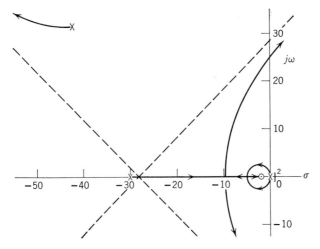

Figure 7-3   Root locus for the roll stabilization system shown in Figure 7-2.

An alternate control system could consist of an integrating gyro with its input axis parallel to the longitudinal axis of the missile. This would provide the same quality of control as the attitude reference system as long as the drift rate of the gyro were low enough. The root locus for this system would be the same as Figure 7-3.

7-3   Control of Aerodynamic Missiles

Figure 7-4 is a sketch of a typical aerodynamic missile showing the orientation of the axis system. By using a body axis system the product of inertia term J_{xz} is zero and $I_z = I_y$. Thus for $P = 0$ there is no coupling between the longitudinal and lateral equations. Control can be accomplished by either conventional control surfaces with the canards stationary or absent or by use of the canards with no control surfaces on the main lifting surfaces.

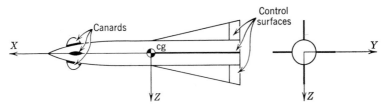

Figure 7-4   Sketch of an aerodynamic missile and axis system.

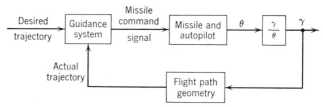

Figure 7-5   Generalized block diagram for missile control system.

For dynamic analysis of the longitudinal modes, the short-period approximation can be used along with the body axis stability derivatives. Depending on the configuration of the missile, there may not be any downwash effect, thus $C_{m\dot{\alpha}}$ would be zero. This would be true of the missile configuration shown in Figure 7-4 with or without the canards. For the lateral modes, the one-degree-of-freedom Dutch roll approximation can be used. It will usually be found that the stability derivatives for missiles are determined using the body axis system.

The missile autopilot need not be complicated. In general, rate feedback will probably be necessary for both the pitch and yaw channel. The rest of the autopilot will be determined by the mission of the missile and the type of guidance used. For control of this type of missile the actuating signal would come from the guidance system. If command guidance is used, the actuating signal would be sent from the ground in the form of a pitch or yaw command. For a beam rider, proportional navigation, or a homing guidance* system, the command signal would be generated internally in the missile. For all these systems, the command signals result from the fact that the missile trajectory was in error. Figure 7-5 is a generalized block diagram that is applicable to any type of system. The comparison of the actual and desired trajectory would be done either in the missile or by the launch controller. The γ/θ relation would be the same as the one derived in Chapter 2, for the glide slope analysis.

7-4   Transfer Function for a Ballistic-Type Missile

Because all detailed information on ballistic missiles is classified, the Vanguard missile is used here to illustrate the control problems associated with ballistic missiles. The stability derivatives for the rigid missile, the structural transfer function, and the block diagram of the autopilot were furnished through the courtesy of the Martin Company, Baltimore, Maryland.

* These pursuit courses are described in detail in Ref. 1.

To derive the transfer function of the missile, it is necessary to orient the missile axis system. The trajectory of ballistic-type missiles is planned to maintain the missile at a zero angle of attack. This is normally attempted by programming the pitch attitude or pitch rate to yield a zero "g" trajectory. This, of course, assumes a certain velocity profile which may or may not be realized due to variations in the fuel flow rate. This condition plus the presence of gusts results in the angle of attack not being zero at all times; however, the angle of attack must be kept small to avoid excessive loading of the structure. Therefore the assumption of zero angle of attack for the equilibrium condition is quite valid, with any changes in angle of attack being considered perturbations from the equilibrium condition. With this assumption, if the X axis of the missile is along the longitudinal axis of the missile, this body axis system becomes a stability axis system. By placing the Z axis in the trajectory plane, all pitch motion is about the Y axis (see Figure 7-6). Basically the axis system is the same as the one used for the pitching motion for the aircraft

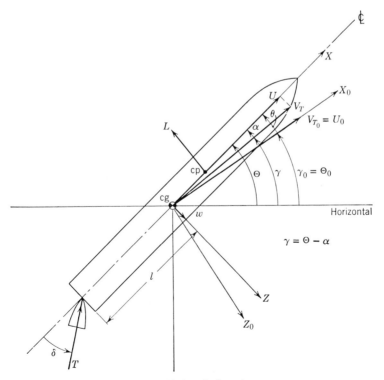

Figure 7-6 Ballistic missile axis system.

(see Figure 1-3). Assuming that the angle of attack is zero for the equilibrium condition, then α for the missile is the same as $'\alpha$ for the aircraft. For the missile to fly a zero angle of attack trajectory, the centrifugal force resulting from the curved trajectory $(V_T\dot{\gamma})$ must exactly balance the component of weight normal to the X axis, which establishes the equilibrium condition. The dynamic analysis to follow is based on perturbations from this equilibrium condition. It should be noted from Figure 7-6, that if an angle of attack is generated, the lift vector acting at the center of pressure (cp) has a destabilizing effect. This factor is one of the major control problems of the missile.

In the derivation of the aircraft's equations of motion, certain assumptions were made, which are reviewed here for their applicability to the missile control problem. These assumptions are listed on p. 21, and are repeated here for reference with the word missile replacing the word aircraft. The assumptions are

1. The X and Z axes lie in the plane of symmetry, and the origin of the axis system is located at the center of gravity of the missile. (Actually the X and Y axes also lie in a plane of symmetry.)
2. The mass of the missile is constant.
3. The missile is a rigid body.
4. The earth is an inertial reference.
5. The perturbations from equilibrium are small.

All these assumptions are still valid. Although the missile is consuming fuel at a terrific rate, if the instantaneous mass is used, the mass may be assumed constant during the period of analysis. The ballistic missile in general cannot be considered a rigid body; however, the missile is first analyzed as a rigid body and then the effect of body bending is studied in Chapter 8.

The assumption that the earth is an inertial reference is satisfactory for the analysis of the control system but not of the inertial guidance system that would be sending the command signals to the control system. The assumption of small perturbations is even more valid for the ballistic missile than for the aircraft. Because the perturbations have been assumed small, and as the duration of the disturbance is short, the velocity can be assumed constant during the period of the dynamic analysis. Thus the equations for the short-period approximation of the aircraft may be used for the longitudinal analysis. From Eq. 1-128 these equations are

$$\left(\frac{mU}{Sq}s - C_{z_\alpha}\right)\alpha(s) + \left[-\frac{mU}{Sq}s - C_w(\sin\Theta)\right]\theta(s) = C_{z_\delta}\delta$$

$$\left(-\frac{d}{2U}C_{m_{\dot\alpha}}s - C_{m_\alpha}\right)\alpha(s) + \left(\frac{I_y}{Sqd}s^2 - \frac{d}{2U}C_{m_q}s\right)\theta(s) = C_{m_\delta}\delta \quad (7\text{-}2)$$

where c has been replaced by d, the diameter of the missile. In the case of the aircraft, $C_{m_{\dot{\alpha}}}$ arose from the time lag for the downwash created by the wing to reach the tail. As there is neither a wing nor a horizontal stabilizer for this missile, $C_{m_{\dot{\alpha}}} = 0$. The actuating signal δ is the deflection of the thrust chamber, with the same sign convention as was used for the conventional aircraft. The component of thrust normal to the X axis is proportional to the sine δ; however, if δ is small, the sine δ can be replaced by δ in radians. The values of the rest of the quantities in Eq. 7-2 are given for the time of maximum dynamic pressure which occurred at 75 sec after launch, at an altitude of 36,000 ft, a velocity of 1285 ft/sec, and a mass of 445 slugs. Then

$$C_{m_\delta} = -\frac{Tl}{Sqd} = -34.25$$

$$C_{z_\delta} = -\frac{T}{Sq} = -4.63$$

$$C_{z_\alpha} = -3.13$$

$$C_{m_\alpha} = +11.27 \text{ (as the cp is ahead of the cg)}$$

$$l = 27 \text{ ft}$$

$$\Theta_0 = 68.5°$$

$$d = 3.75 \text{ ft}$$

$$q = 585 \text{ lb/sq ft}$$

$$S = 11.04 \text{ sq ft}$$

$$\frac{d}{2U} C_{m_q} = -0.321 \text{ sec}$$

$$\frac{mU}{Sq} = 88.5 \text{ sec}$$

$$\frac{mg}{Sq} = -C_w = 2.22$$

$$I_y = 115,000 \text{ slug ft}^2$$

$$\frac{I_y}{Sqd} = 4.75 \text{ sec}^2$$

Substituting these values, Eq. 7-2 becomes

$$(88.5s + 3.13)\alpha(s) + [-88.5s + 2.06]\theta(s) = -4.63\delta(s)$$
$$-11.27\alpha(s) + (4.75s^2 + 0.321s)\theta(s) = -34.25\delta(s) \qquad (7\text{-}3)$$

Then the transfer function for δ input to θ out is

$$\frac{\theta(s)}{\delta(s)} = \frac{-7.21(s + 0.0526)}{(s + 1.6)(s - 1.48)(s - 0.023)} \qquad (7\text{-}4)$$

Equation 7-4 shows that the missile is unstable with a pole at $+1.48$. The pole at $+0.023$ arises from the fact that θ_0 is not zero.

7-5   Vanguard Control System (Rigid Missile)

One of the main functions of the control system is to stabilize the otherwise unstable missile. To accomplish this, the Martin Company added a lead compensator to the forward loop. Figure 7-7 is the block diagram for the Vanguard rigid body control system. The transfer function of the servo is

$$[TF]_{(\text{servo})[\delta_i;\delta]} = \frac{2750}{s^2 + 42.3s + 2750} \qquad (7\text{-}5)$$

After canceling the zero at -0.0526, with the pole at 0.023, the missile transfer function is

$$[TF]_{(\text{missile})[\delta;\theta]} \simeq \frac{-7.21}{(s + 1.6)(s - 1.48)} \qquad (7\text{-}6)$$

The root locus for the Vanguard control system is plotted in Figure 7-8. The stabilizing action of the lead network is apparent. The student is probably wondering about the pole and zero that were canceled in the missile transfer function. Figure 7-9 is an exploded view of the root locus in the vicinity of the origin. From Figure 7-9 it can be seen that only the portion of the root locus near the origin is modified by the addition of the canceled pole and zero. The angle contribution of the canceled pole and zero to the rest of the root locus would be negligible. Actually the missile pole at -1.6 can be canceled by the zero of the lead compensator without significantly changing the root locus. Another control system that is capable of controlling the Vanguard or similar missiles is discussed in Section 7-6.

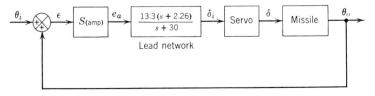

Figure 7-7   Vanguard control system (rigid missile).

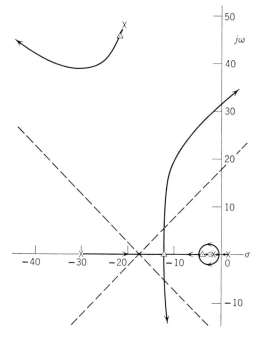

△ $S_{(amp)}$ = 2.3 volt/volt

Figure 7-8   Root locus for the Vanguard rigid body
control system.

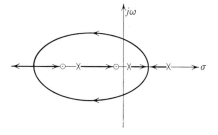

Figure 7-9   Root locus in vicinity of the
origin with the complete missile transfer
function.   *Note:* This figure is not to scale.

7-6   Alternate Misisle Control System (Rigid Missile)

In Chapter 2 it was found that the pitch orientational control system
was capable of controlling an aircraft that was unstable in pitch.   This
same basic autopilot then should be capable of controlling a ballistic

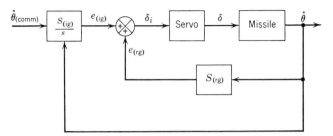

Figure 7-10   Alternate missile control system (rigid missile).

missile.   Some guidance systems also require that the control system be capable of accepting a commanded pitch rate.   Thus a system of the same configuration as the pitch orientational control system should be very useful for ballistic missiles.   Figure 7-10 is the block diagram for this system.   The transfer function for the Vanguard is again used, and the same servo with the damping ratio increased to 0.8.   Thus

$$[TF]_{(\text{servo})[\delta_i:\delta]} = \frac{2750}{s^2 + 84s + 2750} \tag{7-7}$$

Figure 7-11 is the block diagram for the inner loop showing the transfer functions of the servo and missile, and Figure 7-12 is the root locus for the inner loop.   The location of the closed loop poles of the inner loop are shown for a $S_{(\text{rg})} = 1.58$ volt/rad/sec.   The rate gyro sensitivity is selected so that the missile pole at -1.6 will be driven well into the left-half plane without excessively decreasing the damping of the servo.

The need for a higher value of the initial damping ratio of the servo is now evident.   Figure 7-13 is the block diagram for the outer loop of the alternate control system using the closed loop poles from Figure 7-12. The root locus for the outer loop is shown in Figure 7-14 with the location of the closed loop poles for $S_{(\text{ig})} = 10.2$ volt/rad/sec.   From Figure 7-14 it can be seen that this system is capable of controlling the instability of the ballistic type missile.

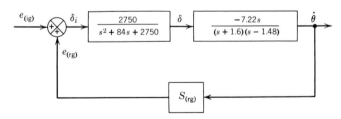

Figure 7-11   Block diagram of inner loop of the alternate control system.

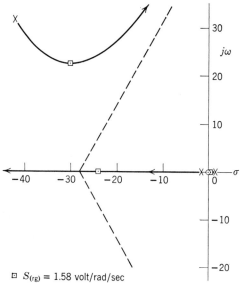

▫ $S_{(rg)}$ = 1.58 volt/rad/sec

Figure 7-12   Root locus of the inner loop of the
alternate control system.

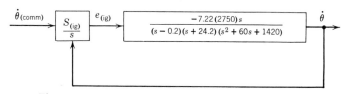

$\dot{\theta}$ (comm) $\dfrac{S_{(ig)}}{s}$ $e_{(ig)}$ $\dfrac{-7.22\,(2750)\,s}{(s-0.2)(s+24.2)(s^2+60s+1420)}$ $\dot{\theta}$

Figure 7-13   Block diagram for outer loop root locus.

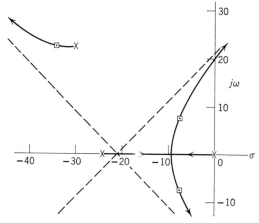

▫ $S_{(ig)}$ = 10.2 volt/rad/sec

Figure 7-14   Root locus of the outer loop of the
alternate control system.

Table 7-1 Aerodynamic Coefficients for the Vanguard Missile (Rigid Missile)

t (sec)	h (ft)	V (ft sec)	M	q (lb/ft²)	m (slugs)	Θ (deg)	C_{m_α}	C_{z_α}	$\dfrac{d}{2U}C_{m_q}$ (sec)	C_{m_δ}	C_{z_δ}	I_y (slug ft²)
48	12,200	604	0.57	300	547	81.7	10.55	−3.08	−0.683	−60.83	−8.69	121,400
75	36,000	1285	1.26	585	445	68.5	11.27	−3.13	−0.321	−34.25	−4.63	115,000
100.4	76,000	2400	2.41	300	354	57.2	16.59	−2.88	−0.172	−70.15	−9.17	104,000
*139	185,000	5600	5.02	12	218	46.5	8.32	−2.3	−0.0736	−2140	−229	75,400

* First stage burn out, $l = 27$ ft; $S = 11.04$ ft²; $d = 3.75$ ft.

In order for the student to determine the control requirements for other portions of the power trajectory, Table 7-1 lists the aerodynamic coefficients for three other flight conditions in addition to the one already used.

7-7 Summary

In this chapter the control of both aerodynamic and ballistic-type missiles has been discussed. It was shown that the analysis of missiles is simpler than the analysis of aircraft, since there is no coupling between the lateral and longitudinal modes, because the missiles are roll stabilized and turning is accomplished by yawing the missile. Thus the short-period approximation can be used for analyzing the longitudinal dynamics, and the one-degree-of-freedom Dutch roll approximation can be used for the lateral analysis. Two techniques for stabilizing ballistic type missiles were discussed.

References

1. A. S. Locke, *Guidance*, D. Van Nostrand Company, Princeton, New Jersey, 1955.
2. C. Broxmeyer, *Inertial Navigation Systems*, McGraw-Hill Book Co., New York, 1964.
3. C. S. Draper, W. Wrigley, and J. Hovorka, *Inertial Guidance*, Pergamon Press, New York, 1960.
4. M. Fernandez, and G. R. Macomber, *Inertial Guidance Engineering*, Prentice-Hall, Englewood Cliffs, New Jersey, 1962.
5. R. H. Parvin, *Inertial Navigation*, D. Van Nostrand Co., Princeton, New Jersey, 1962.
6. G. R. Pitman, Jr., *Inertial Guidance*, John Wiley and Sons, New York, 1962.

8

Structural Flexibility

8-1 Introduction

With the introduction of long slender missiles such as the Vanguard, the Redstone, and the various ballistic missiles, the problem of structural flexibility became acute. Due to the limited thrust available from our rocket engines, these missiles had to be as light as possible. This meant a sacrifice in structural rigidity. Missile flexure causes additional aerodynamic loads which in turn cause additional flexure, etc. Also coupling occurs between the elastic modes and the control system as the control system gyros sense the flexure motion and the rigid body motion.

In this chapter a method for determining the natural frequencies and mode shapes of the body bending modes is presented, followed by the derivation of the uncoupled body bending equations in normalized coordinates. Next, the transfer function for the flexible missile including the "tail-wags-dog" zero is derived, followed by the derivation of the rigid body transfer function, including the propellant sloshing mode. The chapter concludes with a discussion of the compensation required for body bending to assure stability.

8-2 Lagrange's Equation

Lagrange's equation can be derived from Hamilton's principle or the principle of virtual work,[1,2,3] The most general form of Lagrange's equation is

$$\frac{d}{dt}\left(\frac{\partial T}{\partial \dot{q}_j}\right) - \frac{\partial T}{\partial q_j} + \frac{\partial F}{\partial \dot{q}_j} + \frac{\partial U}{\partial q_j} = Q_j \quad (j = 1, 2, \ldots, n) \qquad (8\text{-}1)$$

where T is the kinetic energy of the system, U is the potential energy including strain energy, F is the dissipation function equal to one-half the rate at which energy is dissipated, Q_j is the generalized external force acting on the jth station, and q_j is the generalized coordinate of the jth station.

To illustrate the application of Lagrange's equation, the differential equations for two simple mechanical systems are derived. The first system discussed is the simple mass, spring, and damper system shown in Figure 8-1. From Figure 8-1, $q_1 = x$ and $\dot{q}_1 = \dot{x}$, then $T = \frac{1}{2}m\dot{x}^2$; the dissipation function $F = \frac{1}{2}C\dot{x}^2$ and the potential energy stored in the spring is $U = \int_0^x Kx\,dx = Kx^2/2$, and $Q_1 = F(t)$. As T is only a function of $\dot{x}$, then $\partial T/\partial q_1 = \partial T/\partial x = 0$; the $\partial T/\partial \dot{q}_1 = \partial T/\partial \dot{x} = (\partial/\partial \dot{x})(m\dot{x}^2/2) = m\dot{x}$ and $(d/dt)(\partial T/\partial \dot{q}_1) = (d/dt)m\dot{x} = m\ddot{x}$. The dissipation function term is $\partial F/\partial \dot{q}_1 = \partial F/\partial \dot{x} = (\partial/\partial \dot{x})(C\dot{x}^2/2) = C\dot{x}$, and the potential energy term is $\partial U/\partial q_1 = \partial U/\partial x = (\partial/\partial x)(Kx^2/2) = Kx$. Substituting these values into Eq. 8-1, it becomes

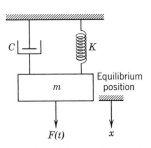

Figure 8-1　Mass, spring, and damper system.

$$m\ddot{x} + C\dot{x} + Kx = F(t) \tag{8-2}$$

the standard second-order equation for the mass, spring, and damper system.

The second example is the derivation of the equations of motion for the double pendulum shown in Figure 8-2. For this system, the two generalized coordinates are ϕ_1 and ϕ_2. The lengths of the two pendulums are l_1 and l_2. The expressions for the kinetic and potential energies are expressed in rectangular coordinates and are then transformed into the generalized coordinates. Thus, from Figure 8-2, $x_1 = l_1 \sin \phi_1$; $x_2 = l_1 \sin \phi_1 + l_2 \sin \phi_2$; $y_1 = l_1 \cos \phi_1$, and $y_2 = l_1 \cos \phi_1 + l_2 \cos \phi_2$. Differentiating the displacements yields the components of velocity of the two masses; thus $\dot{x}_1 = l_1\dot{\phi}_1 \cos \phi_1$; $\dot{x}_2 = l_1\dot{\phi}_1 \cos \phi_1 + l_2\dot{\phi}_2 \cos \phi_2$, and $\dot{y}_1 = -l_1\dot{\phi}_1 \sin \phi_1$; $\dot{y}_2 = -l_1\dot{\phi}_1 \sin \phi_1 - l_2\dot{\phi}_2 \sin \phi_2$. Then the total kinetic energy of the system is

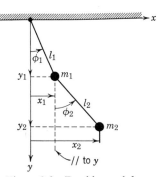

Figure 8-2　Double pendulum.

$$T = \frac{m_1}{2}(\dot{x}_1{}^2 + \dot{y}_1{}^2) + \frac{m_2}{2}(\dot{x}_2{}^2 + \dot{y}_2{}^2) \tag{8-3}$$

Substituting the values of $\dot{x}_1{}^2$, $\dot{x}_2{}^2$, $\dot{y}_1{}^2$, and $\dot{y}_2{}^2$ into Eq. 8-3, and then simplifying, yields

$$T = \frac{m_1}{2} l_1{}^2 \dot{\phi}_1{}^2 + \frac{m_2}{2} [l_1{}^2 \dot{\phi}_1{}^2 + l_2{}^2 \dot{\phi}_2{}^2 + 2l_1 l_2 \dot{\phi}_1 \dot{\phi}_2 \cos(\phi_2 - \phi_1)] \quad (8\text{-}4)$$

The potential energy can be obtained by determining the amount of vertical rise of m_1 and m_2 caused by a rotation ϕ_1 and ϕ_2; thus

$$U = m_1 g(l_1 - y_1) + m_2 g(l_1 + l_2 - y_2) \quad (8\text{-}5)$$

Substituting for y_1 and y_2 in Eq. 8-5, and simplifying, yields

$$U = (m_1 + m_2)g l_1 (1 - \cos\phi_1) + m_2 g l_2 (1 - \cos\phi_2) \quad (8\text{-}6)$$

Taking the partial derivative of T, Eq. 8-4, with respect to $\dot{q}_j$ yields

$$\frac{\partial T}{\partial \dot{q}_1} = \frac{\partial T}{\partial \dot{\phi}_1} = m_1 l_1{}^2 \dot{\phi}_1 + m_2 l_1{}^2 \dot{\phi}_1 + m_2 l_1 l_2 \dot{\phi}_2 \cos(\phi_2 - \phi_1)$$

and

$$\frac{\partial T}{\partial \dot{q}_2} = \frac{\partial T}{\partial \dot{\phi}_2} = m_2 l_2{}^2 \dot{\phi}_2 + m_2 l_1 l_2 \dot{\phi}_1 \cos(\phi_2 - \phi_1)$$

Differentiating with respect to time,

$$\frac{d}{dt}\left(\frac{\partial T}{\partial \dot{\phi}_1}\right) = (m_1 + m_2) l_1{}^2 \ddot{\phi}_1$$
$$+ m_2 l_1 l_2 [\ddot{\phi}_2 \cos(\phi_2 - \phi_1) - \dot{\phi}_2(\dot{\phi}_2 - \dot{\phi}_1)\sin(\phi_2 - \phi_1)] \quad (8\text{-}7)$$

and

$$\frac{d}{dt}\left(\frac{\partial T}{\partial \dot{\phi}_2}\right) = m_2 l_2{}^2 \ddot{\phi}_2$$
$$+ m_2 l_1 l_2 [\ddot{\phi}_1 \cos(\phi_2 - \phi_1) - \dot{\phi}_1(\dot{\phi}_2 - \dot{\phi}_1)\sin(\phi_2 - \phi_1)] \quad (8\text{-}8)$$

The partial derivative of T with respect to q_j yields

$$\frac{\partial T}{\partial q_1} = \frac{\partial T}{\partial \phi_1} = m_2 l_1 l_2 \dot{\phi}_1 \dot{\phi}_2 \sin(\phi_2 - \phi_1) \quad (8\text{-}9)$$

and

$$\frac{\partial T}{\partial q_2} = \frac{\partial T}{\partial \phi_2} = -m_2 l_1 l_2 \dot{\phi}_1 \dot{\phi}_2 \sin(\phi_2 - \phi_1) \quad (8\text{-}10)$$

Taking the partial derivative of U, Eq. 8-6, with respect to q_j, yields

$$\frac{\partial U}{\partial q_1} = \frac{\partial U}{\partial \phi_1} = (m_1 + m_2)g l_1 \sin\phi_1 \quad (8\text{-}11)$$

and

$$\frac{\partial U}{\partial q_2} = \frac{\partial U}{\partial \phi_2} = m_2 g l_2 \sin \phi_2 \qquad (8\text{-}12)$$

Substituting Eqs. 8-7 through 8-12 into Lagrange's equation, Eq. 8-1 provides the equations of motion for the double pendulum, thus

$$(m_1 + m_2)l_1{}^2\ddot{\phi}_1 + m_2 l_1 l_2 \ddot{\phi}_2 \cos{(\phi_2 - \phi_1)} - m_2 l_1 l_2 \dot{\phi}_2{}^2 \sin{(\phi_2 - \phi_1)}$$
$$+ (m_1 + m_2)g l_1 \sin \phi_1 = 0 \quad (8\text{-}13)$$
and

$$m_2 l_1 l_2 \ddot{\phi}_1 \cos{(\phi_2 - \phi_1)} + m_2 l_2{}^2 \ddot{\phi}_2 + m_2 l_1 l_2 \dot{\phi}_1{}^2 \sin{(\phi_2 - \phi_1)}$$
$$+ m_2 g l_2 \sin \phi_2 = 0 \quad (8\text{-}14)$$

8-3 Lagrange's Equation Applied to a System of Lumped Parameters

In actuality, a flexible aircraft or missile structure has an infinite number of degrees of freedom making an exact analysis almost impossible. However, a very accurate approximate analysis can be obtained by reducing the system to one of a finite number of degrees of freedom; the greater the number of degrees of freedom, the greater the accuracy of analysis. To reduce the number of degrees of freedom from an infinite number, the structure is broken down into discrete masses. Thus the total mass of the vehicle is replaced by a number of rigid masses connected by weightless connectors. The connectors are assumed to have the same elastic properties as the physical structure that they replace. The application of this principle is shown in Figure 8-3, which shows the wing of an aircraft reduced to two discrete masses. Also included are the generalized coordinates providing for rigid body motion as well as flexure. Notice

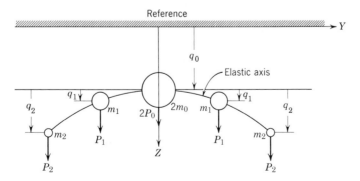

Figure 8-3 Aircraft wing represented by lumped parameters.

that the flexure motion is measured relative to the body so that the total motion of each element of the wing is the sum of rigid body motion and bending motion. It should also be noted that the moment of inertia effect of an angular rotation of the discrete masses has been neglected, since they are small in comparison to the translation inertia terms. Including them would have doubled the number of degrees of freedom. The forces represented by the P's are the forces due to gravity and the aerodynamic loads; these forces can be combined as the generalized forces Q_j. To demonstrate this, the P's are broken down into P_g's and P_a's, the gravitational and aerodynamic forces. The gravitational forces are included in the potential energy terms, and the aerodynamic forces in the generalized forces Q_j. As the system is symmetrical, only half of it need be considered for the analysis. The equation for the kinetic energy of the system can be obtained by taking half the product of each mass in question and the square of its velocity with respect to the reference. Then

$$T = \tfrac{1}{2}(m_0 \dot{z}_0{}^2 + m_1 \dot{z}_1{}^2 + m_2 \dot{z}_2{}^2) \tag{8-15}$$

but $z_0 = q_0$, $z_1 = q_0 + q_1$, and $z_2 = q_0 + q_2$. Substituting the values of z into Eq. 8-15 and simplifying, the equation for the kinetic energy becomes

$$T = \tfrac{1}{2}[m_T \dot{q}_0{}^2 + m_1 \dot{q}_1{}^2 + 2m_1 \dot{q}_1 \dot{q}_0 + 2m_2 \dot{q}_0 \dot{q}_2 + m_2 \dot{q}_2{}^2] \tag{8-16}$$

where $m_T = m_0 + m_1 + m_2$. The potential energy is made up of two parts: the internal strain energy U_i, and the potential energy resulting from the vertical displacement U_g. Thus (see Ref. 3, p. 133)

$$U_i = \tfrac{1}{2} \sum_{i=1}^{2} \sum_{j=1}^{2} k_{ij} q_i q_j = \tfrac{1}{2}(k_{11} q_1{}^2 + k_{12} q_1 q_2 + k_{21} q_2 q_1 + k_{22} q_2{}^2) \tag{8-17}$$

where k_{ij}'s are the influence coefficients to be discussed later in this chapter. The gravitational potential energy is

$$U_g = -(P_{g_0})q_0 - P_{g_1}(q_0 + q_1) - P_{g_2}(q_0 + q_2) \tag{8-18}$$

Finally, the aerodynamic forces

$$Q_0 = P_{a_0} + P_{a_1} + P_{a_2}, \; Q_1 = P_{a_1}, \; Q_2 = P_{a_2} \tag{8-19}$$

Taking the partial derivative of Eq. 8-16, with respect to $\dot{q}_0$, yields $\partial T / \partial \dot{q}_0 = m_T \dot{q}_0 + m_1 \dot{q}_1 + m_2 \dot{q}_2$; differentiating with respect to time

$$\frac{d}{dt}\left(\frac{\partial T}{\partial \dot{q}_0}\right) = m_T \ddot{q}_0 + m_1 \ddot{q}_1 + m_2 \ddot{q}_2 \tag{8-20}$$

The partial derivatives of Eqs. 8-16 and 8-17 with respect to q_0 are zero; however, the partial derivative of Eq. 8-18 yields

$$\frac{\partial U_g}{\partial q_0} = -(P_{g_0} + P_{g_1} + P_{g_2}) \tag{8-21}$$

Substituting the Q_0 term of Eqs. 8-19, 8-20, and 8-21 into Lagrange's equation yields

$$m_T \ddot{q}_0 + m_1 \ddot{q}_1 + m_2 \ddot{q}_2 = P_{a_0} + P_{g_0} + P_{a_1} + P_{g_1} + P_{a_2} + P_{g_2}$$
$$= P_0 + P_1 + P_2 \tag{8-22}$$

In like manner,

$$\frac{d}{dt}\left(\frac{\partial T}{\partial \dot{q}_1}\right) = m_1 \ddot{q}_0 + m_1 \ddot{q}_1 \tag{8-23}$$

However, the partial derivative of Eq. 8-17 with respect to q_1 is not zero, thus

$$\frac{\partial U_i}{\partial q_1} = k_{11}q_1 + \tfrac{1}{2}k_{12}q_2 + \tfrac{1}{2}k_{21}q_2 = k_{11}q_1 + k_{12}q_2 \tag{8-24}$$

as $k_{12} = k_{21}$. The partial derivative of Eq. 8-16 with respect to q_1 is zero; therefore, substituting Eqs. 8-23 and 8-24 into Lagrange's equation and combining with the partial derivative of Eq. 8-18 with respect to q_1, and adding Q_1 from Eq. 8-19 yields

$$m_1 \ddot{q}_0 + m_1 \ddot{q}_1 + k_{11}q_1 + k_{12}q_2 = P_1 \tag{8-25}$$

The third equation is obtained in the same manner as Eq. 8-25; thus the three equations of motion are

$$m_T \ddot{q}_0 + m_1 \ddot{q}_1 + m_2 \ddot{q}_2 \qquad\qquad\qquad = P_0 + P_1 + P_2$$

$$m_1 \ddot{q}_0 + m_1 \ddot{q}_1 \qquad + k_{11}q_1 + k_{12}q_2 = P_1$$

$$m_2 \ddot{q}_0 \qquad + m_2 \ddot{q}_2 + k_{21}q_1 + k_{22}q_2 = P_2 \tag{8-26}$$

Equation 8-26 summarizes the three differential equations which describe the behavior of the system shown in Figure 8-3. As just mentioned, the k_{ij}'s are the influence coefficients and can be written in the form of a matrix. The influence coefficient matrix is equal to the inverse, or reciprocal, of the stiffness matrix or $[k_{ij}] = [C_{ij}]^{-1}$ where the C's are the deflections at point i due to a unit load at point j (see Appendix E for a discussion of matrix algebra). Again the stiffness matrix is symmetrical.

8-4 Mode Shapes and Frequencies

Before looking at the solution of the simultaneous vibration equations, the technique for determining the shapes and natural frequencies of the vibratory modes is given. First, the vibratory equations are written in matrix form for ease of handling. In matrix form, Eqs. 8-26, or any set of simultaneous equations, can be expressed as

$$[m_{ij}]\{\ddot{q}_j\} + [k_{ij}]\{q_j\} = \{Q_j\} \tag{8-27}$$

where $[m_{ij}]$ is the inertial matrix and $[k_{ij}]$ is the stiffness matrix and

$$\{\ddot{q}_j\} = \begin{bmatrix} \ddot{q}_1 \\ \ddot{q}_2 \\ \vdots \\ \ddot{q}_n \end{bmatrix} \qquad \{q_j\} = \begin{bmatrix} q_1 \\ q_2 \\ \vdots \\ q_n \end{bmatrix} \qquad \{Q_j\} = \begin{bmatrix} Q_1 \\ Q_2 \\ \vdots \\ Q_n \end{bmatrix}$$

and n is equal to the number of degrees of freedom. Both the inertial and stiffness matrices are symmetrical but some of the terms may be zero.

To illustrate, Eq. 8-26 is written in matrix form; thus

$$\begin{bmatrix} m_T & m_1 & m_2 \\ m_1 & m_1 & 0 \\ m_2 & 0 & m_2 \end{bmatrix} \begin{bmatrix} \ddot{q}_0 \\ \ddot{q}_1 \\ \ddot{q}_2 \end{bmatrix} + \begin{bmatrix} 0 & 0 & 0 \\ 0 & k_{11} & k_{12} \\ 0 & k_{21} & k_{22} \end{bmatrix} \begin{bmatrix} q_0 \\ q_1 \\ q_2 \end{bmatrix} = \begin{bmatrix} P_0 + P_1 + P_2 \\ P_1 \\ P_2 \end{bmatrix} \tag{8-28}$$

Note that if the rigid body motion is neglected, $q_0 = 0$, and Eq. 8-28 reduces to

$$\begin{bmatrix} m_1 & 0 \\ 0 & m_2 \end{bmatrix} \begin{bmatrix} \ddot{q}_1 \\ \ddot{q}_2 \end{bmatrix} + \begin{bmatrix} k_{11} & k_{12} \\ k_{21} & k_{22} \end{bmatrix} \begin{bmatrix} q_1 \\ q_2 \end{bmatrix} = \begin{bmatrix} P_1 \\ P_2 \end{bmatrix} \tag{8-29}$$

Returning to the determination of the mode shapes and frequencies, Eq. 8-27 is written in general terms for a system with three degrees of freedom; thus

$$\begin{bmatrix} m_{11} & m_{12} & m_{13} \\ m_{21} & m_{22} & m_{23} \\ m_{31} & m_{32} & m_{33} \end{bmatrix} \begin{bmatrix} \ddot{q}_1 \\ \ddot{q}_2 \\ \ddot{q}_3 \end{bmatrix} + \begin{bmatrix} k_{11} & k_{12} & k_{13} \\ k_{21} & k_{22} & k_{23} \\ k_{31} & k_{32} & k_{33} \end{bmatrix} \begin{bmatrix} q_1 \\ q_2 \\ q_3 \end{bmatrix} = \begin{bmatrix} Q_1 \\ Q_2 \\ Q_3 \end{bmatrix} \tag{8-30}$$

The solution of Eq. 8-30 is assumed to be of the form $q_j = A_j \sin(\omega t + \psi)$; then $\ddot{q}_j = -A_j \omega^2 \sin(\omega t + \psi)$. Substituting these values into Eq. 8-30, and using only the characteristic equation, yields

$$- \begin{bmatrix} m_{11} & m_{12} & m_{13} \\ m_{21} & m_{22} & m_{23} \\ m_{31} & m_{32} & m_{33} \end{bmatrix} \begin{bmatrix} A_1 \omega^2 \\ A_2 \omega^2 \\ A_3 \omega^2 \end{bmatrix} + \begin{bmatrix} k_{11} & k_{12} & k_{13} \\ k_{21} & k_{22} & k_{23} \\ k_{31} & k_{32} & k_{33} \end{bmatrix} \begin{bmatrix} A_1 \\ A_2 \\ A_3 \end{bmatrix} = 0 \tag{8-31}$$

after dividing through by $\sin(\omega t + \psi)$.

Simplifying,

$$\begin{bmatrix} (k_{11} - m_{11}\omega^2) & (k_{12} - m_{12}\omega^2) & (k_{13} - m_{13}\omega^2) \\ (k_{21} - m_{21}\omega^2) & (k_{22} - m_{22}\omega^2) & (k_{23} - m_{23}\omega^2) \\ (k_{31} - m_{31}\omega^2) & (k_{32} - m_{32}\omega^2) & (k_{33} - m_{33}\omega^2) \end{bmatrix} \begin{bmatrix} A_1 \\ A_2 \\ A_3 \end{bmatrix} = 0 \quad (8\text{-}32)$$

As in the case of the equations of motion of the aircraft, for the solution of Eq. 8-32 to be nontrivial, the determinant of the coefficients of A_j must equal zero. The expansion of the determinant formed by the coefficient matrix of A_j will yield an equation with the highest power of ω equal to $2n$. The factoring of this equation will yield the natural frequencies of the oscillatory modes. These are designated by ω_r with $r = 1, 2, \ldots, n$ and $\omega_1 < \omega_2 < \ldots \omega_n$. The next step is the determination of the mode shapes. To illustrate the required manipulation of the matrix equation, Eq. 8-32 first is expanded, which yields

$$(k_{11} - m_{11}\omega^2)A_1 + (k_{12} - m_{12}\omega^2)A_2 + (k_{13} - m_{13}\omega^2)A_3 = 0$$

$$(k_{21} - m_{21}\omega^2)A_1 + (k_{22} - m_{22}\omega^2)A_2 + (k_{23} - m_{23}\omega^2)A_3 = 0$$

$$(k_{31} - m_{31}\omega^2)A_1 + (k_{32} - m_{32}\omega^2)A_2 + (k_{33} - m_{33}\omega^2)A_3 = 0$$

$$(8\text{-}33)$$

The first two equations of Eq. 8-33 can be written as

$$(k_{11} - m_{11}\omega^2)A_1 + (k_{12} - m_{12}\omega^2)A_2 = -(k_{13} - m_{13}\omega^2)A_3$$

$$(k_{21} - m_{21}\omega^2)A_1 + (k_{22} - m_{22}\omega^2)A_2 = -(k_{23} - m_{23}\omega^2)A_3 \quad (8\text{-}34)$$

Returning to matrix form and dividing both sides by A_3 yields

$$\begin{bmatrix} (k_{11} - m_{11}\omega^2) & (k_{12} - m_{12}\omega^2) \\ (k_{21} - m_{21}\omega^2) & (k_{22} - m_{22}\omega^2) \end{bmatrix} \begin{bmatrix} A_1/A_3 \\ A_2/A_3 \end{bmatrix} = - \begin{bmatrix} k_{13} - m_{13}\omega^2 \\ k_{23} - m_{23}\omega^2 \end{bmatrix} \quad (8\text{-}35)$$

The ratios A_1/A_3 and A_2/A_3 are defined as the mode shapes and designated ϕ_j. Previously, the natural frequencies for the n modes of oscillation were determined; thus Eq. 8-35 can be evaluated for each ω_r yielding a ϕ_j for each natural frequency; these are designated $\phi_j^{(r)}$. Introducing the r notation, Eq. 8-35 becomes

$$\begin{bmatrix} (k_{11} - m_{11}\omega_r^2) & (k_{12} - m_{12}\omega_r^2) \\ (k_{21} - m_{21}\omega_r^2) & (k_{22} - m_{22}\omega_r^2) \end{bmatrix} \begin{bmatrix} \phi_1^{(r)} \\ \phi_2^{(r)} \end{bmatrix} = - \begin{bmatrix} k_{13} - m_{13}\omega_r^2 \\ k_{23} - m_{23}\omega_r^2 \end{bmatrix} \quad (8\text{-}36)$$

and $\phi_3^{(r)} = A_3/A_3 = 1$. The mode shapes $\phi_j^{(r)}$ yield the maximum amplitude of each mode of vibration (r) for each station or discreet mass j, as shown in Figure 8-4 for the first mode shape ($r = 1$). Normally the steps between Eq. 8-33 and Eq. 8-36 can be eliminated, and Eq. 8-36 can be obtained directly from Eq. 8-33. Equation 8-36 then can be solved for the required mode shapes.

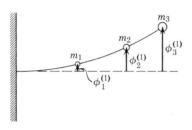

To better illustrate this procedure, a simple numerical example is worked. The system analyzed is the same as that shown in Figure 8-3, which is redrawn in Figure 8-5 showing only half of the aircraft. The differential equation for this system is given in Eq. 8-28. In comparing Eq. 8-28 to Eq. 8-30, it can be seen that the

Figure 8-4   Illustration of mode shapes for the first mode.

m's and k's of Eq. 8-28 can be equated to the m's and k's of Eq. 8-30. A prime is used with the k's from Eq. 8-28 to distinguish them from the k's of Eq. 8-30; thus

$$m_{11} = m_T = 103.3 \, \frac{\text{lb-sec}^2}{\text{in.}}$$

$$m_{12} = m_{21} = m_{22} = m_1 = 25.8 \, \frac{\text{lb-sec}^2}{\text{in.}}$$

$$m_{13} = m_{31} = m_{33} = m_2 = 12.9 \, \frac{\text{lb-sec}^2}{\text{in.}}$$

$$m_{23} = m_{32} = 0$$

$$k_{11} = k_{12} = k_{13} = k_{21} = k_{31} = 0$$

$$k_{22} = k'_{11}, \quad k_{23} = k_{32} = k'_{12}, \quad k_{33} = k'_{22}$$

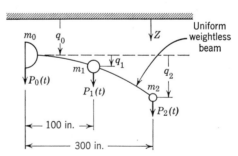

Figure 8-5   System for sample problem.

$EI = 3.6 \times 10^{10}$ lb-in.^{2} $m_2 = 12.9 \, \frac{\text{lb-sec}^2}{\text{in.}}$

$m_0 = 64.6 \, \frac{\text{lb-sec}^2}{\text{in.}}$ $m_T = 103.3 \, \frac{\text{lb-sec}^2}{\text{in.}}$

$m_1 = 25.8 \, \frac{\text{lb-sec}^2}{\text{in.}}$ $g = 387$ in/sec^2

It is now necessary to determine k'_{11}, k'_{12}, and k'_{22}. Since $[k_{ij}] = [C_{ij}]^{-1}$, the stiffness matrix must be determined first. The deflection at point i due to a unit load applied at point j is given by $C_{ij} = \dfrac{x_i^2}{6EI}(3x_j - x_i)$ for $x_i \le x_j^{(\text{Ref. 1})}$, where x_i and x_j are the distances from the centerline to the points i and j, respectively, as shown in Figure 8-6. For C_{11}, $x_i = x_j = 100$ in., then

$$C_{11} = \frac{(100)^2(300 - 100)}{6(3.6 \times 10^{10})}$$
$$= 0.925 \times 10^{-5}\ \text{in/lb}$$

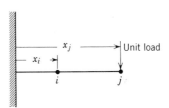

Figure 8-6

For C_{12}, $x_i = 100$ in., $x_j = 300$ in., then

$$C_{12} = C_{21} = \frac{(100)^2(900 - 100)}{6(3.6 \times 10^{10})} = 3.7 \times 10^{-5}\ \text{in/lb}$$

For C_{22}, $x_i = x_j = 300$ in., then

$$C_{22} = \frac{(300)^2(900 - 300)}{6(3.6 \times 10^{10})} = 25 \times 10^{-5}\ \text{in/lb}$$

Then,

$$[C_{ij}] = \begin{bmatrix} 0.925 & 3.7 \\ 3.7 & 25 \end{bmatrix} \times 10^{-5}\ \text{in/lb}$$

But,

$$[k'_{ij}] = [C_{ij}]^{-1} = \frac{(-1)^{i+j}\begin{vmatrix} \text{Minor of} \\ [C_{ji}] \end{vmatrix}}{|C_{ij}|}$$

Then,

$$k'_{11} = \frac{(-1)^2(25 \times 10^{-5})}{\begin{vmatrix} 0.925 \times 10^{-5} & 3.7 \times 10^{-5} \\ 3.7 \times 10^{-5} & 25 \times 10^{-5} \end{vmatrix}} = \frac{25 \times 10^{-5}}{9.5 \times 10^{-10}} = 2.63 \times 10^5\ \text{lb/in.}$$

Similarly,

$$k'_{12} = k'_{21} = \frac{(-1)^3(3.7 \times 10^{-5})}{9.5 \times 10^{-10}} = -0.39 \times 10^5\ \text{lb/in.}$$

and

$$k'_{22} = \frac{0.925 \times 10^{-5}}{9.5 \times 10^{-10}} = 0.0975 \times 10^5\ \text{lb/in.}$$

Therefore,

$$[k'_{ij}] = \begin{bmatrix} 2.63 & -0.39 \\ -0.39 & 0.0975 \end{bmatrix} \times 10^5 \text{ lb/in.}$$

The differential equations describing the motion of the wing and fuselage can now be written in matrix form; thus

$$\begin{bmatrix} 103.3 & 25.8 & 12.9 \\ 25.8 & 25.8 & 0 \\ 12.9 & 0 & 12.9 \end{bmatrix} \begin{bmatrix} \ddot{q}_0 \\ \ddot{q}_1 \\ \ddot{q}_2 \end{bmatrix} + \begin{bmatrix} 0 & 0 & 0 \\ 0 & 2.63 & -0.39 \\ 0 & -0.39 & 0.0975 \end{bmatrix} \begin{bmatrix} q_0 \\ q_1 \\ q_2 \end{bmatrix} \times 10^5$$

$$= \begin{bmatrix} P_0(t) + P_1(t) + P_2(t) \\ P_1(t) \\ P_2(t) \end{bmatrix}$$

Next it is necessary to determine the natural frequencies of oscillation, which can be accomplished by substituting the values of m_{ij} and k_{ij} into Eq. 8-32, which yields

$$\begin{bmatrix} (0 - 103.3\omega^2) & (0 - 25.8\omega^2) & (0 - 12.9\omega^2) \\ (0 - 25.8\omega^2) & (2.63 + 10^5 - 25.8\omega^2) & (-0.39 \times 10^5 - 0) \\ (0 - 12.9\omega^2) & (-0.39 \times 10^5 - 0) & (0.0975 \times 10^5 - 12.9\omega^2) \end{bmatrix}$$

$$\times \begin{bmatrix} A_1 \\ A_2 \\ A_3 \end{bmatrix} = 0$$

Expanding the determinant and simplifying yields

$$\omega^2(\omega^4 - 13.35 \times 10^3\omega^2 + 41.6 \times 10^5) = 0$$

Factoring,

$$\omega^2(\omega^2 - 13 \times 10^3)(\omega^2 - 320) = 0$$

Therefore the natural frequencies of the modes are $\omega_0 = 0$, $\omega_1 = 17.9$ rad/sec, $\omega_2 = 114$ rad/sec. The ω_0 frequency is the rigid body mode.

The mode shapes corresponding to the three frequencies of vibration are now evaluated. As the mode shapes represent the eigenfunctions of the solution of the homogeneous equation, each mode shape represents only relative displacements of the portions of the beam; thus the mode shape for the rigid body mode can be obtained by inspection. If the relative displacement of the fuselage is taken as unity, the mode shape corresponding to ω_0 for the fuselage, $\phi_0^{(0)} = 1$. The mode shapes for the other two stations, m_1 and m_2, are zero since the displacements of these two stations are measured with respect to the horizontal reference that moves with the fuselage; thus $\phi_1^{(0)} = \phi_2^{(0)} = 0$. The remaining mode shapes can be evaluated using Eq. 8-36. Substituting for $\omega_1{}^2$ in Eq. 8-36, it becomes

$$
\begin{bmatrix} -(103.3)(320) & -(25.8)(320) \\ -(25.8)(320) & 2.63 \times 10^5 - (25.8)(320) \end{bmatrix} \begin{bmatrix} \phi_0^{(1)} \\ \phi_1^{(1)} \end{bmatrix} = - \begin{bmatrix} -(12.9)(320) \\ -0.39 \times 10^5 \end{bmatrix}
$$

Multiplying through by the inverse of the coefficient matrix yields

$$
\begin{bmatrix} \phi_0^{(1)} \\ \phi_1^{(1)} \end{bmatrix} = \begin{bmatrix} -0.03 & -9.7 \times 10^{-4} \\ -9.7 \times 10^{-4} & 0.389 \times 10^{-2} \end{bmatrix} \begin{bmatrix} 4.13 \\ 39 \end{bmatrix}
$$

Solving yields the mode shapes for ω_1, which are $\phi_0^{(1)} = -0.161$, $\phi_1^{(1)} = 0.148$, and $\phi_2^{(1)} = 1$. In like manner, substituting for ω_2 in Eq. 8-36, and simplifying, yields

$$
\begin{bmatrix} -134.2 & -33.5 \\ -33.5 & -7.2 \end{bmatrix} \begin{bmatrix} \phi_0^{(2)} \\ \phi_1^{(2)} \end{bmatrix} = \begin{bmatrix} 16.75 \\ 3.9 \end{bmatrix}
$$

Solving for the mode shapes for ω_2 yields $\phi_0^{(2)} = -0.0598$, $\phi_1^{(2)} = -0.264$, and $\phi_2^{(2)} = 1$. Thus the natural frequencies and the mode shapes have been determined.

8-5 Normal Coordinates

So far, the natural frequencies (the eigenvalues) and the mode shapes (the eigenfunctions) for the coupled homogeneous equations have been determined; however, the response of the system to the forcing functions has not been determined. To obtain the so called particular solution, use is made of the normal coordinates ξ_r, which yield a set of uncoupled second-order equations of the form

$$
M_r \ddot{\xi}_r + M_r \omega_r{}^2 \xi_r = \Xi_r \tag{8-37}
$$

where $M_r = \lfloor \phi_j^{(r)} \rfloor [m_{ij}] \{\phi_j^{(r)}\}$ are the generalized masses and

$$
\Xi_r = \lfloor \phi_j^{(r)} \rfloor \{Q_j\}
$$

are the normalized forces. The ω_r's are the natural frequencies of the mode shapes already determined. After solving the uncoupled equations for the ξ_r's, the solution of the original equations in terms of the generalized coordinates, q_j, can be obtained by using the following equation:

$$\{q_j\} = [\phi_j^{(r)}]\{\xi_r\} \tag{8-38}$$

The numerical example, started in Section 8-3, is continued here to obtain the uncoupled equations in normal coordinates. Substituting the appropriate values into Eq. 8-37 yields

$$M_0 = [1 \ 0 \ 0] \begin{bmatrix} 103.3 & 25.8 & 12.9 \\ 25.8 & 25.8 & 0 \\ 12.9 & 0 & 12.9 \end{bmatrix} \begin{bmatrix} 1 \\ 0 \\ 0 \end{bmatrix}$$

$$= [103.3 \ 25.8 \ 12.9] \begin{bmatrix} 1 \\ 0 \\ 0 \end{bmatrix} = 103.3 \frac{\text{lb-sec}^2}{\text{in.}}$$

Note that the first operation consists of multiplying the square matrix by the row matrix. In like manner

$$M_1 = [-0.161 \ \ 0.148 \ \ 1] \begin{bmatrix} 103.3 & 25.8 & 12.9 \\ 25.8 & 25.8 & 0 \\ 12.9 & 0 & 12.9 \end{bmatrix} \begin{bmatrix} -0.161 \\ 0.148 \\ 1 \end{bmatrix} = 9.76 \frac{\text{lb-sec}^2}{\text{in.}}$$

and

$$M_2 = [-0.0598 \ \ -0.264 \ \ 1] \begin{bmatrix} 103.3 & 25.8 & 12.9 \\ 25.8 & 25.8 & 0 \\ 12.9 & 0 & 12.9 \end{bmatrix}$$

$$\times \begin{bmatrix} -0.0598 \\ -0.264 \\ 1 \end{bmatrix} = 14.33 \frac{\text{lb-sec}^2}{\text{in.}}$$

Using the equation for the normalized forces,

$$\Xi_0 = [1 \ 0 \ 0] \begin{bmatrix} P_0(t) + P_1(t) + P_2(t) \\ P_1(t) \\ P_2(t) \end{bmatrix} = P_0(t) + P_1(t) + P_2(t) \text{ lb}$$

$$\Xi_1 = [-0.161 \quad 0.148 \quad 1] \begin{bmatrix} P_0(t) + P_1(t) + P_2(t) \\ P_1(t) \\ P_2(t) \end{bmatrix}$$

$$= -0.161P_0(t) - 0.013P_1(t) + 0.839P_2(t) \text{ lb}$$

and

$$\Xi_2 = [-0.0598 \quad -0.264 \quad 1] \begin{bmatrix} P_0(t) + P_1(t) + P_2(t) \\ P_1(t) \\ P_2(t) \end{bmatrix}$$

$$= -0.0598P_0(t) - 0.3238P_1(t) + 0.9402P_2(t) \text{ lb}$$

Therefore the uncoupled equations in normal coordinates are

$$103.3\ddot{\xi}_0 = P_0(t) + P_1(t) + P_2(t)$$

$$9.76\ddot{\xi}_1 + 3120\xi_1 = -0.161P_0(t) - 0.013P_1(t) + 0.839P_2(t)$$

$$14.33\ddot{\xi}_2 + 18.6 \times 10^4\xi_2 = -0.0598P_0(t) - 0.3238P_1(t) + 0.9402P_2(t)$$

8-6 System Transfer Function, Including Body Bending

The body bending equations have been derived, but to determine the effects of body bending on the control system, the body bending equations must be combined with the rigid body equations. This coupling between the rigid body and body bending arises from the fact that the attitude and rate gyros sense both the rigid body attitude changes and the body bending motion. By proper combination of the body bending equations with the rigid body equations, modified as necessary, the transfer function for the combined rigid body and bending modes can be obtained.

For the body bending, the body bending equations in normal coordinate form are used (see Eq. 8-37). The solution of the body bending equations in generalized coordinates (Eq. 8-38) is not generally required; however, as is shown later in this section, the actual displacement of specific stations of the vehicle and the actual slope of the elastic axis at specific points, θ_{B_j}, may be required. These values can be obtained from the following equation:

$$\{\theta_{B_j}\} = [\sigma_j^{(r)}]\{\xi_r\} \tag{8-39}$$

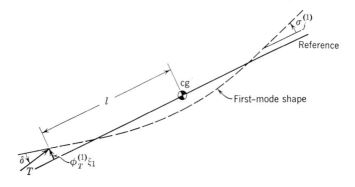

Figure 8-7   First-mode shape for a missile showing normalized mode slope.

where $\sigma_j^{(r)}$ is the slope of the rth mode shape at the jth station, and is referred to as the normalized mode slope (rad/ft).   See Figure 8-7.

To illustrate the simultaneous solution of the modified rigid body equations and the body bending equations, the equations for the ballistic missile modified for the effects of body bending at the tail, are used. This procedure neglects the aerodynamic effects associated with the body bending which is generally satisfactory for vehicles of this type.[4]   Equation 7-2 is rewritten to include the effects of body bending; thus

$$\left(\frac{mU}{Sq}s - C_{z_\alpha}\right)\alpha(s) + \left[-\frac{mU}{Sq}s - C_w\sin\Theta\right]\theta(s) = C_{z_\delta}[\delta(s) + \sigma_T^{(r)}\xi_r(s)]$$

$$-C_{m_\alpha}\alpha(s) + \left(\frac{I_y}{Sqd}s^2 - \frac{d}{2U}C_{m_q}s\right)\theta(s) = C_{m_\delta}[\delta(s) + \sigma_T^{(r)}\xi_r(s)] + \frac{C_{z_\delta}}{d}\phi_T^{(r)}\xi_r(s)$$

$$\left(\frac{M_r}{Sq}s^2 + \frac{M_r\omega_r^2}{Sq}\right)\xi_r(s) = C_{z_\delta}\phi_T^{(r)}\delta(s) \qquad (8\text{-}40)$$

The expression $[\delta(s) + \sigma_T^{(r)}\xi_r(s)]$ gives the effective direction of the thrust vector relative to the reference, which is the centerline of the rigid missile; the normalized mode slope is negative at the tail of the missile.   The term $\dfrac{C_{z_\delta}}{d}\phi_T^{(r)}\xi_r(s)$ gives the torque resulting from the lateral displacement of the rocket motor gimbal from the reference axis. C_{z_δ} must be divided by the characteristic length to keep the equation nondimensional.   The damping of the body bending modes has been neglected.   Equation 8-40 is rewritten considering only the first two body bending modes and transcribing the body bending terms in the rigid body equations to the left-hand side; thus

$$A\alpha(s) + B\theta(s) - C_{z_\delta}\sigma_T^{(1)}\xi_1(s) - C_{z_\delta}\sigma_T^{(2)}\xi_2(s) = C_{z_\delta}\delta(s)$$

$$C\alpha(s) + D\theta(s) - \left(C_{m_\delta}\sigma_T^{(1)} + \frac{C_{z_\delta}}{d}\phi_T^{(1)}\right)\xi_1(s) - \left(C_{m_\delta}\sigma_T^{(2)} + \frac{C_{z_\delta}}{d}\phi_T^{(2)}\right)\xi_2(s)$$
$$= C_{m_\delta}\delta(s)$$

$$\left(\frac{M_1}{Sq}s^2 + \frac{M_1\omega_1^2}{Sq}\right)\xi_1(s) = C_{z_\delta}\phi_T^{(1)}\delta(s)$$

$$\left(\frac{M_2}{Sq}s^2 + \frac{M_2}{Sq}\omega_2^2\right)\xi_2(s) = C_{z_\delta}\phi_T^{(2)}\delta(s) \qquad (8\text{-}41)$$

where $A = \left(\dfrac{mU}{Sq}s - C_{z_\alpha}\right)$; $B = \left(-\dfrac{mU}{Sq}s - C_w \sin \Theta\right)$; $C = -C_{m_\alpha}$; and

$D = \left(\dfrac{I_y}{Sqd}s^2 - \dfrac{d}{2U}C_{m_q}s\right)$.

The characteristic equation can be obtained by setting the determinant of the coefficients of Eq. 8-41 to zero. Denoting this by ∇ yields

$$\nabla = (AD - BC)\left(\frac{M_1}{Sq}s^2 + \frac{M_1}{Sq}\omega_1^2\right)\left(\frac{M_2}{Sq}s^2 + \frac{M_2}{Sq}\omega_2^2\right) = 0 \quad (8\text{-}42)$$

The term $(AD - BC)$ of Eq. 8-42 is equal to the expansion of the determinant formed by the coefficients of Eq. 7-2, and is equivalent to the open loop poles of the rigid body transfer function. Since Eq. 8-42 forms the denominator of the missile transfer functions, including body bending, the roots of Eq. 8-42 yield the poles of the flexible missile transfer function. From Eq. 8-42 it can be seen that the poles of the flexible missile transfer function consist of the rigid body poles and the poles from each of the body bending modes. To obtain the overall flexible

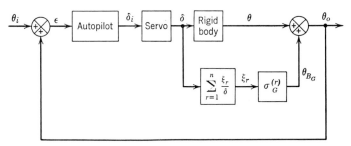

Figure 8-8 Block diagram for the flexible missile. *Note:* $\sigma_G^{(r)}$ is the normalized mode slope at the gyro or stable platform, used to sense the missile attitude.

missile transfer function, it is first necessary to look at the block diagram of the flexible missile, as shown in Figure 8-8. From Figure 8-8 it can be seen that θ_0, which is the pitch attitude sensed by the attitude gyro or stable platform, consists of θ from the rigid missile plus the attitude resulting from the missile flexure. Thus, the overall transfer function is

$$\frac{\theta_0(s)}{\delta(s)} = \frac{\theta(s)}{\delta(s)} + \sigma_G^{(1)} \frac{\xi_1(s)}{\delta(s)} + \sigma_G^{(2)} \frac{\xi_2(s)}{\delta(s)} \tag{8-43}$$

The individual transfer functions can be obtained from Eq. 8-41; they are

$$\frac{\theta(s)}{\delta(s)} = \frac{1}{\nabla}\left[(AC_{m_\delta} - CC_{z_\delta})\left(\frac{M_1 s^2}{Sq} + \frac{M_1\omega_1^2}{Sq}\right)\left(\frac{M_2 s^2}{Sq} + \frac{M_2\omega_2^2}{Sq}\right)\right.$$

$$- \phi_T^{(1)} C_{z_\delta}\left\{A\left(C_{m_\delta}\sigma_T^{(1)} + \frac{C_{z_\delta}}{d}\phi_T^{(1)}\right) + CC_{z_\delta}\sigma_T^{(1)}\right\}\left(\frac{M_2 s^2}{Sq} + \frac{M_2\omega_2^2}{Sq}\right)$$

$$\left. - \phi_T^{(2)} C_{z_\delta}\left\{A\left(C_{m_\delta}\sigma_T^{(2)} + \frac{C_{z_\delta}}{d}\phi_T^{(2)}\right) + CC_{z_\delta}\sigma_T^{(2)}\right\}\left(\frac{M_1 s^2}{Sq} + \frac{M_1\omega_1^2}{Sq}\right)\right]$$

$$\frac{\sigma_G^{(1)}\xi_1(s)}{\delta(s)} = \frac{\sigma_G^{(1)} C_{z_\delta}\phi_T^{(1)}(AD - BC)\left(\frac{M_2 s^2}{Sq} + \frac{M_2\omega_2^2}{Sq}\right)}{\nabla}$$

and

$$\frac{\sigma_G^{(2)}\xi_2(s)}{\delta(s)} = \frac{\sigma_G^{(2)} C_{z_\delta}\phi_T^{(2)}(AD - BC)\left(\frac{M_1 s^2}{Sq} + \frac{M_1\omega_1^2}{Sq}\right)}{\nabla} \tag{8-44}$$

As mentioned earlier in this paragraph the poles of the flexible missile transfer function are the rigid body poles plus the body bending poles. The zeros of the flexible transfer function in general consist of zeros only slightly different from the rigid body zeros, plus complex zeros near each of the body bending poles. This pattern is normally found when analyzing the flexible missile or aircraft;[5] however, gyro location influences the location of the zeros, and as will be seen in Section 8-9, other zero locations are possible.

Contained in Ref. 4 is a diagram of the stability boundaries for several flight times. This plot resulted from analog studies and gives the relation between gyro location and the maximum allowable gain for stability. This figure is represented in Figure 8-9. From this figure it can be seen that the best location for the sensors (gyros) is just to the rear of the center of the missile. This location puts the sensors near the antinode of the first body bending mode.

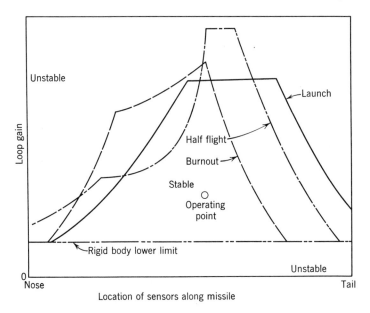

Figure 8-9   Typical stability boundary plot for several flight times.

8-7   The "Tail-Wags-Dog" Zero

Another factor that must be considered in the missile transfer function is the so-called "tail-wags-dog" (TWD) zero. This zero, actually a pair of complex zeros, occurs at the frequency at which the inertial forces resulting from the gimbaling of the rocket engine cancel the component of thrust normal to the missile axis due to deflection of the motor chamber. There are two inertial torques, one resulting from the angular acceleration of the rocket engine about its gimbal pivot, and the other resulting from the translation of the center of gravity of the rocket engine, as shown in Figure 8-10. Due to an angular acceleration of the rocket chamber there will be an inertial reaction torque at the gimbal. This torque can be transferred to the center of gravity of the missile and will be in the direction shown and equal to $I_y\ddot{\delta}$. As a result of the lateral translation of the center of gravity of the chamber, there will be an inertial reaction force equal to $m_R l_R\ddot{\delta}$, where m_R is the mass of the chamber and $l_R\ddot{\delta}$ is the tangential acceleration of the center of mass of the rocket motor. This inertial reaction force multiplied by l gives a torque acting about the missile center of gravity in the direction shown in Figure 8-10. The torque due to engine deflection is equal to $Tl\delta$ as before; the summation of the applied moments is then

$$\sum \text{applied moments} = -(Tl\delta + m_R l_R l\ddot{\delta} + I_y\ddot{\delta}) \qquad (8\text{-}45)$$

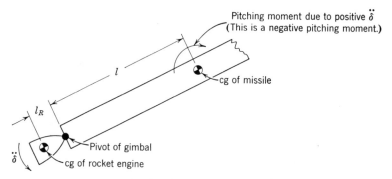

Figure 8-10   Sketch showing inertial torques resulting from gimbal deflections.

Before Eq. 8-45 can be introduced into the moment equation of Eq. 8-41, it must be nondimensionalized by dividing by Sqd. For the force equation, Eq. 8-45 must be divided by l then nondimensionalized by dividing by Sq. The right-hand side of Eq. 8-41 then becomes

For force equations:
$$\left[C_{z_\delta} - \frac{1}{Sq} \left(m_R l_R + \frac{I_y}{l} \right) s^2 \right] \delta(s)$$

Moment equation:
$$\left[C_{m_\delta} - \frac{1}{Sqd} (m_R l_R l + I_y) s^2 \right] \delta(s)$$

Body bending equations:
$$\phi_T^{(r)} \left[C_{z_\delta} - \frac{1}{Sq} \left(m_R l_R + \frac{I_y}{l} \right) s^2 \right] \delta(s)$$

(as C_{z_δ} and C_{m_δ} are negative)

The result of including these inertial terms is to add a pair of complex zeros to the overall flexible missile transfer function. These zeros will be on the $j\omega$ axis at a value of $\omega_n = \left(\dfrac{T}{m_R l_R + I_y/l} \right)^{1/2}$ rad/sec, if the convention is used that $\phi_T^{(r)} = 1$. This is true since the right-hand side of the moment equation can be written as $(l/d)[C_{z_\delta} - (1/Sq)(m_R l_R + I_y/l)s^2]\delta(s)$ and $C_{z_\delta} = -T/Sq$.

8-8   Effects of Propellant Sloshing

Another important factor that affects the stability of large liquid-fuel missiles is the sloshing of the propellants. The so-called sloshing modes may couple with the rigid body, and even the bending modes, and cause instability. The problem is further aggravated because the damping of the propellant modes is very light in large vehicles, varying inversely with tank diameter.[4]

In Appendix A of Ref. 4 it is shown that for cylindrical tanks the effects of propellant sloshing can be approximated by a series of pendulums or spring masses (one for each sloshing mode). This mechanical analog must be repeated for each tank. Normally it is sufficient to consider only the first propellant mode,[4] as the forces and moments produced by the higher order modes are negligible. In the analysis to follow, a simple pendulum is used to derive the transfer function for a rigid missile with one tank. The analysis can be extended to include multiple tanks, but is too unwieldy to be presented here. Before proceeding with the derivation, a short discussion of the behavior of a pendulum under the influence of pivot acceleration is presented.

To illustrate this behavior, a simple pendulum, the pivot of which is being accelerated linearly with an acceleration equal to $\mathbf{a}$, is considered. This is shown in Figure 8-11 where $\mathbf{F} = m_p\mathbf{a}$ and m_p is the mass of the pendulum, and two forces $\mathbf{F}_1$ and $\mathbf{F}_2$ have been adding acting through the cg of the pendulum with $\mathbf{F}_1 = \mathbf{F} = -\mathbf{F}_2$. The addition of $\mathbf{F}_1$ and $\mathbf{F}_2$ has no effect on the total summation of forces and moments acting on the pendulum. $\mathbf{F}_1$ acting on the cg of the pendulum causes the pendulum to accelerate in the direction of $\mathbf{F}_1$, but causes no rotation. However, the two forces $\mathbf{F}$ and $\mathbf{F}_2$ form a couple and impart an angular acceleration to the pendulum. The angular acceleration resulting from the couple is the phenomenon of particular interest. The moment resulting from the couple can be obtained by summing torques about any point in the body; for simplicity the torque produced by the couple is summed about the cg. Thus

$$\sum \mathbf{M}_{(cg)} = \sum \mathbf{M}_{(pivot)} = \mathbf{F} \times l = m_p\mathbf{a} \times l \qquad (8\text{-}46)$$

From this discussion it can be seen that the results of accelerating the pivot of a pendulum are twofold. First, it causes the whole pendulum to accelerate in the direction of the pivot acceleration, and second, it causes a moment to be produced about the pivot. Having examined the effects of accelerating the pivot of a pendulum, the analysis of the effects of propellant sloshing are presented next.

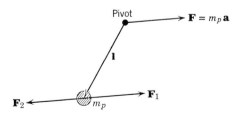

Figure 8-11 Simple pendulum under the influence of pivot acceleration.

The parameters to be used are shown in Figure 8-12, and equations for relating them to the tank properties are given in Appendix A of Ref. 4. From Figure 8-12, the accelerations acting on the pivot of the pendulum are the thrust acceleration (a_T), a tangential acceleration due to pitching angular accelerations ($l_p\ddot{\theta}$), and the Z component of vehicle acceleration, $\dot{w}$. The centripetal acceleration caused by a pitch angular velocity ($l_p\dot{\theta}^2$) which is very small in comparison to a_T is neglected. The pivot acceleration expressed in the missile axis system is then

$$\mathbf{a}_{(\text{pivot})} = \mathbf{i}a_T + \mathbf{k}(\dot{w} - l_p\ddot{\theta}) \tag{8-47}$$

and

$$\mathbf{L}_p = -\mathbf{i}L_p + \mathbf{k}L_p\theta_p \tag{8-48}$$

using the small angle assumption for θ_p. The moment resulting from these accelerations can be obtained by substituting Eqs. 8-47 and 8-48 into Eq. 8-46. Thus

$$\sum \mathbf{M}_{(\text{pivot})} = m_p\mathbf{a}_{(\text{pivot})} \times \mathbf{L}_p = \mathbf{j}[m_pL_p(l_p\ddot{\theta} - \dot{w}) - m_pa_TL_p\theta_p] \tag{8-49}$$

Now $\sum \mathbf{M}_{(\text{pivot})}$ is equal to the moment of inertia of the pendulum (I_p) times the angular acceleration of the pendulum with respect to inertial space, which can be approximated by the angular acceleration of the

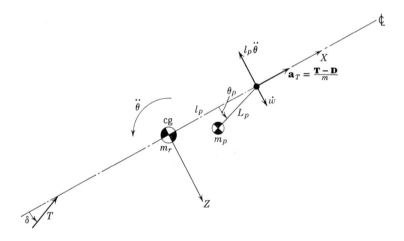

Figure 8-12   Propellant sloshing model for one tank where m_p is the mass of the propellant in the tank; $m_r = m - m_p$ is the mass of the missile less the propellant mass; L_p is the length of the equivalent pendulum representing the first sloshing mode of one tank; l_p is the distance between the (cg) of the missile, less propellant, and the pivot of the pendulum.

pendulum with respect to the earth which is $\ddot{\theta}_p + \ddot{\theta}$. Substituting for $\sum \mathbf{M}_{(pivot)}$ in Eq. 8-49 yields

$$m_p L_p(l_p \ddot{\theta} - \dot{w}) - m_p a_T L_p \theta_p = L_p^2 m_p (\ddot{\theta}_p + \ddot{\theta}) \qquad (8\text{-}50)$$

where $I_p = L_p^2 m_p$. Rearranging and dividing by Sqd to nondimensionalize, Eq. 8-50 becomes

$$\frac{L_p^2 m_p}{Sqd} \ddot{\theta}_p + \frac{m_p a_T L_p \theta_p}{Sqd} + \frac{m_p L_p U \dot{\alpha}}{Sqd} - \frac{L_p m_p}{Sqd} (l_p - L_p) \ddot{\theta} = 0 \qquad (8\text{-}51)$$

which is the equation describing the effects of the sloshing of the liquid propellant on the vehicle dynamics.

To obtain the transfer function for the missile, including propellant sloshing, Eq. 8-51 must be solved simultaneously with the rigid body equations. (The combination of Eq. 8-51 with the body bending equations is discussed later in this section.) This situation requires the addition of one term in each of the Z force and pitching moment equations to account for the displacement of the center of gravity of the propellant from the X axis of the missile. Due to the longitudinal acceleration of the missile there is a tension generated in the rod of the pendulum equal to $m_p a_T$ (assuming θ_p a small angle). This tension generates a force on the missile at the pivot, as shown in Figure 8-13, which results in a Z force equal to $m_p a_T \theta_p$ and a negative pitching moment equal to $m_p a_T l_p \theta_p$. The

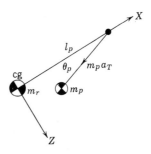

Figure 8-13   Force produced by propellant on the missile.

force and moment after nondimensionalizing must be added to the rigid body equations to complete the equations of motion including propellant sloshing. The final equations in Laplace transform notation for zero initial conditions, are given in Eq. 8-52.

$$\left(\frac{m_r U}{Sq} s - C_{z_\alpha}\right)\alpha(s) - \left(\frac{m_r U}{Sq} s - C_w \sin \Theta\right)\theta(s) - C_{z_p}\theta_p(s) = C_{z_\delta}\delta(s)$$

$$-C_{m_\alpha}\alpha(s) + \left(\frac{I_{y_r}}{Sqd} s^2 - \frac{d}{2U} C_{m_q}\right)\theta(s) + C_{m_p}\theta_p(s) = C_{m_\delta}\delta(s)$$

$$\frac{L_p m_p U}{Sqd} s\alpha(s) - \frac{L_p m_p (l_p - L_p)}{Sqd} s^2\theta(s) + \frac{L_p^2 m_p}{Sqd}(s^2 + \omega_p^2)\theta_p(s) = 0 \qquad (8\text{-}52)$$

Where I_{y_r} is the moment of inertia of the missile less the propellant, $\omega_p{}^2 = a_T/L_p$; $C_{z_p} = m_p a_T/Sq$; and $C_{m_p} = m_p a_T l_p/Sqd$. Using the determinants to obtain the $\theta(s)/\delta(s)$ transfer function, and if $\theta(s)/\delta(s) = N/\nabla$ then

$$N = \frac{L_p{}^2 m_p}{Sqd} \left\{ \left[C_{m_\delta}\left(\frac{m_r U}{Sq}s - C_{z_\alpha}\right) + C_{m_\alpha} C_{z_\delta}\right](s^2 + \omega_p{}^2) \right.$$
$$\left. + \frac{U}{L_p}(C_{z_\delta}C_{m_p} + C_{z_p}C_{m_\delta})s \right\} \quad (8\text{-}53)$$

and

$$\nabla = \frac{L_p{}^2 m_p}{Sqd} \left\{ \left[\left(\frac{m_r U}{Sq}s - C_{z_\alpha}\right)\left(\frac{I_{y_r}}{Sqd}s^2 - \frac{d}{2U}C_{m_q}s\right) \right.\right.$$
$$\left. - C_{m_\alpha}\left(\frac{m_r U}{Sq}s - C_w \sin\Theta\right)\right]$$
$$\times (s^2 + \omega_p{}^2) + \left[C_{m_p}\left(\frac{m_r U}{Sq}s - C_{z_\alpha}\right) + C_{m_\alpha}C_{z_p}\right]\frac{(l_p - L_p)}{L_p}s^2$$
$$\left. + \left[-C_{m_p}\left(\frac{m_r U}{Sq}s - C_w \sin\Theta\right) + C_{z_p}\left(\frac{I_{y_r}}{Sqd}s^2 - \frac{d}{2U}C_{m_q}s\right)\right]\frac{U}{L_p}s \right\}$$
$$(8\text{-}54)$$

The first term of Eq. 8-53 contains the rigid body zero and a pair of complex zeros on the imaginary axis. The actual location of these zeros will be slightly modified by the last term. Similarly, the first term of the denominator contains the rigid body poles and the complex poles from the propellant sloshing mode, again modified by the other terms. In this

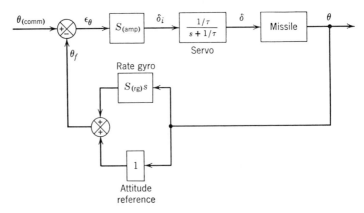

Figure 8-14   Basic control system to show effects of propellant sloshing.

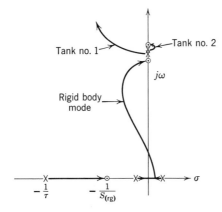

Figure 8-15   Typical root locus for ballistic
missile with two propellant tanks (Ref. 4).

case only one tank was considered to illustrate the technique; if two tanks
had been considered, there would have been two equivalent pendulum
equations and a total of four simultaneous equations, etc.   The case of
two propellant tanks is considered in Ref. 4 including the rigid body
degrees of freedom, with a typical root locus for the control system shown
in Figure 8-14.   The form of the root locus is shown in Figure 8-15 which
was taken from Ref. 4, p. 19.

From Figure 8-15 it can be seen that the root locus for the second tank is
slightly unstable; however, this problem can be alleviated by the addition
of mechanical baffles to provide damping.

Thus far only the rigid body modes have been considered in connection
with the propellant sloshing; however, it is possible that there may be
adverse coupling between the flexible modes and the propellant sloshing.
In deriving Eq. 8-50 the inputs were considered the acceleration of the
pivot of the pendulum representing the sloshing mode.   The flexible
modes would also yield an acceleration in the direction of the Z axis.   The
location of the pivot of the pendulum with respect to the antinode of
the flexible missile, and the frequency of the bending mode relative to the
propellant sloshing frequency, determine the amount of coupling.

8-9   Compensation Required for Body Bending

Thus far the transfer function for the flexible missile has been derived
to include the "tail-wags-dog" zero.   The effects of propellant slosh

were also discussed. Instabilities resulting from the sloshing modes normally can be controlled through the use of baffles; however, the correction of instabilities arising from the body bending modes requires some form of compensation.

As the flexible missile poles lie on or near the imaginary axis, the angle of departure of the root locus from these poles is very important. Very often the angle of departure, with no compensation for body bending, carries these poles into the right-half plane (see Figure 8-19). To prevent this, a lag network of the form $K/[s + (1/\tau)]$ is usually added, with the pole adjusted to provide the best angle of departure for the low-frequency modes (phase stabilize) and still maintain stability of the rigid body poles. This normally takes care of the low-frequency modes but may not completely stabilize the higher frequency modes.

If phase stabilization (controlling the angle of departure) of the higher frequency modes is impossible or impractical, the use of so-called "notched filters" is resorted to. The notched filter[6] is designed to attenuate frequencies associated with the higher frequency modes so that the portion of the signal produced by the sensors at these frequencies will be sufficiently well attenuated so as to cause no stability problems. This is sometimes referred to as "gain stabilization" of the higher frequency modes.

To illustrate the effects of compensation, the adaptive control system for advance booster systems, developed by Minneapolis-Honeywell Regulator Company is used.[7] This control system employs some interesting concepts that are discussed along with the compensation for the body bending poles. Minneapolis-Honeywell used the fourth NASA Scout configuration, which is aerodynamically stable, to analyze their control system. The block diagram of the control system, with the details of the gain changer eliminated, is shown in Figure 8-16. Before simplifying the block diagram for the root locus plot, the purpose and operation of the rate gyro blender is discussed.

The main purpose of the rate gyro blender is to insure favorable first bending mode zero locations. It was essential that the first-mode zero have a higher natural frequency than the first-mode pole for proper first-mode phase stability for the body bending filters used in the forward loop. As shown in Eq. 8-44, the numerator of the transfer function is influenced by the amount of body bending sensed by the gyro. For this application one gyro is placed forward and the other aft of the antinode of the first body bending mode. The attenuated output of each gyro is passed through a band pass filter centered on the first body bending frequency. The attenuation factor K is automatically adjusted so that the output of the absolute value circuits from each gyro will be equal (the output of the

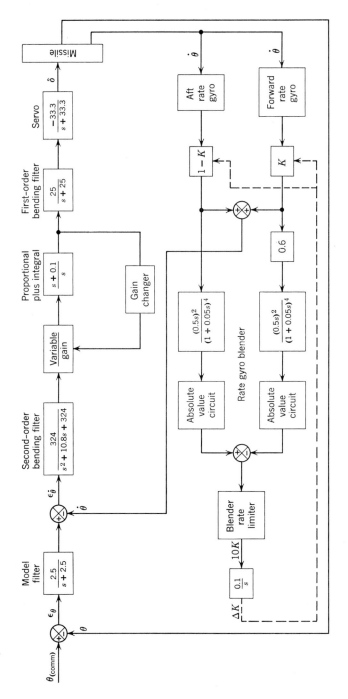

Figure 8-16   Block diagram of control system for an advanced booster system.

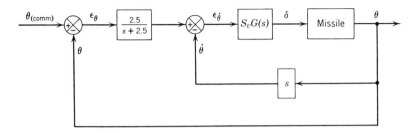

Figure 8-17   Simplified block diagram of missile control system, where S_v is the variable gain,

$$G(s) = \frac{-(33.3)(25)(324)(s + 0.1)}{s(s^2 + 10.8s + 324)(s + 25)(s + 33.3)}$$

absolute value circuits is always positive). As the output of the forward rate gyro is attenuated by a fixed factor of 0.6, besides the variable factor K, the feedback signal to the control system from the forward gyro is greater than the signal from the aft gyro. This favoring of the forward gyro provides the favorable first-mode zero location. Since the output of each gyro is summed, after being attenuated by a factor of K and $1 - K$, effectively there is unity feedback of $\dot{\theta}$. The block diagram can then be redrawn as shown in Figure 8-17.

The effect of the rate gyro blender is accounted for in the transfer function of the missile. From Figure 8-17,

$$\epsilon_{\dot{\theta}} = \frac{2.5(\theta_{(comm)} - \theta)}{s + 2.5} - s\theta$$

Simplifying, $$\epsilon_{\dot{\theta}} = \frac{2.5\theta_{(comm)} - 2.5\theta - s^2\theta - 2.5s\theta}{s + 2.5}$$

or

$$\epsilon_{\dot{\theta}} = \frac{2.5}{s + 2.5} \theta_{(comm)} - \left(\frac{s^2 + 2.5s + 2.5}{s + 2.5} \right) \theta \qquad (8\text{-}55)$$

Using Eq. 8-55 an equivalent block diagram may be drawn for the control system as shown in Figure 8-18. From this figure it can be seen that the effect of passing the attitude error signal through the model filter, and summing that with the $\dot{\theta}$ feedback, is the same as feeding θ through a feedback filter to be summed with the signal from the model filter. The time constant of the model filter determines the zeros of the feedback filter which, as can be seen from Figure 8-19*b*, determines the approximate

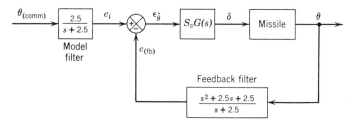

Figure 8-18   Equivalent block diagram of the missile control system.

location of the closed loop rigid body poles, and therefore serves as the model for the rigid body poles.

Before looking at the final root locus, the purpose of the two bending filters is discussed.   Figure 8-19*a* shows the angle of departure from the body bending poles for the first three body bending modes for no compensation, the first-order bending filter only, and the second-order bending filter only.   From Figure 8-19*a* it can be seen that neither filter alone will phase stabilize all three bending modes.   In fact, all three modes are

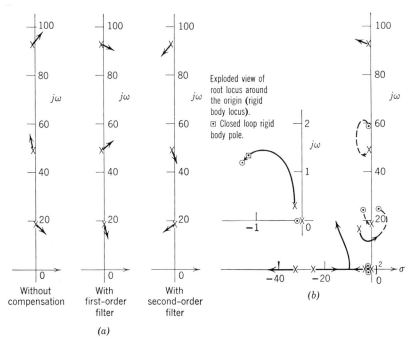

Figure 8-19   (*a*) Angle of departure for the body bending poles with and without compensation.   (*b*) Root locus of the complete system, maximum dynamic pressure.

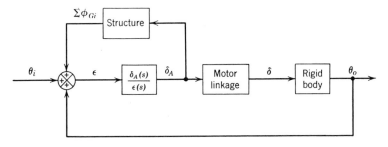

Figure 8-20   Block diagram of Vanguard control system showing flexible and rigid body loops.

unstable with only the first-order filter.   The second-order filter used with the first-order filter provides phase stability for all the body bending modes.   The second-order filter also serves another role.

As this is a self-adaptive control system using a principle similar to the MH 90 control system discussed in Chapter 6, a limit cycle must be generated by a pair of poles crossing the imaginary axis.   The second-order filter provides these poles.   The frequency at which the filter poles cross the imaginary axis is governed mainly by the open loop location of the filter poles and the first body bending poles and zeros.   The location shown provided almost a constant frequency for this imaginary axis crossing and therefore for the limit cycle.   With the second-order filter

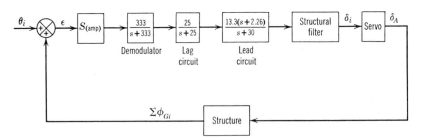

Figure 8-21   Block diagram for flexible missile including compensators for body bending.

$$\frac{\Sigma\phi_{Gi}}{\delta_A} = \frac{-0.686(s + 53)(s - 53)(s^2 - 152.2s + 14{,}500)(s^2 + 153.8s + 14{,}500)}{(s^2 + s + 605)(s^2 + 45.5s + 2660)(s^2 + 2.51s + 3900)(s^2 + 3.99s + 22{,}980)}$$

$$[TF]_{(\text{servo})[\delta_i;\delta_A]} = \frac{2750}{s^2 + 42.2s + 2750}$$

$$[TF]_{(\text{structural filter})} = \frac{(s^2 + 12s + 5810)(s^2 + 22s + 12{,}800)}{(s^2 + 13s + 3520)(s^2 + 20.8s + 21{,}000)}$$

poles on the imaginary axis the closed loop rigid body poles were very near the zeros of the feedback filter (actual location shown on Figure 8-19*b*), thus providing favorable rigid body response. The final root locus is shown in Figure 8-19*b* (the dotted portion of the root locus is assumed to be the approximate locus).

For this particular missile it was possible to phase stabilize all three body bending modes; however, this is not true in every case, as can be seen by examining the compensation required for the Vanguard missile. The Martin Company considered the first three body bending modes but handled the analysis differently, as shown in Figure 8-20. For the subsequent analysis only the structure loop is considered, which is shown in more detail in Figure 8-21. The natural frequencies and the damping ratios of the first three body bending modes are

First Mode: $\zeta = 0.02$, $\omega_n = 24.6$ rad/sec
Second Mode: $\zeta = 0.02$, $\omega_n = 62.5$ rad/sec
Third Mode: $\zeta = 8.7 \times 10^{-5}$, $\omega_n = 151.5$ rad/sec

The demodulator is part of the rigid body control system but was not included in the rigid body analysis since the contribution of the pole at -333 to the rigid body root locus would have been negligible. The lag circuit was added to phase stabilize the first body bending mode. But as can be seen from the root locus (Figure 8-22), it also prevents system instability which would have resulted from the third body bending mode. The root locus for the flexible missile is shown in Figure 8-22; also shown

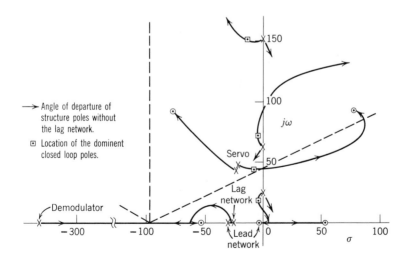

Figure 8-22 Root locus of the flexible missile with compensation. *Note:* Poles and zeros not identified are from the structure.

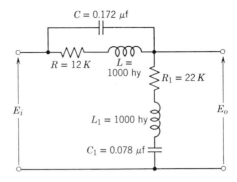

Figure 8-23 Double-notch filter for second body bending mode.

$$\frac{E_o(s)}{E_i(s)} = \frac{(LCs^2 + RCs + 1)(L_1C_1s^2 + R_1C_1s + 1)}{(LCs^2 + RCs + 1)(L_1C_1s^2 + R_1C_1s + 1) + LC_1s^2 + RC_1s}$$

are the angles of departure from the body bending poles for the system without the lag circuit. The effect of the lag circuit is very evident from the root locus. However, as can be seen from the root locus, the damping of the second mode pole is not increased noticeably. The Martin Company found that no simple compensator would improve the damping of the second-mode pole without driving the rigid missile unstable. Their solution was to use a double-notch filter (see Figure 8-23), which provided attenuation between 60 and 100 rad/sec. The double-notch filter then attenuated any signals resulting from the second body mode. The poles and zeros of the double-notch filter are close enough to each other so that the filter does not appreciably alter the root locus.

8-10 Summary

In this chapter a method for determining the transfer function of a flexible missile has been presented, as well as the rigid missile transfer function including propellant sloshing. In deriving the flexible missile transfer function it was assumed that natural frequencies and shapes of the body bending modes were known. The usefulness of the flexible missile transfer functions depends upon the accuracy of the mode shapes and frequencies used. In some cases actual tests on a full-size missile have been made to obtain a comparison between the theoretical and actual mode shapes and frequencies.[4] Experience has shown that the frequencies are more accurately predicted than the mode shapes.

There has been no attempt in this chapter to make a structures expert out of a control system engineer, but only to provide him with some idea of how to use the data that he can expect to receive from the structures people.

The methods of compensation were discussed, but like any other control system problem, the actual compensation required will vary from system to system. The determination of the required compensation may become quite involved and the compensator very complex.

References

1. R. L. Bisplinghoff, H. Ashley, and R. L. Halfman, *Aeroelasticity*, Addison-Wesley Publishing Co., Cambridge, Massachusetts, 1955.
2. R. H. Scanlan, and R. Rosenbaum, *Introduction to the Study of Aircraft Vibration and Flutter*, MacMillan Co., New York, 1951.
3. J. W. Mar, *Structural Problems in Weapons*, 16.48 Notes, Massachusetts Institute of Technology (unpublished), 1956.
4. D. R. Lukens, A. F. Schmitt, and G. T. Broucek, "Approximate Transfer Functions for Flexible-Booster-and-Autopilot Analysis," WADD TR-61-93, April 1961.
5. R. C. Seamans, F. A. Barnes, T. B. Garber, and V. W. Howard, "Recent Developments in Aircraft Control," *Journal of the Aeronautical Sciences*, March 1955.
6. M. W. Reed, J. A. Wolfe, and D. L. Mellen, "Advanced Flight Vehicle Self-Adaptive Flight Control System, Part IV—Notch Filter Development," WADD-TR-60-651, Part IV, June 1962.
7. L. T. Prince, "Design, Development, and Flight Research of an Experimental Adaptive Control Technique for Advanced Booster Systems," ASD-TDR-62-178, Part I, Nov. 1962.

9

Application of Statistical Design Principles

9-1 Introduction

In many design problems the use of sinusoidal, step, or impulse functions as system inputs is inadequate. The designer may design an autopilot that gives extremely good response to a step or a pulse input, but how will this system behave under the influence of atmospheric turbulence? If the system responds rapidly, which means a high gain, then in the presence of turbulence excessive "g" loads may be imposed on the airframe or the crew. Another application would be in the design of an automatic terrain avoidance system which would enable the aircraft to fly automatically at low levels over rough terrain. The fire-control problem also has emphasized the need to evaluate a system under the influence of noise. Here the problem is generated by the desire to detect the target, compute a solution to the fire control problem, and launch the interceptor missile at the maximum range possible. This requires the system to operate accurately under the most adverse conditions, that is, extremely low signal-to-noise ratio.

All these examples have one thing in common, that is, the response, or lack of response, of the system to a nonanalytical type of input. The atmospheric turbulence, terrain elevation, and radar noise would form a random-type input that could not in general be reproduced analytically. True, the response of the terrain avoidance system can be simulated by use of an analog computer employing some typical terrain profile as an input, and the behavior of the autopilot and the fire-control system under the influence of their noise inputs (turbulence and radar noise) can be simulated using a noise generator to furnish the disturbing inputs; however, it would be useful to have an analytical means for predicting the performance of the system to these random inputs during the design phase. This then is the purpose of this chapter, to provide the control system engineer with an analytical means to predict the behavior of a system to random inputs.

268

To accomplish this it is necessary to obtain the power spectral density of the random input. For this, the input must have certain statistical properties. Thus the first portion of this chapter deals with the basic statistical concepts; next a suitable error criterion is selected, followed by the definition of the correlation function, and finally the power spectral density. The rigorous mathematical derivations of the various statistical parameters are not attempted in this book. For these the reader is referred to Ref. 1 and Chapters 7 and 8 of Ref. 2.

9-2 Random Processes

Before discussing the characteristics of the random process some of the aspects of probability theory are discussed.

In any experiment, the output of which is dependent upon chance, the individual outcome is defined as a "sample point." Thus, in the tossing of a coin, the two outputs, a head or a tail, would be the two sample points; in the throwing of a single die there would be six sample points. All the sample points of a particular experiment make up what is called the "sample space," which consists of all the possible outcomes of the experiment. It is convenient to define a variable whose values over a sample space are determined by the various sample points of a sample space. This variable is referred to as a "random" or "stochastic" variable. In this textbook the term random variable is used. Thus for the throwing of one die, if X is considered the random variable, then X may take on values from 1 to 6 depending upon the outcome of each throw. Associated with each sample point in the sample space there is a "probability" of any event occurring; thus the probability is a measure of the likelihood that an event will occur. The probability is always a real number varying from 0 to +1 inclusive. The probability can be expressed mathematically as the ratio of the number of successful outcomes to the total number of outcomes if each outcome is equally likely[3]; thus

$$P(E) = \frac{S}{S + F} \tag{9-1}$$

where S = number of successful outcomes, F = number of failures, and $S + F$ = total number of outcomes, or the number of sample points in the sample space. Using Eq. 9-1 the probability of throwing a 7 on one throw of two dice is calculated. For this experiment $S = 6$ (the number of combinations making 7) and $S + F = 36$ (the total combinations). Therefore $P(E) = 6/36 = 1/6$.

Thus far in the discussion of probability, the sample space has been considered to have contained a finite number of sample points. This is not generally true in the physical world; for instance, radar noise in a fire-control system or atmospheric noise in the form of turbulence. Considering the turbulence further, if a measure were made of the velocity of the vertical gusts for a given flight path, the magnitude of the gusts would be of a random nature and the probability of measuring a gust of exactly 10 ft/sec normally would be considered zero. However, there would be some probability of experiencing a gust greater than some specified value, and the probability of being subjected to a gust of some velocity other than zero would be one (a certainty). If the random variable X is defined as the instantaneous values of the gusts measured, there will be a certain probability density or distribution function, $f(x)$, such that

$$P(x < X < x + dx) = f(x)\,dx \tag{9-2}$$

Equation 9-2 states that the probability that the instantaneous velocity of the gust, as denoted by the random variable X, lies within the range of x to $x + dx$ is equal to the probability density function times the range of values, dx. If it is assumed that the velocity of the vertical gusts follows a so-called normal or Gaussian distribution,[1] then

$$f(x) = \frac{1}{\sigma\sqrt{2\pi}}\,e^{-(x-m)^2/2\sigma^2} \tag{9-3}$$

where m = mean or average value of X, σ = standard deviation, and σ^2 = the variance and is equal to the average of $(x - m)^2$.

A plot of Eq. 9-3 for various values of σ is shown in Figure 9-1. The probability density function $f(x)$ must satisfy two conditions:

$$f(x) \geq 0$$

$$\int_{-\infty}^{\infty} f(x)\,dx = 1$$

Returning to the discussion of the vertical wind gusts, if it is assumed that the average value of the gusts over a long flight path is zero, then $m = 0$,

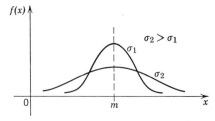

Figure 9-1   Plot of normal probability density function for two different standard deviations.

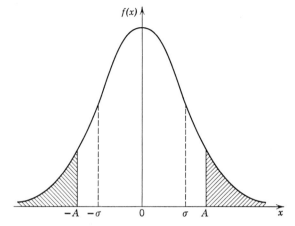

Figure 9-2   Normal probability density function for $m = 0$.

and the probability density function will be symmetrical about the $x = 0$ axis, as shown in Figure 9-2.   If it is assumed that Figure 9-2 represents the probability density function of the gusts, to find the probability that the magnitude of the velocity of the gusts is less than some value A, that is $|X| < A$, it is only necessary to find the area of the unshaded portion under the curve, or mathematically,

$$P(|X| < A) = 2 \int_0^A f(x)\, dx \tag{9-4}$$

$$P(|X| < A) = 2 \int_0^A \frac{1}{\sigma\sqrt{2\pi}}\, e^{-(x^2/2\sigma^2)}\, dx \tag{9-5}$$

Let $u = x/\sigma\sqrt{2}$, then $x = u\sigma\sqrt{2}$ and $dx = \sigma\sqrt{2}\, du$, for the upper limit when $x = A$; then $u = A/\sigma\sqrt{2}$.   Substituting these values into Eq. 9-5 yields

$$P(|X| < A) = 2 \int_0^{A/\sigma\sqrt{2}} \frac{1}{\sigma\sqrt{2\pi}}\, e^{-u^2}(\sigma\sqrt{2}\, du) \tag{9-6}$$

Simplifying,

$$P(|X| < A) = \frac{2}{\sqrt{\pi}} \int_0^{A/\sigma\sqrt{2}} e^{-u^2}\, du \tag{9-7}$$

Equation 9-7 is referred to as the "error integral."   The evaluation of this integral is tabulated in various handbooks.[4]   A sketch of the error integral as a function of the upper limit of the integral is shown in Figure 9-3.   As the total area under the curve of Figure 9-2 is 1, then the $P(|X| > A) = 1 - P(|X| < A)$.   The function representing the $P(|X| > A)$ is called the "complementary error function."

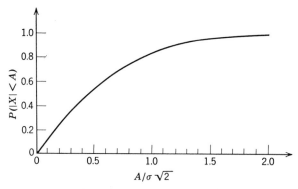

Figure 9-3   The error integral.

Having defined some of the basic probability concepts it is now time to define the "random process." A random process is a sample space, or ensemble of time functions, with the individual time functions representing the sample points. Associated with any random process is a series of probability density functions. The so-called "first probability density function" is the probability that the magnitude of any one of the random functions making up the sample space at some time t_1 lies between the two values x_1 and $x_1 + dx_1$. The "second probability density function" is the probability that the magnitude of any one of the random functions at t_1 lies between the two values x_1 and $x_1 + dx_1$, and at time t_2 that it also lies between the two values x_2 and $x_2 + dx_2$. The probability density functions can be continued on to even higher orders. A complete set of these probability density functions completely defines the statistical characteristics of the random process. It should be noted that the individual functions of a random process need not be random; for example, the expression $A \cos (\omega t + \phi)$ could represent the sample space with the randomnicity arising from the selection of A, ω, and ϕ.

A random process of particular interest is the "stationary random process," one for which the statistical characteristics are independent of the time origin. Thus, the first probability density function is the probability that one of the functions lies between x_1 and $x_1 + dx_1$ at any time; the second probability density function then depends only on the difference between t_1 and t_2, etc. Another property of the stationary random process is the "ergodic hypothesis," which assumes that the ensemble averages equal the time averages of a single representative function of the sample space or ensemble. This means that any function making up the random process can be used to represent the ensemble if studied for a sufficient length of time. It must be remembered that the ergodic hypothesis applies only to stationary random processes and in the work to follow, it is assumed that the random process is a stationary ergodic random process.

9-3   Mean-Square Error

Having defined the random process it is now important to select a measure of the error to be minimized, as this determines which statistical characteristics of the input are important.   The measure of the error chosen by Wiener is the mean-square error which is defined as

$$\overline{e^2} = \lim_{T \to \infty} \frac{1}{2T} \int_{-T}^{T} [f_0(t) - f_d(t)]^2 \, dt \qquad (9\text{-}8)$$

where $f_0(t)$ is the actual output and $f_d(t)$ is the desired output.   For most autopilots the desired output would be the command input.   The decision to use the mean-square error as the error criterion is not meant to imply that this is the optimum error measure to use.   In fact, as the square of the error is used it places more emphasis on large errors than on small errors. This point is discussed in considerable detail in Ref. 2, pp. 413–415. However, as the mean-square error can be expressed easily in terms of certain statistical characteristics, such as the correlation function or the power spectral density, this error criterion is used in spite of the short-comings mentioned.

9-4   Autocorrelation Functions

An important statistical characteristic of any signal is the autocorrelation function which is a measure of the predictability of the signal at some future time based upon the present knowledge of the signal.   The auto-correlation function is defined as (Ref. 2, p. 429)

$$\phi_{xx}(\tau) = \lim_{T \to \infty} \frac{1}{2T} \int_{-T}^{T} f_x(t) f_x(t + \tau) \, dt$$

or

$$\phi_{xx}(\tau) = \lim_{T \to \infty} \frac{1}{2T} \int_{-T}^{T} f_x(t - \tau) f_x(t) \, dt \qquad (9\text{-}9)$$

which is simply the time average of the products of the values of the function, τ sec apart, as τ is allowed to vary from zero to some large value. The first expression of Eq. 9-9 indicates that the function is measured at some time t and at some later time $t + \tau$.   The second expression signifies that the values of the function are taken at some time t and at an earlier time $t - \tau$; in both cases the values of the function are taken τ sec apart, thus the equivalence of the two expressions.

An examination of Eq. 9-9 indicates that if $\tau = 0$, then

$$\phi_{xx}(0) = \lim_{T \to \infty} \frac{1}{2T} \int_{-T}^{T} [f_x(t)]^2 \, dt$$

which is the mean-square value of the function as defined by Eq. 9-8. Also, if the function contains no *DC* or periodic component, then, in the limit, as τ approaches ∞, $\phi_{xx}(\tau)$ approaches zero; that is to say that for τ large there is no correlation between the present value of the signal and its future value. Finally, it can be shown mathematically (Ref. 2, p. 431) that $\phi_{xx}(0) \geq \phi_{xx}(\tau)$.

To further illustrate some of the properties of the autocorrelation function, the correlation function of a sine wave and a rectangular pulse is evaluated.

Example I.   A Sine Wave

Let

$$f_\psi(t) = A \sin (\omega t + \psi) \tag{9-10}$$

where ψ is a random phase angle.   Then the autocorrelation function for $f_\psi(t)$ is

$$\phi_{\psi\psi}(\tau) = \lim_{T \to \infty} \frac{1}{2T} \int_{-T}^{T} A \sin (\omega t + \psi) \cdot A \sin [\omega(t + \tau) + \psi] \, dt$$

Since the signal is periodic, it is only necessary to average over one period; thus

$$\phi_{\psi\psi}(\tau) = \frac{\omega A^2}{2\pi} \int_{0}^{2\pi/\omega} \sin (\omega t + \psi) \sin (\omega t + \psi + \omega \tau) \, dt \tag{9-11}$$

To simplify the integral, let $u = \omega t + \psi$, then $du = \omega \, dt$ or $dt = du/\omega$. The limits of integration are now $u = \psi$ for $t = 0$ and $u = 2\pi + \psi$ for $t = 2\pi/\omega$.   With these substitutions Eq. 9-11 becomes

$$\phi_{\psi\psi}(\tau) = \frac{A^2}{2\pi} \int_{\psi}^{2\pi + \psi} \sin u \sin (u + \omega \tau) \, du \tag{9-12}$$

Expanding $\sin (u + \omega \tau)$ and combining yields

$$\phi_{\psi\psi}(\tau) = \frac{A^2}{2\pi} \int_{\psi}^{2\pi + \psi} (\sin^2 u \cos \omega \tau + \sin u \cos u \sin \omega \tau) \, du.$$

The $\sin u \cos u$ can be replaced by $\frac{1}{2} \sin 2u$; making this substitution and integrating yields

$$\phi_{\psi\psi}(\tau) = \frac{A^2}{2\pi} \left[\cos \omega \tau \left(\frac{u}{2} - \frac{\sin 2u}{4} \right) \Big|_{\psi}^{2\pi + \psi} - \frac{1}{4} \sin \omega \tau \cos 2u \Big|_{\psi}^{2\pi + \psi} \right]$$

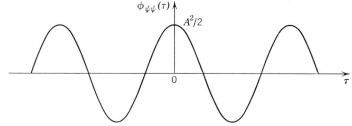

Figure 9-4 Autocorrelation function of a sinusoidal signal.

Substitution of the limits and simplifying yields

$$\phi_{\psi\psi}(\tau) = \frac{A^2}{2} \cos \omega\tau \qquad (9\text{-}13)$$

A sketch of Eq. 9-13 is presented in Figure 9-4.

As mentioned earlier in this section, the value of the correlation function for $\tau = 0$ is the mean-square value of the sine wave.

Example 2. A Rectangular Pulse

The autocorrelation function of the rectangular pulse shown in Figure 9-5 is determined here. The autocorrelation function is given by

$$\phi_{aa}(\tau) = \lim_{T \to \infty} \frac{1}{2T} \int_{-T}^{T} f_a(t)f_a(t + \tau)\, dt \qquad (9\text{-}14)$$

As the function is zero for $|t| > b$, the averaging only need be performed over the interval of $-b$ to b. Also the integrand is zero except where the two functions $f_a(t)$ and $f_a(t + \tau)$ overlap; therefore Eq. 9-14 becomes

$$\phi_{aa}(\tau) = \frac{1}{2b} \int_{-b}^{b-|\tau|} a^2\, dt = \frac{a^2}{2b} t \Big|_{-b}^{b-|\tau|}$$

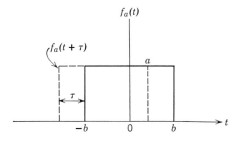

Figure 9-5 Rectangular pulse.

$$f_a(t) = \begin{cases} a & -b < t < b \\ 0 & |t| > b \end{cases}$$

Substituting the limits yields

$$\phi_{aa}(\tau) = \frac{a^2}{2b}\,[b - |\tau| + b] = a^2\!\left(1 - \frac{|\tau|}{2b}\right) \qquad (9\text{-}15)$$

The correlation function given by Eq. 9-15 is shown in Figure 9-6. The autocorrelation function shown in Figure 9-6 corresponds to several other time functions. Two of them are shown in Figure 9-7. The wave in Figure 9-7*a* has a constant period of 2*b* but a random amplitude. The value of the autocorrelation function for $\tau = 0$ is dependent only on, and equal to, the variance of the assumed amplitude distribution function (Ref. 2, pp. 433–435). The wave shown in Figure 9-7*b* has a constant amplitude of $\pm a$ but a random period of $n(2b)$ where n is a purely random variable.

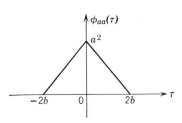

Figure 9-6   Autocorrelation function for the rectangular pulse shown in Figure 9-5.

These two examples illustrate some additional properties of the autocorrelation function which are:

(*a*) The autocorrelation function is an even function.

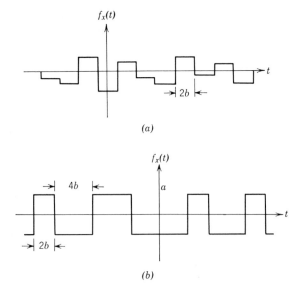

(*a*)

(*b*)

Figure 9-7   Random square waves with autocorrelation function shown in Figure 9-6.

(*b*) The autocorrelation function will contain the same periodic components as the basic signal.

One additional property not illustrated by the examples is that if the basic signal contains a *DC* component, the autocorrelation function approaches a constant nonzero value, as τ approaches infinity.

9-5 Cross-Correlation Function

Thus far the autocorrelation function of a single signal has been considered; however, in practice the input may consist of the desired signal plus noise. If the input is designated as $f_i(t) = f_s(t) + f_n(t)$, where $f_s(t)$ and $f_n(t)$ represent the signal and noise components of the input respectively, the autocorrelation function of the input can be obtained from Eq. 9-9. Thus

$$\phi_{ii}(\tau) = \lim_{T \to \infty} \frac{1}{2T} \int_{-T}^{T} f_i(t) f_i(t + \tau)\, dt \qquad (9\text{-}16)$$

Substituting for $f_i(t)$ and $f_i(t + \tau)$ yields

$$\phi_{ii}(\tau) = \lim_{T \to \infty} \frac{1}{2T} \int_{-T}^{T} [f_s(t) + f_n(t)][f_s(t + \tau) + f_n(t + \tau)]\, dt \quad (9\text{-}17)$$

Expanding the integrand yields

$$\phi_{ii}(\tau) = \lim_{T \to \infty} \frac{1}{2T} \int_{-T}^{T} f_s(t) f_s(t + \tau)\, dt + \lim_{T \to \infty} \frac{1}{2T} \int_{-T}^{T} f_s(t) f_n(t + \tau)\, dt$$

$$+ \lim_{T \to \infty} \frac{1}{2T} \int_{-T}^{T} f_n(t) f_s(t + \tau)\, dt + \lim_{T \to \infty} \frac{1}{2T} \int_{-T}^{T} f_n(t) f_n(t + \tau)\, dt$$

$$(9\text{-}18)$$

The first and fourth terms of Eq. 9-18 are the autocorrelation functions of the signal and the noise, respectively. The other two terms are the cross-correlation functions of the signal and the noise. The cross-correlation function is defined as

$$\phi_{sn}(\tau) = \lim_{T \to \infty} \frac{1}{2T} \int_{-T}^{T} f_s(t) f_n(t + \tau)\, dt$$

or

$$\phi_{ns}(\tau) = \lim_{T \to \infty} \frac{1}{2T} \int_{-T}^{T} f_n(t) f_s(t + \tau)\, dt \qquad (9\text{-}19)$$

The cross-correlation function is a measure of the dependence of one signal upon the other. If the two signals are from independent sources, the cross-correlation functions are zero, that is, $[\phi_{sn}(\tau) = \phi_{ns}(\tau) = 0]$, and the signals are said to be "uncorrelated." Thus the correlation function of the sum of two uncorrelated signals is simply the sum of the autocorrelation functions for each signal.

9-6   Power Spectral Density

The correlation function describes the statistical properties of the signal in the time domain. It would be useful, as will be seen in this section, to be able to work in the frequency domain. The Fourier transform normally is used to go from the time domain to the frequency domain. Taking the Fourier transform of the correlation function yields the power spectral density denoted by

$$\Phi_{xx}(\omega) = \int_{-\infty}^{\infty} \phi_{xx}(\tau) e^{-j\omega\tau} \, d\tau \tag{9-20}$$

The autocorrelation function can then be obtained by taking the inverse Fourier transform of the power spectral density. Then

$$\phi_{xx}(\tau) = \frac{1}{2\pi} \int_{-\infty}^{\infty} \Phi_{xx}(\omega) e^{+j\omega\tau} \, d\omega \tag{9-21}$$

Now the autocorrelation function for $\tau = 0$ has been shown to yield the mean-square value. If the signal is a voltage, the mean-square value of the voltage is proportional to the average power. For $\tau = 0$, Eq. 9-21 reduces to

$$\phi_{xx}(0) = \frac{1}{2\pi} \int_{-\infty}^{\infty} \Phi_{xx}(\omega) \, d\omega \tag{9-22}$$

If it is assumed that the voltage is dissipated across a 1-ohm resistor, the average power in the signal is given by

$$P_{\text{(avg)}} = \frac{1}{2\pi} \int_{-\infty}^{\infty} \Phi_{xx}(\omega) \, d\omega \tag{9-23}$$

which is the integral of the area under the power spectral density curve; thus the power spectral density is a measure of the power contained in the signal as a function of frequency. Before looking at some examples of the power spectral density, Eq. 9-20 is simplified. Now

$$e^{-j\omega\tau} = \cos \omega\tau - j \sin \omega\tau \tag{9-24}$$

Substituting Eq. 9-24 into Eq. 9-20 yields

$$\Phi_{xx}(\omega) = \int_{-\infty}^{\infty} \phi_{xx}(\tau)(\cos \omega\tau - j \sin \omega\tau) \, d\tau$$

Expanding the integrand

$$\Phi_{xx}(\omega) = \int_{-\infty}^{\infty} \phi_{xx}(\tau) \cos \omega\tau \, d\tau - j \int_{-\infty}^{\infty} \phi_{xx}(\tau) \sin \omega\tau \, d\tau \tag{9-25}$$

As shown in Section 9-4, the autocorrelation function is an even function, the sine function is an odd function; therefore, the second integral of Eq. 9-25 is the integral of an odd function from minus infinity to plus infinity, and is zero. Equation 9-20 then reduces to

$$\Phi_{xx}(\omega) = \int_{-\infty}^{\infty} \phi_{xx}(\tau) \cos \omega\tau \, d\tau \tag{9-26}$$

The power spectral densities of two different signals are derived in the following examples.

Example 1.

A common noise used to evaluate the performance of a system under the influence of noise is "white noise." By definition white noise has a constant power spectral density, or $\Phi_{nn}(\omega) = $ constant. It is impossible to obtain a noise that has a constant frequency spectrum from minus infinity to plus infinity, because the signal would contain an infinite amount of power. However, it is possible to obtain a noise signal with a flat frequency spectrum over a range of frequencies wider than the bandwidth of the system being analyzed. This then is considered white noise for the particular application. Having defined the power spectral density of white noise, let it be assumed the autocorrelation function for white noise is an impulse at $\tau = 0$. Then $\phi_{nn}(\tau) = u_0(t)$. The power spectral density is then

$$\Phi_{nn}(\omega) = \int_{-\infty}^{\infty} \phi_{nn}(\tau) e^{-j\omega\tau} \, d\tau \tag{9-27}$$

but if $\phi_{nn}(\tau)$ is an impulse at $\tau = 0$ and at 0 for all other values of τ, then

$$\Phi_{nn}(\omega) = \int_{-0}^{+0} \phi_{nn}(\tau) \, d\tau$$

Since $\Phi_{nn}(\omega)$, the area under the impulse, is equal to a constant as already stated, the autocorrelation function for white noise is indeed an impulse at $\tau = 0$ as assumed.

Example 2.

This example is an illustration of how to obtain the power spectral density of the square wave for which the autocorrelation function was found earlier in this chapter. From Eq. 9-15

$$\phi_{aa}(\tau) = a^2 \left(1 - \frac{|\tau|}{2b} \right) \quad \text{for} \quad -2b < \tau < 2b$$

and zero for other values of τ. Substituting into Eq. 9-26 yields

$$\Phi_{aa}(\omega) = \int_{-2b}^{2b} a^2 \left(1 - \frac{|\tau|}{2b} \right) \cos \omega\tau \, d\tau \tag{9-28}$$

Simplifying and dropping the absolute magnitude signs yields

$$\Phi_{aa}(\omega) = \frac{a^2}{b} \int_0^{2b} (2b - \tau) \cos \omega\tau \, d\tau \tag{9-29}$$

Expanding,

$$\Phi_{aa}(\omega) = \frac{a^2}{b} \left[\int_0^{2b} 2b \cos \omega\tau \, d\tau - \int_0^{2b} \tau \cos \omega\tau \, d\tau \right]$$

Integrating,

$$\Phi_{aa}(\omega) = \frac{a^2}{b} \left[\frac{2b}{\omega} \sin \omega\tau \, \Big|_0^{2b} - \left(\frac{1}{\omega^2} \cos \omega\tau + \frac{\tau}{\omega} \sin \omega\tau \right) \Big|_0^{2b} \right]$$

Substituting the limits,

$$\Phi_{aa}(\omega) = \frac{a^2}{b} \left(\frac{2b}{\omega} \sin 2\omega b - \frac{\cos 2\omega b - 1}{\omega^2} - \frac{2b}{\omega} \sin 2\omega b \right)$$

or

$$\Phi_{aa}(\omega) = \frac{2a^2}{b} \left(\frac{1 - \cos 2\omega b}{2\omega^2} \right) \tag{9-30}$$

But $\sin^2 \dfrac{\alpha}{2} = \dfrac{1 - \cos \alpha}{2}$; let $\alpha = 2\omega b$, then Eq. 9-30 becomes

$$\Phi_{aa}(\omega) = \frac{2a^2}{b} \left(\frac{\sin^2 \omega b}{\omega^2} \right) = 2a^2 b \left(\frac{\sin^2 \omega b}{\omega^2 b^2} \right)$$

Therefore,

$$\Phi_{aa}(\omega) = 2a^2 b \left(\frac{\sin \omega b}{\omega b} \right)^2 \tag{9-31}$$

As a check, the average power contained in the rectangular pulse is determined. Then

$$P_{(avg)} = \frac{1}{2\pi} \int_{-\infty}^{\infty} 2a^2 b \left(\frac{\sin \omega b}{\omega b} \right)^2 d\omega \tag{9-32}$$

Let $x = \omega b$, then $dx = b \, d\omega$. Substituting, Eq. 9-32 becomes

$$P_{(avg)} = \frac{a^2}{\pi} \int_{-\infty}^{\infty} \frac{\sin^2 x}{x^2} \, dx = \frac{2a^2}{\pi} \int_0^{\infty} \frac{\sin^2 x}{x^2} \, dx$$

but

$$\int_0^{\infty} \frac{\sin^2 x}{x^2} \, dx = \frac{\pi}{2}$$

Substituting,

$$P_{(avg)} = \frac{2a^2}{\pi} \left(\frac{\pi}{2} \right) = a^2$$

as before.

9-7 Application of Statistical Design Principles

The goal of this chapter is the determination of the behavior of a physical system to random inputs. This section is devoted to the determination of the mean-square error if the power spectral density of the input and the system transfer function are known.

First, let $G(s)$ denote the transfer function between the input and the output; then if s is replaced by $j\omega$, and this denoted by $G(j\omega)$, and if the input and output are voltages, then $e_{(out)} = e_{(in)}G(j\omega)$. Squaring, $e_{(out)}^2 = e_{(in)}^2|G(j\omega)|^2$, where $|G(j\omega)|^2$ signifies the product of the complex conjugates of $G(j\omega)$. Now e^2 is proportional to power; therefore the power spectral density of the output can be related to the power spectral density of the input in the same manner. Thus

$$\Phi_{oo}(\omega) = \Phi_{ii}(\omega)|G(j\omega)|^2 \tag{9-33}$$

where $\Phi_{oo}(\omega)$ is the power spectral density of the output and $\Phi_{ii}(\omega)$ is the power spectral density of the input. The input may be the desired input or a noise input. From Eq. 9-23 $\phi_{xx}(0) = 1/2\pi \int_{-\infty}^{\infty} \Phi_{xx}(\omega)\, d\omega$. But $\phi_{xx}(0)$ is the mean-square value of the signal; therefore

$$\overline{e_o^2} = \frac{1}{2\pi} \int_{-\infty}^{\infty} \Phi_{oo}(\omega)\, d\omega \tag{9-34}$$

Substituting for $\Phi_{oo}(\omega)$ from Eq. 9-33 yields

$$\overline{e_o^2} = \frac{1}{2\pi} \int_{-\infty}^{\infty} \Phi_{ii}(\omega)|G(j\omega)|^2\, d\omega \tag{9-35}$$

where $\overline{e_o^2}$ is the mean-square value of the output.

The mean-square value of the pitch attitude for an aircraft with a simple control system under the influence of turbulence is now determined using Eq. 9-35. The control system is shown in Figure 9-8. The aircraft is

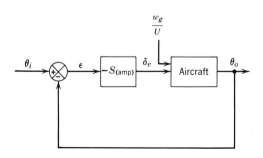

Figure 9-8 Simplified pitch control system.

represented by the short-period approximation. The elevator servo dynamics have been eliminated for simplicity, and a negative gain amplifier is used as the variable gain. The turbulence input is shown as w_g/U where w_g is the magnitude of the vertical gust. The effect of the gust is to change the angle of attack of the aircraft; and since an updraft causes a positive change in the angle of attack, $'\alpha_g = -w_g/U$, where w_g is considered negative for an updraft. The required transfer functions can be obtained from Eq. 1-128 after adding $C_{z_\alpha}'\alpha_g$ and $C_{m_\alpha}'\alpha_g$ to the right-hand sides of the force and moment equations, respectively. The equations of motion then become

$$(13.78s + 4.46)'\alpha(s) - 13.78s\theta(s) = -0.246\delta_e(s) - 4.46'\alpha_g(s)$$
$$(0.055s + 0.619)'\alpha(s) + (0.514s^2 + 0.4s)\theta(s)$$
$$= -0.71\delta_e(s) - 0.619'\alpha_g(s) \quad (9\text{-}36)$$

The transfer function then for θ out and δ_e input is

$$\frac{\theta_o(s)}{\delta_e(s)} = \frac{-1.38(s + 0.309)}{s(s^2 + 1.21s + 1.46)} \quad (9\text{-}37)$$

and for θ out and $'\alpha_g$ input

$$\frac{\theta_o(s)}{'\alpha_g(s)} = \frac{-1.03}{s^2 + 1.21s + 1.46} \quad (9\text{-}38)$$

The closed loop transfer function for $'\alpha_g$ input to θ out is of the form

$$\frac{\theta_o(s)}{'\alpha_g(s)}\bigg]_{CL} = \frac{\theta_o(s)/'\alpha_g(s)}{1 + S_{(amp)}[\theta_o(s)/\delta_e(s)]} \quad (9\text{-}39)$$

Substituting the appropriate transfer functions and simplifying yields

$$\frac{\theta_o(s)}{'\alpha_g(s)}\bigg]_{CL} = \frac{-1.03s}{s^3 + 1.21s^2 + (1.46 + 1.38S)s + 0.426S} \quad (9\text{-}40)$$

where S signifies $S_{(amp)}$. From Eq. 9-35,

$$\overline{\theta_o^2} = \frac{1}{2\pi} \int_{-\infty}^{\infty} \Phi_{gg}(\omega) \left|\frac{\theta_o(j\omega)}{'\alpha_g(j\omega)}\right|^2 d\omega \quad (9\text{-}41)$$

where $\Phi_{gg}(\omega)$ is the power spectral density of the gusts. The representation of the power spectral density was obtained from Ref. 5, p. 28 and is given in Eq. 9-42

$$\Phi_{gg}(\omega) = \frac{2\tau_g \, \overline{'\alpha_g^2}}{1 + \tau_g^2\omega^2} \quad (9\text{-}42)$$

where $\overline{'\alpha_g^2}$ is the mean-square value of the gusts and τ_g the time constant associated with the gusts. A τ_g of 2 sec is used for this analysis; then Eq. 9-42 reduces to

$$\Phi_{gg}(\omega) = \frac{\overline{'\alpha_g^2}}{\omega^2 + 0.25} \quad (9\text{-}43)$$

Substituting Eqs. 9-40 and 9-43 into Eq. 9-41, after replacing s by $j\omega$, yields

$$\frac{\overline{\theta_o^2}}{\overline{\alpha_g^2}} = \frac{1}{2\pi} \int_{-\infty}^{\infty}$$

$$\times \frac{1.06\omega^2 \, d\omega}{(\omega^2 + 0.25) \, | - j\omega^3 - 1.21\omega^2 + (1.46 + 1.38S)j\omega + 0.426S|^2} \tag{9-44}$$

Breaking $\omega^2 + 0.25$ into its two complex conjugates and multiplying the $j\omega + 0.5$ term by the rest of the denominator yields

$$\frac{\overline{\theta_o^2}}{\overline{\alpha_g^2}} = \frac{1.06}{2\pi} \int_{-\infty}^{\infty}$$

$$\times \frac{\omega^2 \, d\omega}{|\omega^4 - 1.71j\omega^3 - (2.07 + 1.38S)\omega^2 + (0.73 + 1.116S)j\omega + 0.213S|^2} \tag{9-45}$$

The method for evaluating the integral contained in Eq. 9-45 can be found in many references, and is repeated here (see Ref. 1, pp. 395–397). It should be noted in the following equations that $g_n(\omega)$ contains only even powers of ω and that the denominator is written as the product of the complex conjugates; also that $h_n(\omega)$ is the complex conjugate made up of $+j\omega$. If

$$I_n = \frac{1}{2\pi j} \int_{-\infty}^{\infty} \frac{g_n(\omega) \, d\omega}{h_n(\omega)h_n(-\omega)} \tag{9-46}$$

$$h_n(\omega) = a_0\omega^n + a_1\omega^{n-1} + \cdots + a_n \tag{9-47}$$

where

$$g_n(\omega) = b_0\omega^{2n-2} + b_1\omega^{2n-4} + \cdots + b_{n-1} \tag{9-48}$$

and $n = \frac{1}{2}$ the highest power of the denominator (n for h_n and g_n is the same for a particular problem). Then

$$I_n = \frac{(-1)^{n+1}}{2a_0} \frac{N_n}{D_n} \tag{9-49}$$

where

$$D_n = \begin{vmatrix} a_1 & a_0 & 0 & \cdots & 0 \\ a_3 & a_2 & a_1 & a_0 & 0 \\ a_5 & a_4 & a_3 & \cdots & 0 \\ \vdots & \vdots & \vdots & \vdots & \vdots \end{vmatrix} \qquad \begin{matrix} \text{This must be a square} \\ n\text{th-order determinant.} \end{matrix}$$

and, $N_n = D_n$ with the first column $(a_1, a_3, a_5, \cdots)$ replaced by $b_0, b_1,$ $b_2, \cdots, b_{n-1}$. But

$$\overline{e^2} = \frac{1}{2\pi} \int_{-\infty}^{\infty} \frac{g_n(\omega) \, d\omega}{h_n(\omega)h_n(-\omega)}$$

Therefore,

$$\overline{e^2} = jI_n$$

For this problem $n = 4$ and comparing the denominator of Eq. 9-45 with Eq. 9-47,

$a_0 = 1$

$a_1 = -1.71j$

$a_2 = -(2.07 + 1.38S)$

$a_3 = (0.73 + 1.116S)j$

$a_4 = 0.213S$

Forming D_n,

$$D_n =$$

$$\begin{vmatrix} -1.71j & 1 & 0 & 0 \\ (0.73 + 1.116S)j & -(2.07 + 1.38S) & -1.71j & 1 \\ 0 & 0.213S & (0.73 + 1.116S)j & -(2.07 + 1.38S) \\ 0 & 0 & 0 & 0.213S \end{vmatrix}$$

Expanding and simplifying yields

$$D_n = -0.213S(1.39S^2 + 3.41S + 2.05) \tag{9-50}$$

Comparing the numerator of Eq. 9-45 with Eq. 9-48,

$b_0 = 0$

$b_1 = 0$

$b_2 = 1$

$b_3 = 0$

Forming N_n,

$$N_n = \begin{vmatrix} 0 & 1 & 0 & 0 \\ 0 & -(2.07 + 1.38S) & -1.71j & 1 \\ 1 & 0.213S & (0.73 + 1.116S)j & -(2.07 + 1.38S) \\ 0 & 0 & 0 & 0.213S \end{vmatrix}$$

Expanding and simplifying yields

$$N_n = -1.71j(0.213S) \tag{9-51}$$

Substituting Eqs. 9-50 and 9-51 into Eq. 9-49 yields

$$I_n = \frac{(-1)^5(+1.71j)}{2(1.39S^2 + 3.41S + 2.05)} \qquad (9\text{-}52)$$

Therefore,

$$\frac{\overline{\theta_o{}^2}}{\overline{\alpha_g{}^2}} = \frac{0.907}{1.39S^2 + 3.41S + 2.05} \qquad (9\text{-}53)$$

which gives the ratio of the mean-square error of the output to the mean-square value of the vertical gusts. From Eq. 9-53 it can be seen that increasing the amplifier gain reduces the response of the aircraft to the gusts. However, increasing the gain decreases the damping. This is one disadvantage of using the mean-square as an error criterion; it tends to yield a lightly damped oscillatory system due to the dependence upon large errors.

9-8 Additional Applications of Statistical Design Principles

Another important application of statistical design principles is the use of cross-correlation techniques to experimentally determine the transfer function of a control system (Ref. 2, pp. 437–438). This would solve the identification problem inherent in some types of self-adaptive control systems. To accomplish this white noise is added to the input of the system, then the output of the system is cross-correlated with the white noise. A block diagram of the system is shown in Figure 9-9. The cross-correlation between the white noise input and the system output is

$$\phi_{io}(\tau) = \lim_{T \to \infty} \frac{1}{2T} \int_{-T}^{T} f_{(ni)}(t - \tau) f_{(no)}(t) \, dt \qquad (9\text{-}54)$$

as the cross-correlation between the input noise and the system output from the signal is zero (the signal and the noise are uncorrelated).

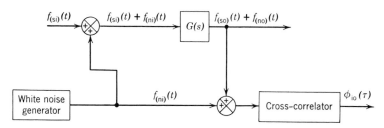

Figure 9-9 Measurement of the system transfer function using cross-correlation.

Now the output of a system can be related to its input in the time domain by use of the convolution integral and $g(t)$ the impulse response of the system—$g(t)$ is the inverse Laplace transform of $G(s)$.   Then

$$f_{(no)}(t) = \int_{-\infty}^{\infty} g(x)f_{(ni)}(t - x)\, dx \qquad (9\text{-}55)$$

where x is a dummy variable of integration.   Substituting Eq. 9-55 into Eq. 9-54 yields

$$\phi_{io}(\tau) = \lim_{T \to \infty} \frac{1}{2T} \int_{-T}^{T} f_{(ni)}(t - \tau)\, dt \int_{-\infty}^{\infty} g(x)f_{(ni)}(t - x)\, dx \qquad (9\text{-}56)$$

Interchanging the order of integration yields

$$\phi_{io}(\tau) = \int_{-\infty}^{\infty} g(x)\, dx \left[\lim_{T \to \infty} \frac{1}{2T} \int_{-T}^{T} f_{(ni)}(t - \tau)f_{(ni)}(t - x)\, dt \right] \qquad (9\text{-}57)$$

The term in the brackets of Eq. 9-57 is the autocorrelation function of $f_{(ni)}(t)$ with argument $(\tau - x)$, or

$$\phi_{nn}(\tau - x) = \lim_{T \to \infty} \frac{1}{2T} \int_{-T}^{T} f_{(ni)}(t - \tau)f_{(ni)}(t - x)\, dt$$

Thus Eq. 9-57 can be written as

$$\phi_{io}(\tau) = \int_{-\infty}^{\infty} g(x)\phi_{nn}(\tau - x)\, dx \qquad (9\text{-}58)$$

But since the noise input is white noise, $\phi_{nn}(\tau - x)$ is an impulse (see Section 9-6), and if $\phi_{nn}(\tau - x)$ is considered the input, $\phi_{io}(\tau)$ would be the impulse response of the system (see Eq. 9-55).   Then $\phi_{io}(\tau) = Ag(\tau)$ for $\phi_{nn}(\tau) = Au_0(t)$.   Therefore $G(s)$ can be obtained by taking the Laplace transform of $\phi_{io}(\tau)$.

Another application of statistical design principles is the prediction of the reliability of a control system subject to random failures; however, the subject is too broad to be covered here (for information on this the reader is referred to Ref. 6).

9-9 Summary

Some of the more important statistical properties have been discussed in this chapter; the main purpose has been to present a method for determining the mean-square error or output for a given condition.   The material here is primarily a condensation of the material given in the references.   The reader who wishes to pursue the subject further is referred to the extensive bibliography at the end of Laning and Battin's book (Ref. 1, pp. 427–429).   However, the material presented here should give the reader at least a nodding acquaintance with the theory and application of statistical design.

References

1. J. H. Laning, Jr., and R. H. Battin, *Random Processes in Automatic Control*, McGraw-Hill Book Co., New York, 1956.
2. J. G. Truxall, *Automatic Feedback Control System Synthesis*, McGraw-Hill Book Co., New York, 1955.
3. "Fundamentals of Design of Piloted Aircraft Flight Control Systems," BuAer Report AE-61-4, Sec. 3–4, Chap. 5, Vol. I, 1952.
4. E. Jahnke, and F. Emde, *Tables of Functions*, 4th ed., pp. 24–25, Dover Publications, New York, 1945.
5. R. C. Seamans, F. A. Barnes, T. B. Garber, and V. W. Howard, "Recent Developments in Aircraft Control," *Journal of the Aeronautical Sciences*, March 1955.
6. G. H. Sandler, *System Reliability Engineering*, Prentice-Hall, Englewood Cliffs, New Jersey, 1963.

Appendix *A*

Review of Vector Analysis

A-1   Vector Representation of Linear Quantities

Many of the physical quantities are most conveniently represented by vectors. Some of the linear quantities involved that can be represented by vectors are forces, velocity, acceleration, and position. By definition a vector has magnitude, direction, and sense (that is, plus or minus) in that direction.

Figure A-1 shows several examples of vector quantities, such as velocity, force, and position. For the velocity vector **V** and the force vector **F** the length of the arrow is proportional to the magnitude of the velocity or force, and the direction is indicated by the direction of the arrow. The position vector is a vector representing the distance and direction between two points.

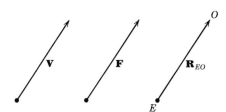

Figure A-1   Vector representation of linear
quantities.

A-2   Vector Representation of Rotational Quantities

In addition to the linear quantities discussed in Section A-1, certain rotational quantities such as angular velocity, angular momentum, and torques can be represented by vectors. Any rotating body rotates about some instantaneous axis. The vector representing this angular velocity, **ω**, lies along the instantaneous axis of rotation. To determine the positive

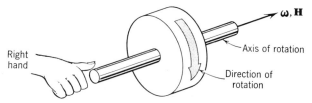

Figure A-2   Vector representation of angular velocity and angular
momentum.

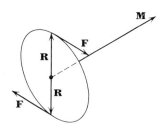

Figure A-3   Vector representation of a torque vector where **M** represents the torque vector; **F** represents the forces of the couple generating the torque; and **R** represents the two lever arms.

sense of the vector use the right-hand rule as illustrated in Figure A-2. As the angular momentum vector is $\mathbf{H} = I_{(sp)}\boldsymbol{\omega}$, where $I_{(sp)}$ is the moment of inertia of the body about the axis of rotation, the angular momentum vector lies along the angular velocity vector. Another important rotational quantity is the vector representation of a torque, which is shown in Figure A-3. The length of the arrow is proportional to the magnitude of the torque, which is equal to the sum of the products of the components of the forces **F** perpendicular to the lever arms **R** times the length of the lever arms. A vector expression for this is given in Section A-7.

A-3   Addition and Subtraction of Vectors

To add vectors it is necessary only to place the tail of one vector coincident with the head of the other. This procedure is continued until all the vectors have been added (see Figure A-4*a*). To subtract the two vectors, the negative of one is added to the other (see Figure A-4*b*).

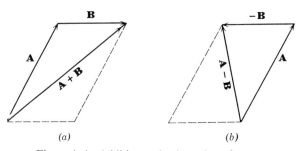

| (a) | (b) |

Figure A-4   Addition and subtraction of vectors.

A-4 Multiplication of a Vector by a Scalar

When a vector is multiplied by a scalar quantity only the magnitude of the vector is altered; the new magnitude is equal to the old magnitude times the scalar quantity. The sign associated with the scalar quantity must be taken into account; if it is negative it reverses the sense of the vector but does not otherwise change the direction of the vector.

A-5 Rectangular Components of Vectors

To obtain the components of a vector, an orthogonal axis system is usually used with the axes signified by X, Y, and Z. In general this system is right-handed, which means that if the positive X axis is rotated about the Z axis toward the positive Y axis through the smallest angle, using the right-hand rule, with the fingers pointed in the direction of rotation, the thumb points in the positive direction of the Z axis.

To represent vectors of unit length which lie along the X, Y, and Z axes the symbols $\mathbf{i}$, $\mathbf{j}$, and $\mathbf{k}$ are used in that order. Thus a vector $\mathbf{R}$ can be expressed in terms of its components along the three axes as $\mathbf{R} = \mathbf{i}R_x + \mathbf{j}R_y + \mathbf{k}R_z$, where R_x, R_y, and R_z are the magnitudes of the projections of the vector $\mathbf{R}$ (including algebraic signs), as illustrated in Figure A-5. The addition and subtraction of vectors that have been written in terms of their rectangular components can be accomplished by adding or subtracting the respective components. Thus if $\mathbf{A} = \mathbf{i}A_x + \mathbf{j}A_y + \mathbf{k}A_z$

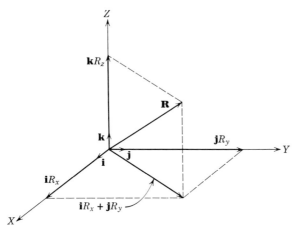

Figure A-5 Representation of a vector in terms of its rectangular components.

and $\mathbf{B} = \mathbf{i}B_x + \mathbf{j}B_y + \mathbf{k}B_z$, then $\mathbf{A} + \mathbf{B} = \mathbf{i}(A_x + B_x) + \mathbf{j}(A_y + B_y) + \mathbf{k}(A_z + B_z)$.

A-6   Scalar or Dot Product

In equation form the dot product of two vectors is written as $\mathbf{A} \cdot \mathbf{B} = AB \cos A_{AB}$, where A_{AB} is the angle between the two vectors. Remember that the result of the dot product is a *scalar* quantity. Since the result of the dot product is dependent upon the angle between the two vectors, if the vectors are perpendicular, their dot product is zero or if they are parallel, the dot product is equal to the product of the magnitudes of the vectors. Thus $\mathbf{i} \cdot \mathbf{j} = \mathbf{j} \cdot \mathbf{k} = \mathbf{k} \cdot \mathbf{i} = 0$ and $\mathbf{i} \cdot \mathbf{i} = \mathbf{j} \cdot \mathbf{j} = \mathbf{k} \cdot \mathbf{k} = 1$, also $\mathbf{A} \cdot \mathbf{A} = A^2$ or $|\mathbf{A}| = \sqrt{\mathbf{A} \cdot \mathbf{A}}$. If the two vectors are written in terms of their rectangular coordinates, $\mathbf{A} \cdot \mathbf{B} = (\mathbf{i}A_x + \mathbf{j}A_y + \mathbf{k}A_z) \cdot (\mathbf{i}B_x + \mathbf{j}B_y + \mathbf{k}B_z) = A_x B_x + A_y B_y + A_z B_z$, as the dot products of the other unit vectors are zero.

A-7   Vector or Cross-Product

The vector or cross-product in equation form is $\mathbf{A} \times \mathbf{B} = \mathbf{1}_{(\mathbf{A} \times \mathbf{B})} AB \sin A_{AB}$, where $\mathbf{1}_{(\mathbf{A} \times \mathbf{B})}$ is a unit vector in the direction of $\mathbf{A} \times \mathbf{B}$. The unit vector by definition is normal to the plane containing the vectors $\mathbf{A}$ and $\mathbf{B}$; the positive direction is determined by use of the right-hand rule as vector $\mathbf{A}$ is rotated toward $\mathbf{B}$ through the smallest angle, as illustrated in Figure A-6. If the two vectors are written in terms of their rectangular components,

$$\mathbf{A} \times \mathbf{B} = (\mathbf{i}A_x + \mathbf{j}A_y + \mathbf{k}A_z) \times (\mathbf{i}B_x + \mathbf{j}B_y + \mathbf{k}B_z)$$

$$= \begin{vmatrix} \mathbf{i} & \mathbf{j} & \mathbf{k} \\ A_x & A_y & A_z \\ B_x & B_y & B_z \end{vmatrix}$$

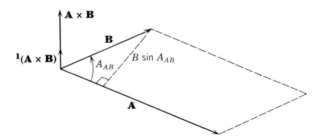

Figure A-6   Vector or cross product of two vectors.

As mentioned in Section A-2, the torque resulting from a force acting on a moment arm can be expressed by a vector equation. Referring to Figure A-3, $\mathbf{M} = 2\mathbf{R} \times \mathbf{F}$.

A-8 Multiple Products

Besides the dot and cross products there are various combinations of these two products. Two of these combinations are of particular interest. One is the so-called "triple scalar product," which in equation form is $\mathbf{A} \cdot (\mathbf{B} \times \mathbf{C})$ and results in a scalar quantity. Since the triple scalar product yields a scalar quantity, the terms may be rotated within the expression; thus $\mathbf{A} \cdot (\mathbf{B} \times \mathbf{C}) = \mathbf{C} \cdot (\mathbf{A} \times \mathbf{B}) = \mathbf{B} \cdot (\mathbf{C} \times \mathbf{A})$. The only requirement is that the order of rotation be maintained. If the three vectors are expressed in terms of their rectangular components,

$$\mathbf{A} \cdot (\mathbf{B} \times \mathbf{C}) = \begin{vmatrix} A_x & A_y & A_z \\ B_x & B_y & B_z \\ C_x & C_y & C_z \end{vmatrix}$$

The other multiple product of interest is the "triple cross-product," which in equation form is $\mathbf{A} \times (\mathbf{B} \times \mathbf{C})$ and results in a vector. The operation in parentheses must be performed first for $\mathbf{A} \times (\mathbf{B} \times \mathbf{C}) \neq (\mathbf{A} \times \mathbf{B}) \times \mathbf{C}$. A useful expression for evaluating the triple cross-product is $\mathbf{A} \times (\mathbf{B} \times \mathbf{C}) = \mathbf{B}(\mathbf{A} \cdot \mathbf{C}) - \mathbf{C}(\mathbf{A} \cdot \mathbf{B})$.

A-9 Differentiation of a Vector

If a vector is time dependent, it may be differentiated with respect to the scalar quantity time. The derivative of a position vector that is changing both in length and direction is illustrated in Figure A-7. The vector $\mathbf{R}$ can be written as $\mathbf{1}_R R$, where $\mathbf{1}_R$ is a unit vector in the direction of $\mathbf{R}$. Then

$$\frac{d\mathbf{R}}{dt} = \frac{d}{dt}(\mathbf{1}_R R) = \frac{d\mathbf{1}_R}{dt} R + \mathbf{1}_R \frac{dR}{dt} \tag{A-1}$$

The second term $\mathbf{1}_R (dR/dt)$ represents the rate of change of the length of the vector as shown in Figure A-8. The first term is the rate of change of $\mathbf{R}$ due to a rotation. If $\boldsymbol{\omega}_R$ denotes the angular velocity of the vector $\mathbf{R}$ and $\mathbf{1}_R$, the tangential velocity of the tip of the vector $\mathbf{R}$ can be expressed as

$$R\frac{d\mathbf{1}_R}{dt} = \boldsymbol{\omega}_R \times \mathbf{R} \tag{A-2}$$

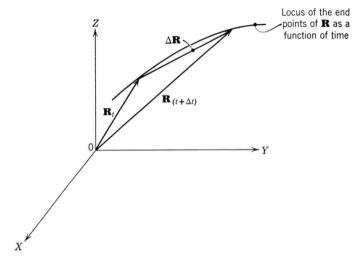

Figure A-7   Variation of a position vector **R** as a function of time.

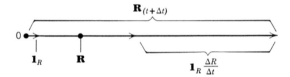

Figure A-8   Rate of change of the length of **R**.

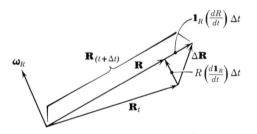

Figure A-9   Total derivative of a position vector.

and Eq. A-1 can now be written as

$$\frac{d\mathbf{R}}{dt} = \mathbf{1}_R \frac{dR}{dt} + \boldsymbol{\omega}_R \times \mathbf{R} \qquad (A\text{-}3)$$

where $\mathbf{1}_R (dR/dt)$ is the linear velocity caused by a change in the length of the position vector and $\boldsymbol{\omega}_R \times \mathbf{R}$ is the tangential velocity of the tip of the position vector due to an angular velocity of the position vector. This expression is illustrated in Figure A-9.

The second derivative of a position vector is also frequently encountered; therefore, the second derivative of the vector $\mathbf{R}$ is now derived. Taking the derivative of Eq. A-3, it becomes

$$\frac{d^2\mathbf{R}}{dt^2} = \frac{d}{dt} \left(\mathbf{1}_R \frac{dR}{dt} \right) + \frac{d}{dt} (\boldsymbol{\omega}_R \times \mathbf{R}) \qquad (A\text{-}4)$$

Expanding,

$$\frac{d^2\mathbf{R}}{dt^2} = \left(\frac{d\mathbf{1}_R}{dt} \right) \frac{dR}{dt} + \mathbf{1}_R \frac{d^2R}{dt^2} + \frac{d\boldsymbol{\omega}_R}{dt} \times \mathbf{R} + \boldsymbol{\omega}_R \times \frac{dR}{dt} \qquad (A\text{-}5)$$

Previously, $d\mathbf{1}_R/dt = \boldsymbol{\omega}_R \times \mathbf{1}_R$. For $d\boldsymbol{\omega}_R/dt \times \mathbf{R}$, the $d\boldsymbol{\omega}_R/dt$ is again the derivative of a vector, but if it is assumed that the angular acceleration vector lies along the same axis as the angular velocity vector, $d\boldsymbol{\omega}_R/dt \times \mathbf{R}$ can be written as $\dot{\boldsymbol{\omega}}_R \times \mathbf{R}$. Finally, if Eq. A-3 is substituted for $d\mathbf{R}/dt$ in the last term, Eq. A-5 becomes

$$\frac{d^2\mathbf{R}}{dt^2} = \mathbf{1}_R \frac{d^2R}{dt^2} + (\boldsymbol{\omega}_R \times \mathbf{1}_R) \frac{dR}{dt} + \dot{\boldsymbol{\omega}}_R \times \mathbf{R} + \boldsymbol{\omega}_R \times \left(\mathbf{1}_R \frac{dR}{dt} + \boldsymbol{\omega}_R \times \mathbf{R} \right)$$
$$(A\text{-}6)$$

Combining like terms, Eq. A-6 becomes

$$\frac{d^2\mathbf{R}}{dt^2} = \mathbf{1}_R \frac{d^2R}{dt^2} + 2(\boldsymbol{\omega}_R \times \mathbf{1}_R) \frac{dR}{dt} + \dot{\boldsymbol{\omega}}_R \times \mathbf{R} + \boldsymbol{\omega}_R \times (\boldsymbol{\omega}_R \times \mathbf{R}) \qquad (A\text{-}7)$$

The significance of each term is discussed in the following paragraph. The first term of Eq. A-7 is the linear acceleration and results from the second rate of change of the length of the vector. The second and third terms are both tangential accelerations. The third term is probably familiar, but the second term may not be. This term is explained most easily by considering a position vector that is rotating at a constant velocity while it is changing in length. As the length of the position vector changes, the tangential velocity of the tip must also change if the angular velocity of the vector is to remain constant. If the tangential velocity changes, there must be a tangential acceleration. The second term should not be confused with the Coriolis acceleration term discussed in Section A-10. The last term is the centripetal acceleration resulting from the rotation of this vector, and is always directed toward the center of rotation.

A-10   The Equation of Coriolis

The need for the equation of Coriolis arises when there are two or more axis systems rotating relative to each other; then the rate of change of a vector is different when viewed from the different axis systems. The equation of Coriolis can be stated in the following manner: The motion of an object as viewed from a reference space is equal to the motion as seen from the moving space, plus the motion resulting from the relative angular velocity of the moving space with respect to the reference space. This situation is illustrated in Figure A-10.   From the vector triangle in Figure A-10, the equation of Coriolis can be written directly; thus

$$[\dot{\mathbf{R}}]_r \, dt = [\dot{\mathbf{R}}]_m \, dt + \boldsymbol{\omega}_{rm} \, dt \times \mathbf{R} \qquad \text{(A-8)}$$

Dividing Eq. A-8 by the dt yields

$$[\dot{\mathbf{R}}]_r = [\dot{\mathbf{R}}]_m + \boldsymbol{\omega}_{rm} \times \mathbf{R} \qquad \text{(A-9)}$$

which is the equation of Coriolis.   It should be noted that $[\dot{\mathbf{R}}]_m$ is the vector velocity of the point in question as seen from the moving space and may contain either or both of the terms given in Eq. A-3.   Another

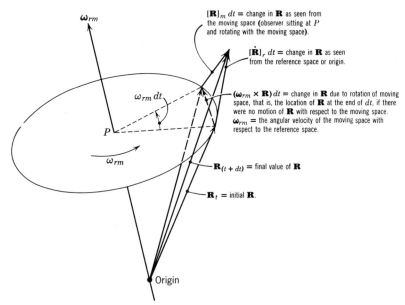

Figure A-10   Representation of the equation of Coriolis.

useful relation can be obtained by taking the time derivative of Eq. A-9 with respect to the reference space. This procedure yields the second equation of Coriolis which is derived here. The time derivative of Eq. A-9 with respect to the reference space yields

$$[\ddot{\mathbf{R}}]_r = \left(\frac{d}{dt}\right)_r [\dot{\mathbf{R}}]_r = \left(\frac{d}{dt}\right)_r [\dot{\mathbf{R}}]_m + \left(\frac{d}{dt}\right)_r (\boldsymbol{\omega}_{rm} \times \mathbf{R}) \qquad \text{(A-10)}$$

The term $(d/dt)_r[\dot{\mathbf{R}}]_m$ is the time derivative with respect to the reference space of a vector defined in the moving space. This requires the application of the first equation of Coriolis, which yields

$$\left(\frac{d}{dt}\right)_r [\dot{\mathbf{R}}]_m = \left(\frac{d}{dt}\right)_m [\dot{\mathbf{R}}]_m + \boldsymbol{\omega}_{rm} \times [\dot{\mathbf{R}}]_m \qquad \text{(A-11)}$$

which can be written as

$$\left(\frac{d}{dt}\right)_r [\dot{\mathbf{R}}]_m = [\ddot{\mathbf{R}}]_m + \boldsymbol{\omega}_{rm} \times [\dot{\mathbf{R}}]_m \qquad \text{(A-12)}$$

The last term of Eq. A-10 expanded is

$$\left(\frac{d}{dt}\right)_r (\boldsymbol{\omega}_{rm} \times \mathbf{R}) = \left(\frac{d}{dt}\right)_r \boldsymbol{\omega}_{rm} \times \mathbf{R} + \boldsymbol{\omega}_{rm} \times \left(\frac{d}{dt}\right)_r \mathbf{R} \qquad \text{(A-13)}$$

As $\boldsymbol{\omega}_{rm}$ is measured with respect to the reference space, $(d/dt)_r \boldsymbol{\omega}_{rm} = [\dot{\boldsymbol{\omega}}_{rm}]_r$ and the equation of Coriolis need not be applied. However, Eq. A-9 must be substituted for $(d/dt)_r \mathbf{R}$ in Eq. A-13. Thus

$$\left(\frac{d}{dt}\right)_r (\boldsymbol{\omega}_{rm} \times \mathbf{R}) = [\dot{\boldsymbol{\omega}}_{rm}]_r \times \mathbf{R} + \boldsymbol{\omega}_{rm} \times [\dot{\mathbf{R}}]_m + \boldsymbol{\omega}_{rm} \times (\boldsymbol{\omega}_{rm} \times \mathbf{R}) \qquad \text{(A-14)}$$

Substituting Eqs. A-12 and A-14 into Eq. A-10 yields

$$[\ddot{\mathbf{R}}]_r = [\ddot{\mathbf{R}}]_m + [\dot{\boldsymbol{\omega}}_{rm}]_r \times \mathbf{R} + 2\boldsymbol{\omega}_{rm} \times [\dot{\mathbf{R}}]_m + \boldsymbol{\omega}_{rm} \times (\boldsymbol{\omega}_{rm} \times \mathbf{R}) \qquad \text{(A-15)}$$

The significance of each term is discussed in the following. The first term, $[\ddot{\mathbf{R}}]_m$, is the vector acceleration of the point in question as seen from the moving space, and it may contain any or all the terms given in Eq. A-7. The second term is the tangential acceleration resulting from an angular acceleration of the moving frame with respect to the reference frame. The third term is the Coriolis acceleration; the similar looking term in Eq. A-7 should not be confused with this term. The last term is the centripetal acceleration resulting from the rotation of the moving space with respect to the reference space. The centripetal acceleration vector always points toward and is normal to the axis of rotation.

Appendix B

Some Gyroscopic Theory

B-l   The Law of the Gyro[1]

The law of the gyro can be derived from Newton's Law in rotational form which states: The time rate of change with respect to inertial space of the angular momentum of a body about its center of gravity is equal to the applied torque.   This can be written in equation form as

$$\mathbf{M}_{(app)} = [\dot{\mathbf{H}}]_I \tag{B-1}$$

As Eq. B-1 involves the time derivative of a vector with respect to inertial space, the equation of Coriolis must be applied.   Using the earth as the moving space,

$$[\dot{\mathbf{H}}]_I = [\dot{\mathbf{H}}]_E + \boldsymbol{\omega}_{IE} \times \mathbf{H} \tag{B-2}$$

where $\boldsymbol{\omega}_{IE}$ is the angular velocity of the earth with respect to inertial space $(0.07292115 \times 10^{-3}$ rad/sec).   The gyroscope itself can be mounted on a base (aircraft, missile, etc.) that is moving with respect to the earth.   Then

$$[\dot{\mathbf{H}}]_E = [\dot{\mathbf{H}}]_B + \boldsymbol{\omega}_{EB} \times \mathbf{H} \tag{B-3}$$

Also the case (Ca) of the gyroscope can be mounted on a platform so that it can rotate relative to the base.   This yields

$$[\dot{\mathbf{H}}]_B = [\dot{\mathbf{H}}]_{Ca} + \boldsymbol{\omega}_{B,Ca} \times \mathbf{H} \tag{B-4}$$

Finally the inner gimbal (Gi) can rotate relative to the case, thus

$$[\dot{\mathbf{H}}]_{Ca} = [\dot{\mathbf{H}}]_{Gi} + \boldsymbol{\omega}_{Ca,Gi} \times \mathbf{H} \tag{B-5}$$

Substituting Eqs. B-2, B-3, B-4, and B-5 into Eq. B-1, it becomes

$$\mathbf{M}_{(app)} = [\dot{\mathbf{H}}]_{Gi} + (\boldsymbol{\omega}_{Ca,Gi} + \boldsymbol{\omega}_{B,Ca} + \boldsymbol{\omega}_{EB} + \boldsymbol{\omega}_{IE}) \times \mathbf{H} \tag{B-6}$$

But

$$\boldsymbol{\omega}_{Ca,Gi} + \boldsymbol{\omega}_{B,Ca} + \boldsymbol{\omega}_{EB} + \boldsymbol{\omega}_{IE} = \boldsymbol{\omega}_{I,Gi} \tag{B-7}$$

Then, Eq. B-6 becomes

$$\mathbf{M}_{(app)} = [\dot{\mathbf{H}}]_{Gi} + \boldsymbol{\omega}_{I,Gi} \times \mathbf{H} \qquad \text{(B-8)}$$

From Eq. A-3

$$[\dot{\mathbf{H}}]_{Gi} = \mathbf{1}_H \frac{dH}{dt} + \boldsymbol{\omega}_{Gi,H} \times \mathbf{H} \qquad \text{(B-9)}$$

In Eq. B-9, $\boldsymbol{\omega}_{Gi,H}$ is the angular velocity of the $\mathbf{H}$ vector with respect to the inner gimbal, or an angular velocity of the spin axis with respect to the inner gimbal. This term is made to be zero by the construction of the gyroscope, that is, no gimbal or bearing compliance. The term $\mathbf{1}_H(dH/dt)$ of Eq. B-9 is constrained to be zero by maintaining the spin angular velocity of the rotor constant. This is accomplished by making the rotor part of a hysteresis-type synchronous motor and closely controlling the frequency of the excitation voltage. This procedure reduces Eq. B-9 to zero; thus, Eq. B-8 becomes

$$\mathbf{M}_{(app)} = \boldsymbol{\omega}_{I,Gi} \times \mathbf{H} \qquad \text{(B-10)}$$

Equation B-10 is the Law of the Gyro, which states that if a torque is applied to a gyroscope, $\mathbf{M}_{(app)}$, the inner gimbal precesses with respect to inertial space, $\boldsymbol{\omega}_{I,Gi}$, such that Eq. B-10 is satisfied. The precession is of such a direction that the $\mathbf{H}$ vector attempts to align itself with the applied torque vector.

B-2 Dynamic Equation of the Single-Degree-of-Freedom Gyro[1]

Figure B-1 shows the orientation of the gyroscopic element of a single-degree-of-freedom gyro relative to the case. In the single-degree-of-freedom gyro the outer gimbal of the two-degree-of-freedom gyro is fixed with respect to the case. The X, Y, and Z axes of the case are designated the output axis OA, the spin reference axis SRA, and the input axis IA, respectively. The output axis is synonymous with the inner gimbal axis of the two-degree-of-freedom gyro. The designation of the IA, OA, and SRA form a right-handed system such that $\mathbf{1}_{(IA)} \times \mathbf{1}_{(OA)} = \mathbf{1}_{(SRA)}$. The angle A_g, usually referred to as the gimbal angle, locates the spin axis of the gyroscopic element relative to the SRA, and is negative as shown in Figure B-1. As an angular velocity is normally considered the input for a single-degree-of-freedom gyro; its dynamics are determined by analyzing the gyroscopic output torque resulting from an arbitrary angular velocity of the $\mathbf{H}$ vector with respect to inertial space. The gyroscopic output torque is the reaction torque developed by the angular

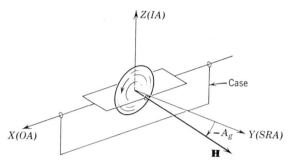

Figure B-1   Orientation of the gyroscopic element relative
to the case for a single-degree-of-freedom gyro.

velocity of precession of the **H** vector and opposes the applied torque;
thus, if the gyroscopic output torque is denoted by $\mathbf{M}_g$,

$$\mathbf{M}_g = -\mathbf{M}_{(\text{app})} = \mathbf{H} \times \boldsymbol{\omega}_{(I,Gi)} \qquad \text{(B-11)}$$

Now

$$\boldsymbol{\omega}_{(I,Gi)} = \boldsymbol{\omega}_{(Ca,Gi)} + \boldsymbol{\omega}_{(I,Ca)} \qquad \text{(B-12)}$$

and $\boldsymbol{\omega}_{(Ca,Gi)} = \mathbf{i}_{Ca}\dot{A}_g$ as the gyro has only one-degree-of-freedom. The
angular velocity of the case can be represented by its X, Y and Z com-
ponents; thus

$$\boldsymbol{\omega}_{I,Ca} = \mathbf{i}_{Ca}\omega_{(I,Ca)_X} + \mathbf{j}_{Ca}\omega_{(I,Ca)_Y} + \mathbf{k}_{Ca}\omega_{(I,Ca)_Z} \qquad \text{(B-13)}$$

Substituting into Eq. B-12

$$\boldsymbol{\omega}_{(I,Gi)} = \mathbf{i}_{Ca}(\dot{A}_g + \omega_{(I,Ca)_X}) + \mathbf{j}_{Ca}\omega_{(I,Ca)_Y} + \mathbf{k}_{Ca}\omega_{(I,Ca)_Z} \qquad \text{(B-14)}$$

The **H** vector can be resolved into the case axis system through the gimbal
angle; thus

$$\mathbf{H} = \mathbf{j}_{Ca}H \cos A_g + \mathbf{k}_{Ca}H \sin A_g \qquad \text{(B-15)}$$

Substituting Eqs. B-14 and B-15 into Eq. B-11, and evaluating the cross-
product, the expression for the gyroscopic output torque becomes

$$\mathbf{M}_g = \mathbf{i}_{Ca}[H\omega_{(I,Ca)_Z} \cos A_g - H\omega_{(I,Ca)_Y} \sin A_g]$$
$$+ \mathbf{j}_{Ca}[H(\dot{A}_g + \omega_{(I,Ca)_X}) \sin A_g] - \mathbf{k}_{Ca}[H(\dot{A}_g + \omega_{(I,Ca)_X}) \cos A_g] \qquad \text{(B-16)}$$

The $\mathbf{j}_{Ca}$ and $\mathbf{k}_{Ca}$ components of Eq. B-16 represent the angular velocity of
the inner gimbal about the output axis and result in a gyroscopic output
torque about the Y and Z axes.   The Y and Z components of the gyroscopic output torque are absorbed by the gimbal bearings and are thus of

no particular interest in the analysis to follow. Taking the $\mathbf{i}_{Ca}$ component of Eq. B-16, then

$$M_{gx} = H(\omega_{(I,Ca)_Z} \cos A_g - \omega_{(I,Ca)_Y} \sin A_g) \qquad \text{(B-17)}$$

Since the input axis of the gyro lies along the Z axis of the case, $\omega_{(I,Ca)_Z}$ is the angular velocity of the case with respect to inertial space about the input axis of the gyro and is denoted by ω_{IA}; also $\omega_{(I,Ca)_Y}$ is the angular velocity of the case with respect to inertial space about the SRA and is denoted by ω_{SRA}; then Eq. B-17 can be rewritten as

$$M_{gx} = H\omega_{IA}\left[\cos A_g - \left(\frac{\omega_{SRA}}{\omega_{IA}}\right)\sin A_g\right] \qquad \text{(B-18)}$$

Equation B-18 describes the response of the gyro to the component of the angular velocity of the case with respect to inertial space that lies along the IA and SRA. The second term in the brackets of Eq. B-18 is referred to as the "coupling term." The coupling term describes the response of the gyro to angular velocities of the case with respect to inertial space about the SRA. Normally, A_g is kept small, especially in the integrating gyro (maximum values of A_g of about 2.5° are common); thus, if A_g is small, the coupling term can be neglected and

$$M_{gx} \simeq H\omega_{IA} \qquad \text{(B-19)}$$

If $S_{(g)[\omega;M]}$ is defined as the sensitivity of the gyro for an angular velocity input to a gyroscopic output torque,

$$S_{(g)[\omega;M]} = \frac{M_{gx}}{\omega_{IA}} = H\frac{\text{gm-cm}^2}{\text{sec}} \qquad \text{(B-20)}$$

In the remaining analysis of the single-degree-of-freedom gyro A_g is assumed to be small, and Eq. B-19 is used to describe the gyroscopic output torque resulting from an angular velocity of the case with respect to inertial space about the input axis. The single-degree-of-freedom gyro shown in Figure B-1 would be of no practical value since there is no method provided for absorbing the gyroscopic output torque. If a spring or elastic restraint is provided that yields a torque proportional to A_g, the single-degree-of-freedom gyro becomes a rate gyro. If a viscous damper is added that yields a torque proportional to $\dot{A}_g$, an integrating gyro results. These two important single-degree-of-freedom gyros are discussed in Sections B-3 and B-4.

B-3 The Rate Gyro

Figure B-2 shows the configuration of a typical rate gyro. Here the elastic restraint is provided by a torsion bar fixed to the inner gimbal and

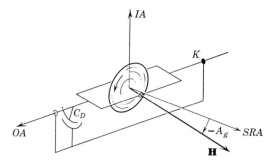

Figure B-2   Configuration of a typical rate gyro.

the case.   The viscous damper is added to provide damping of the transient.   Using Eq. B-19 the differential equation of the rate gyro can be written by summing torques about the output axis.   Then

$$M_{gx} = H\omega_{IA} = I_{OA}\ddot{A}_g + C_D\dot{A}_g + KA_g \tag{B-21}$$

where K is the spring constant of the torsion bar (dyne-cm/rad), C_D is the coefficient of damping of the viscous damper (dyne-cm/rad/sec), and I_{OA} is the moment of inertia of the gimbal and wheel about the output axis. Dividing through by K and taking the Laplace transformation with the initial conditions zero, Eq. B-21 becomes

$$\left(\frac{I_{OA}}{K}s^2 + \frac{C_D}{K}s + 1\right)A_g(s) = \frac{H}{K}\omega_{IA}(s) \tag{B-22}$$

Putting Eq. B-22 in the standard form of the second-order differential equation it becomes

$$\left[\left(\frac{s}{\omega_n}\right)^2 + \frac{2\zeta}{\omega_n}s + 1\right]A_g(s) = \frac{H}{K}\omega_{IA}(s)$$

where
$$\omega_n = \left(\frac{K}{I_{OA}}\right)^{\frac{1}{2}} \quad \text{and} \quad \zeta = \frac{C_D}{2(KI_{OA})^{\frac{1}{2}}} \tag{B-23}$$

For rapid response ω_n should be large, therefore, $K \gg I_{OA}$.   In most commercial rate gyros ζ is made equal to about 0.7 and typical values of ω_n vary between 150 and 400 rad/sec.   The transfer function for the rate gyro for an angular velocity input to a gimbal angle output, can be obtained by solving Eq. B-23 for $A_g(s)/\omega_{IA}(s)$.   Thus

$$[TF]_{(rg)[\omega;A_g]} = \frac{H/K}{\left(\dfrac{s}{\omega_n}\right)^2 + \dfrac{2\zeta}{\omega_n}s + 1} \text{ sec} \tag{B-24}$$

In the analysis of many of the applications of the rate gyro the dynamics can be neglected and the transfer function can be replaced by a sensitivity. Under these conditions, Eq. B-24 becomes

$$S_{(rg)[\omega;A_g]} = \frac{H}{K} \text{ sec} \tag{B-25}$$

For either case the steady-state value of A_g for a step input of ω_{IA} is

$$A_g = \frac{H}{K} \omega_{IA} \tag{B-26}$$

To give the reader some concept of the magnitudes of the various parameters of a typical rate gyro the MIT 10^4 rate gyro, developed by the Instrumentation Lab of MIT, is used as an example. For this gyro

$$H = 10^4 \text{ gm-cm}^2/\text{sec}$$

$$C_D = 5 \times 10^3 \text{ gm-cm}^2/\text{sec}$$

$$K = 3.03 \times 10^5 \text{ gm-cm}^2/\text{sec}^2$$

$$I_{OA} = 34 \text{ gm-cm}^2$$

$$\omega_n = 94.25 \text{ rad/sec}$$

$$\zeta = 0.78$$

These figures substantiate the conclusion drawn earlier that $K \gg I_{OA}$. Also as can be seen from the figures $K \gg C_D \gg I_{OA}$. This relation of the magnitudes of K, C_D, and I_{OA} is common in commercial rate gyros.

Since the output of the rate gyro is a signal proportional to the angular velocity of the case about its input axis with respect to inertial space, the rate gyro has found wide use in flight control systems. In these applications the sensitivity of the rate gyro is such that it cannot detect the component of earth's rate parallel to the input axis of the gyro. Thus, the gyro effectively measures the angular velocity of the case and thus the aircraft with respect to the earth.

B-4 Rate Integrating Gyro

If the elastic restraint is removed from the rate gyro, leaving only the viscous damper, the result is referred to as a "rate integrating gyro" or just an "integrating gyro." The differential equation for the integrating gyro can be obtained from Eq. B-21 by letting $K = 0$, thus

$$M_{g_X} = H\omega_{IA} = I_{OA}\ddot{A}_g + C_D\dot{A}_g \tag{B-27}$$

Taking the Laplace transformation of Eq. B-27 with the initial conditions zero and dividing by C_D yields

$$\left(\frac{I_{OA}}{C_D} s + 1\right) A_g(s) = \frac{H}{C_D} \omega_{IA}(s) \tag{B-28}$$

Writing Eq. B-28 in standard form it becomes

$$(\tau_{(ig)} s + 1) A_g(s) = \frac{H}{C_D} \omega_{IA}(s) \tag{B-29}$$

where $\tau_{(ig)} = I_{OA}/C_D$. For rapid response $\tau_{(ig)}$ must be small; therefore, $C_D \gg I_{OA}$. Typical values of $\tau_{(ig)}$ vary between 0.01 and 0.002 sec. The transfer function for the integrating gyro for ω_{IA} input to a gimbal angle output can be obtained from Eq. B-29. Thus

$$[TF]_{(ig)[\omega;A_g]} = \frac{\dfrac{H}{C_D}}{s(\tau_{(ig)} s + 1)} \text{ sec} \tag{B-30}$$

As in the case of the rate gyro, the dynamics of the integrating gyro often can be neglected. Thus, Eq. B-30 reduces to

$$[TF]_{(ig)[\omega;A_g]} \simeq \frac{\dfrac{H}{C_D}}{s} \tag{B-31}$$

For either case the steady-state value of A_g for a step input of ω_{IA} is

$$A_g = \frac{\dfrac{H}{C_D}}{s} \omega_{IA} = \frac{H}{C_D} \int_0^t \omega_{IA}\, dt = \frac{H}{C_D} A_{IA} \tag{B-32}$$

where $A_{IA} = \int_0^t \omega_{IA}\, dt$. The name, integrating gyro, arises from the fact that the gimbal angle is proportional to the time integral of the input angular velocity. Because the integral of the input angular velocity is the total angle through which the gyro has rotated about its input axis with respect to inertial space A_{IA}, A_g is proportional to this angle. A sensitivity can be defined relating A_g to A_{IA} which is

$$S_{(ig)[A_{IA};A_g]} = \frac{A_g}{A_{IA}} = \frac{H}{C_D} \tag{B-33}$$

The ratio of H/C_D is usually about 1 but may be as large as 2.

The parameters for a couple of integrating gyros are given in Table B-1 for comparison with the MIT 10^4 rate gyro. The two gyros are the HIG-4 and HIG-6 built by Minneapolis-Honeywell. The HIG stands for hermetic integrating gyro. As the gimbal angle of the integrating gyro is proportional to the integral of the input angular velocity, the integrating gyro is always used in a closed loop system. The two gyros are summarized in Table B-2.

Table B-1   Comparison of HIG-4 and HIG-6 Gyros

Parameter	HIG-4	HIG-6
H(gm-cm^2/sec)	10^4	0.725×10^6
C_D(gm-cm^2/sec)	10^4	0.427×10^6
τ(sec)	3.5×10^{-3}	2.5×10^{-3}
I_{OA}(gm-cm^2)	35	1070

Table B-2   Comparison of the Rate and Integrating Gyro

Rate Gyro	Integrating Gyro

Differential Eq.:

$I_{OA}\ddot{A}_g + C_D\dot{A}_g + KA_g = H\omega_{IA}$	$I_{OA}\ddot{A}_g + C_D\dot{A}_g = H\omega_{IA}$
$[TF]_{(rg)[\omega;A_g]} = \dfrac{\dfrac{H}{K}}{\left(\dfrac{s}{\omega_n}\right)^2 + \dfrac{2\zeta}{\omega_n}s + 1}$ sec	$[TF]_{(ig)[\omega;A_g]} = \dfrac{\dfrac{H}{C_D}}{s(\tau_{(ig)}s + 1)}$ sec
$\omega_n = \sqrt{K/I_{OA}} \qquad \zeta = \dfrac{C_D}{2\sqrt{KI_{OA}}}$	$\tau_{(ig)} = \dfrac{I_{OA}}{C_D} \qquad C_D \gg I_{OA}$
$K \gg C_D \gg I_{OA}$	$[TF]_{(ig)[\omega;A_g]} \simeq \dfrac{\dfrac{H}{C_D}}{s}$ sec
$S_{(rg)[\omega;A_g]} = \dfrac{H}{K}$ sec	$S_{(ig)[A_{IA};A_g]} = \dfrac{H}{C_D}$

Steady-state value of A_g for step input of ω_{IA}:

$A_g = \dfrac{H}{K}\omega_{IA}$	$A_g = \dfrac{H}{C_D}\displaystyle\int_0^t \omega_{IA}\,dt$

Reference

1. W. R. Weems, *An Introduction to the Study of Gyroscopic Instruments*, Instrumentation Section, Department of Aeronautical Engineering, Massachusetts Institute of Technology, 1958.

Appendix *C*

Basic Servo Theory

C-I   The Block Diagram

A physical system can be represented by what is called a block diagram. A block diagram is composed of individual blocks that represent the various components of the system. The transfer function, which is the Laplace transform of the ratio of the output over the input with the initial conditions zero, is usually given for each block. The other important element in the block diagram is the summation point or points. The summation point is used for the addition or subtraction of like quantities such as voltages, forces, torques, etc. Some devices used in this text for summation or comparison are the vertical gyro, for comparison of the desired and actual vehicle attitude; the integrating gyro, which actually sums torques about its output axis; and summing amplifiers, for the summation of voltages. The symbol used to represent a summation point is shown in Figure C-1, and the sign associated with the summer indicates the operation being performed. If the quantities being summed are voltages representing the system parameters, for the case shown in Figure C-1a the voltages must be of opposite sign before being summed. Other comparison devices are synchros and potentiometers used for determining the relative positions of shafts. Still others are discussed in the literature and textbooks on the servomechanisms. The derivation of the transfer functions of various electrical, mechanical, and hydraulic components are not covered here as these derivations can be found in such servo texts as those listed at the end of this appendix.

Figure C-1   Representation of summing points.

306

C-2 The Control Ratio

The control ratio is defined as the ratio of the output to the input for a closed loop system. Figure C-2 shows a block diagram of a feedback system using standard symbols. Since the sign at the summer associated with the feedback signal is negative, the system is said to have negative feedback, which is normal.

The control ratio can be derived from the relations given in Figure C-2, which yields

$$\frac{C(s)}{R(s)} = \frac{G(s)}{1 + G(s)H(s)} \tag{C-1}$$

which is the normal form of the control ratio. However, for reasons to be discussed later in the section Eq. C-1 is written here as

$$\frac{C(s)}{R(s)} = \frac{G(s)}{1 - [-G(s)H(s)]} = \frac{G(s)}{1 - [TF]_{OL}} \tag{C-2}$$

where $[TF]_{OL}$ is defined as the open-loop transfer function and consists of the product of the forward transfer function, the feedback transfer function, including the sign associated with each transfer function, and the sign at the summer corresponding to the feedback signal. Thus for Figure C-2 $[TF]_{OL} = -G(s)H(s)$. The control ratio is more general when written as given in Eq. C-2, especially when the transfer functions may be negative, as is the case for many of the transfer functions used in this book. By using Eq. C-2, for the control ratio, the engineer determines the sign needed at the summer to yield the necessary sign for the open-loop transfer function after considering the signs of the various transfer functions. As is shown later in Sections C-5 and C-7, the net sign associated with the open loop transfer function determines the type of root locus to be plotted.

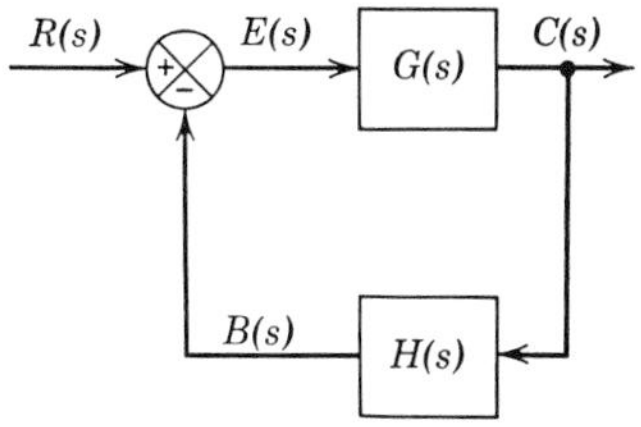

There are two standard forms used in writing any transfer function. One form is obtained by making the coefficients of s all equal to unity as shown in Eq. C-3

Figure C-2   Block diagram of a typical feedback system.

$R(s)$ = Reference input
$E(s)$ = Actuating signal
$G(s)$ = Forward transfer function
$C(s)$ = Controlled variable or output
$H(s)$ = Feedback transfer function (is unity in many cases)
$B(s)$ = Feedback signal
$C(s) = E(s)G(s)$
$B(s) = C(s)H(s)$
$E(s) = R(s) - B(s)$

$$G(s) = \frac{K(s + 1/\tau_1)}{(s + 1/\tau_2)(s^2 + 2\zeta\omega_n s + \omega_n^2)} \tag{C-3}$$

In this form K is defined as the "*static loop sensitivity.*"[1]   This form of the transfer function is employed when using the root locus for analysis. The other form of the transfer function is obtained by making the coefficient of the lowest power of s unity.   In this form Eq. C-3 becomes

$$G(s) = \frac{K'(\tau_1 s + 1)}{(\tau_2 s + 1)\left(\dfrac{s^2}{\omega_n^2} + \dfrac{2\zeta}{\omega_n} s + 1\right)} \tag{C-4}$$

where $K' = \tau_2 K/\tau_1 \omega_n^2$.   The significance of K' is discussed in Section C-3.

C-3 System Types

Closed-loop servo systems with unity feedback are classified according to the number of pure integrations appearing in the forward transfer function, $G(s)$.   Thus a Type-0 system would have no pure integrations in the forward transfer function, a Type 1, one, a Type 2, two, etc.   To illustrate the significance of the various type-systems, some simple examples are used. The unity feedback system shown in Figure C-3 is used for all these examples, with

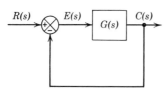

Figure C-3 Unity feedback control system.

$$G(s) = \frac{K'(\tau_1 s + 1)(\tau_2 s + 1)\cdots}{s^n(\tau_a s + 1)(\tau_b s + 1)\cdots} \tag{C-5}$$

and $n = 0, 1, 2$, etc.   Before proceeding with the examples, the $E(s)/R(s)$ transfer function is given.   From the relations given in Figure C-2

$$\frac{E(s)}{R(s)} = \frac{1}{1 + G(s)} \tag{C-6}$$

where $E(s)$ is now the error signal, since $H(s) = 1$.

In the following examples the steady-state value of the error is determined for different inputs for the different types of systems.

Example 1. Type-0 system, n = 0

For a unit step input or $R(s) = 1/s$, then from Eq. C-6

$$E(s) = \frac{1}{s}\left[\frac{1}{1 + G(s)}\right] \tag{C-7}$$

To obtain the steady-state value, that is, the value of $e(t)$ as $t \to \infty$, the final value theorem must be used.   The final value theorem in equation form is[1]

$$\lim_{t \to \infty} e(t) = \lim_{s \to 0} sE(s) \tag{C-8}$$

Then from Eq. C-7

$$e(t)_{ss} = \lim_{s \to 0} \left[\frac{1}{1 + G(s)} \right] = \frac{1}{1 + K'} \qquad \text{(C-9)}$$

as

$$\lim_{s \to 0} \frac{K'(\tau_1 s + 1)(\tau_2 s + 1) \cdots}{(\tau_a s + 1)(\tau_b s + 1) \cdots} = K'$$

Therefore for a Type-0 system for a step input there is a constant error, the value of which depends on the value of K'. For this reason K' is referred to as the error coefficient (see Ref. 1, Sections 6-6, and 6-7, for a discussion of error coefficients). For a unit ramp input or $R(s) = 1/s^2$

$$E(s) = \frac{1}{s^2} \left[\frac{1}{1 + G(s)} \right]$$

Then

$$e(t)_{ss} = \lim_{s \to 0} \frac{1}{s} \left[\frac{1}{1 + G(s)} \right] = \infty$$

Therefore for a Type-0 system for a ramp input the error approaches ∞ as $t \to \infty$.

Example 2. Type-1 system, n = 1

For a unit step input:

$$E(s) = \frac{1}{s} \left[\frac{1}{1 + G(s)} \right]$$

Then

$$e(t)_{ss} = \lim_{s \to 0} \left[\frac{1}{1 + G(s)} \right] = 0$$

as

$$\lim_{s \to 0} \frac{K'(\tau_1 s + 1)(\tau_2 s + 1) \cdots}{s(\tau_a s + 1)(\tau_b s + 1) \cdots} = \infty$$

For a unit ramp input

$$E(s) = \frac{1}{s^2} \left[\frac{1}{1 + G(s)} \right]$$

Then

$$e(t)_{ss} = \lim_{s \to 0} \frac{1}{s} \left[\frac{1}{1 + \dfrac{K'(\tau_1 s + 1)(\tau_2 s + 1) \cdots}{s(\tau_a s + 1)(\tau_b s + 1) \cdots}} \right]$$

or

$$e(t)_{ss} = \lim_{s \to 0} \left[\frac{1}{s + \dfrac{K'(\tau_1 s + 1)(\tau_2 s + 1) \cdots}{(\tau_a s + 1)(\tau_b s + 1) \cdots}} \right] = \frac{1}{K'}$$

In like manner $e(t)_{ss}$ for a parabolic input, $R(s) = 1/s^3$, can be shown to be ∞. The same procedure can be followed for a Type 2 system, etc. The results are tabulated in Table C-1. It must be remembered that the classification of servos into Type 0, 1, 2, etc. is valid only for unity feedback systems; however, the steady-state value of the output or the actuating signal for any aperiodic inputs can be obtained in the manner illustrated here for nonunity feedback systems.

Table C-I   Steady-State Error for Various Inputs for Various Type Servo Systems

Steady-State Error for Inputs Indicated

System Type	Unit Step	Unit Ramp	Unit Parabola
0	$\dfrac{1}{1 + K'}$	∞	∞
1	0	$\dfrac{1}{K'}$	∞
2	0	0	$\dfrac{1}{K'}$
3	0	0	0

C-4 Root Locus

The root locus is a convenient and useful method for determining the effects of varying the gain on the dynamics of a closed-loop servo. The effects of varying the gain on the dynamics of a closed-loop system are shown analytically for a simple system and are then related to the graphical or root-locus method. The system analyzed is shown in Figure C-4. For this system the closed loop transfer function is

$$\frac{C(s)}{R(s)} = \frac{\dfrac{K}{(s + 1)(s + 5)}}{1 + \dfrac{K}{(s + 1)(s + 5)}} \tag{C-10}$$

Simplifying,

$$\frac{C(s)}{R(s)} = \frac{K}{(s + 1)(s + 5) + K} = \frac{K}{s^2 + 6s + K + 5} \tag{C-11}$$

From Eq. C-11 it can be seen that as K is increased from zero, the roots of the quadratic in the denominator vary. Since the denominator of Eq. C-11 is the characteristic equation of the differential equation relating the

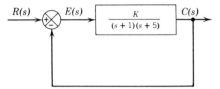

Figure C-4   Simplified control system.

output, $C(s)$, to the input $R(s)$, the roots of the denominator determine the solution of the homogeneous equation, and therefore the characteristics of the transient response of the system.   The roots of the denominator of Eq. C-11 are called the poles of the closed-loop transfer function.   The roots of the characteristic equation can be obtained by using the quadratic formula; thus

$$s_{1,2} = -3 \pm \sqrt{9 - (K + 5)} \qquad \text{(C-12)}$$

From Eq. C-12 it can be seen that for $K \leq 4$ the roots are real, while for $K > 4$ the roots are imaginary.   When the roots of Eq. C-12 are imaginary the general form of the characteristic equation is $s^2 + 2\zeta\omega_n s + \omega_n^2 = 0$, the roots of which are

$$s_{1,2} = -\zeta\omega_n \pm j\omega_n \sqrt{1 - \zeta^2} = -\sigma + j\omega_d$$

The roots given by Eq. C-12 are tabulated for various values of K in Table C-2 and are plotted for all values of K in Figure C-5.   Although K is generally positive, which means that negative feedback is being employed,

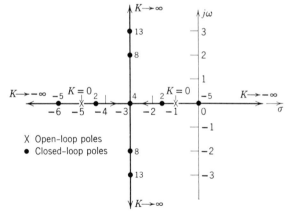

Figure C-5   A plot of all roots of Eq. C-11 for $-\infty < K < \infty$. The values of K for the roots listed in Table C-2 are indicated in the figure.

Table C-2   Tabulation of the Roots of Eq. C-11 for Various Values of K

K	s_1	s_2
0	-1	-5
2	-1.586	-4.414
4	-3	-3
8	$-3 + j2$	$-3 - j2$
13	$-3 + j3$	$-3 - j3$
-5	0	-6

K may be negative.   As can be seen from Figure C-5, the system would be stable for $K > -5$.   The curve shown in the figure is defined as the root-locus plot of Eq. C-11.   Figure C-5 was obtained by substituting various values of K into Eq. C-12 to determine the roots of the characteristic equation.   For this simple example this procedure was not difficult; however, for a more complex system it would be extremely time consuming; therefore, a simpler method is needed.   The rest of this appendix is devoted to the normal procedure for obtaining the root-locus plot and to the interpretation thereof.

Before proceeding with the details of the construction of the root locus, the theory behind the basic rules is discussed.   Figure C-5 was obtained by equating the denominator of Eq. C-11 to zero and solving for the roots for various values of K.   For the root locus it is more useful to use Eq. C-10.   Equating the denominator of Eq. C-10 to zero yields

$$1 + \frac{K}{(s + 1)(s + 5)} = 0$$

or

$$\frac{K}{(s + 1)(s + 5)} = -1 = 1\underline{/180°} \tag{C-13}$$

If some value of s satisfies Eq. C-13, for this value of s

$$\left| \frac{K}{(s + 1)(s + 5)} \right| = 1 \tag{C-14}$$

and

$$\underline{/\frac{K}{(s + 1)(s + 5)}} = \text{an odd multiple of } 180° \tag{C-15}$$

Equation (C-14) is referred to as the "magnitude condition" and is used to determine the value of K for a given value of s, and Eq. C-15 is referred to as the "angle condition" and is used to plot the root locus. The use of Eqs. C-14 and C-15 are first demonstrated by using the value of $s = -3 + j2$ from Figure C-5. This can be done both analytically and graphically. To proceed with the analytic method, the value of s is substituted into Eq. C-13, thus

$$\frac{K}{(-3 + j2 + 1)(-3 + j2 + 5)} = \frac{K}{(-2 + j2)(2 + j2)} = -1 \quad \text{(C-16)}$$

Evaluating, $K/-8 = -1$. For the magnitude condition from Eq. C-14

$$\left|\frac{K}{-8}\right| = \frac{K}{8} = 1$$

Therefore, $K = 8$ as before. For the angle condition

$$\frac{K}{-8} = \frac{K}{8\underline{/180°}} = \frac{K}{8}\underline{/-180°}$$

Therefore the angle is $-180°$, which satisfies Eq. C-18. For the graphical method, which is normally used, examination of the two factors in the denominator of Eq. C-16 shows that they represent the vectors from $s = -1$ and $s = -5$ to the point $s = -3 + j2$. This is illustrated in Figure C-6. Each of the two vectors can be represented by its magnitude and angle. The magnitudes can be obtained by measuring the length of the vectors with a scale. From Eq. C-14, $K = |(s + 1)(s + 5)|$; therefore, K is equal to the product of the magnitudes of the two vectors. Measuring the two vectors yields a magnitude of each of approximately 2.82 units. Squaring this yields 7.97 or approximately 8. Obviously any graphical method is less accurate than an analytic solution. The angles of the two vectors are ϕ_1 and ϕ_2, which can be measured with a protractor. Here $\phi_1 = 135°$ and $\phi_2 = 45°$ and the sum is 180° as before.

Since the whole purpose of the root-locus analysis is to determine the closed-loop poles which determine the behavior of a system for some value of K, the final step is obtaining the closed-loop transfer function

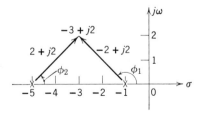

Figure C-6 Representation of the vectors from $s = -1$ and $s = -5$ to $s = -3 + j2$.

from Figure C-5 for a particular value of K. Thus the closed-loop transfer function for $K = 8$ is

$$\frac{C(s)}{R(s)} = \frac{8}{(s + 3 - j2)(s + 3 + j2)} = \frac{8}{s^2 + 6s + 13} \qquad (C-17)$$

With this introduction, Section C-5 deals with the actual construction of a typical root locus and gives all the geometric short cuts.

C-5   Construction of the Root Locus

In the preceding section it was shown that it was necessary to determine the roots of the denominator of the closed-loop transfer function as a function of the variable gain. This means that, in the general case, $1 + G(s)H(s)$ must be made equal to zero. In general, $G(s)H(s)$ will be of the form

$$G(s)H(s) = \frac{K(s + a_1)(s + a_2) \cdots}{(s + b_1)(s + b_2)(s + b_3) \cdots} \qquad (C-18)$$

The roots of the numerator are referred to as the zeros of the transfer function, while the roots of the denominator are called the poles. Due to the physical nature of feedback control systems, the order of the numerator is usually less than or equal to the order of the denominator. From Eq. C-14 the *magnitude condition* can be determined, which yields

$$K = \left| \frac{(s + b_1)(s + b_2)(s + b_3) \cdots}{(s + a_1)(s + a_2) \cdots} \right| \qquad (C-19)$$

Thus K is always equal to the product of the distances from the open-loop poles to a particular closed-loop pole divided by the product of the distances from the open-loop zeros to the same closed-loop pole. For the *angle condition*, from Eq. C-15

$$\left/ \frac{K(s + a_1)(s + a_2) \cdots}{(s + b_1)(s + b_2)(s + b_3) \cdots} \right. = 180(1 + 2m) \qquad (C-20)$$

where $m = 0, 1, 2, \ldots$, which states that the sum of the angles of the numerator terms minus the sum of the angles of the denominator terms is equal to an odd multiple of 180°. The rest of the rules are introduced as required. A sample root locus is drawn to illustrate the procedure.

The system to be analyzed is shown in Figure C-7, which is the same system that is discussed in Section 4-2 with the time constant of the washout circuit adjusted to yield an additional breakaway point and a break-in point. For this system,

$$[TF]_{OL} = G(s)H(s) = \frac{-13.8S_{(yrg)}s(s^2 + 0.05s + 0.066)}{(s + 10)(s - 0.004)(s + 0.2)(s^2 + 0.38s + 1.813)} \qquad (C-21)$$

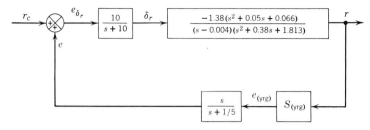

Figure C-7   Block diagram of system to be analyzed.

Then $K = 13.8 S_{(yrg)}$.   The details of the construction are presented here.

1. *Plot the poles and zeros of $G(s)H(s)$:* (see Figure C-8).

$$\text{Zeros: } s = 0$$
$$s = -0.025 \pm j0.244$$
$$\text{Poles: } s = -10$$
$$s = -0.2$$
$$s = -0.19 \pm j1.33$$
$$s = 0.004$$

2. *Real axis locus:* to the right of all poles and zeros the sum of all the angles from the poles and zeros is zero.   After passing to the left of the pole at 0.004, the sum of the angles is 180°, and to the left of the zero at the origin the sum of the angles is 360°, etc.   The extent of the real axis locus is shown in Figure C-8.

3. *Real axis intercept of the asymptotes:* from Eq. C-19, if the value of the pole is equal to any of the zeros, that is, $s = -a$, then $K = \infty$; and

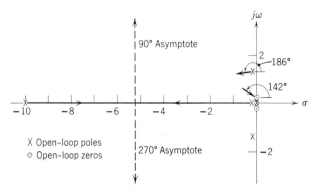

Figure C-8   Sample problem.

if there are more poles than zeros in the open-loop transfer function, for $s = \infty$ in Eq. C-19 $K = \infty$. Therefore, as $K \to \infty$, part of the root locus approaches the open-loop zeros; the remaining branches go to the hypothetical zeros at ∞. The branches that approach infinity approach asymptotes that are readily determined. The intercept of the asymptotes on the real axis is normally designated as σ_0 with

$$\sigma_0 = \frac{\sum \text{Poles} - \sum \text{Zeros}}{\text{Number of Poles} - \text{Number of Zeros}} \qquad \text{(C-22)}$$

The summation of the poles and zeros includes only the real part since the sum of the imaginary parts of the complex conjugates of complex poles are zero. For this problem

$$\sigma_0 = \frac{(-10 - 0.2 - 0.19 - 0.19 + 0.004) - (-0.025 - 0.025)}{5 - 3}$$

or

$$\sigma_0 = \frac{(-10.576) - (-0.05)}{2} = -5.263$$

4. *Angle of the asymptotes:* the angle that the asymptotes make with the real axis is given by Eq. C-23.

$$\text{Angle of the asymptotes} = \frac{(1 + 2m)180°}{\text{Number of Poles} - \text{Number of Zeros}}$$

$$\text{(C-23)}$$

where $m = 0, 1, 2, \ldots$ (number of poles − number of zeros) − 1. For this problem,

$$\text{Angle of asymptotes} = \frac{(1 + 2m)180°}{2}$$

For $m = 0$,

$$\text{Angle of asymptotes} = \frac{180}{2} = 90°$$

For $m = 1$,

$$\text{Angle of asymptotes} = \frac{3(180)}{2} = 270°$$

The asymptotes are also shown in Figure C-8. There is a real axis locus from the pole at 0.004 to the zero at the origin that cannot be shown in Figure C-8 due to the scale.

5. *Angle of departure from the complex pole:* for a point on the root locus very near a complex pole, the angle condition still must be satisfied. If the point being located is very close to the complex pole, the angles from the other poles and zeros to the point can be determined simply by

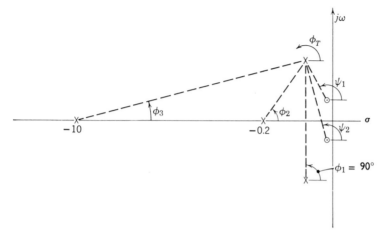

Figure C-9 Determination of the angle of departure from the complex
pole $s = -0.19 + j1.33$. *Note*: This figure is not drawn to scale.

measuring the angles to the complex pole itself.   The angle from the complex pole to the point being located is defined as the "angle of departure,"
and is designated by ϕ_T.   This angle cannot be measured because the test
point is too close to the complex pole; however, since the angle condition
must be satisfied, ϕ_T can be determined by using Eq. C-20.   This procedure
is illustrated by redrawing Figure C-8 to show these angles (see Figure
C-9).   For simplicity the figure is not drawn to scale; the pole at 0.004 and
the zero at the origin are neglected because their contribution to the angle
condition is negligible.   From Eq. C-20, $\lfloor numerator - \lfloor denominator =$
180 $(1 + 2m)°$ or from Figure C-9

$$\psi_1 + \psi_2 - (\phi_1 + \phi_2 + \phi_3 + \phi_T) = 180(1 + 2m)° \qquad (C\text{-}24)$$

Measuring these angles from Figure C-8 and substituting into Eq. C-24
yields $(97 + 95) - (90 + 88 + 8 + \phi_T) = 180(1 + 2m)°$ or $\phi_T = -174° =$
186° (see Figure C-8).

6. *Angle of arrival at a complex zero:* the same procedure can be applied
to determine the angle of arrival at a complex zero (ψ_T), which is illustrated
in Figure C-10.   Again from Eq. C-20,

$$\psi_1 + \psi_T - (\phi_1 + \phi_2 + \phi_3 + \phi_4) = 180(1 + 2m)° \qquad (C\text{-}25)$$

Substituting the values of the angles as measured from Figure (C-8)
yields $90 + \psi_T - (277 + 85 + 48 + 2) = 180(1 + 2m)°$ or $\psi_T = 180 +$
$322 = 502°$, which is equivalent to $502 - 360$ or 142° (see Figure C-8).

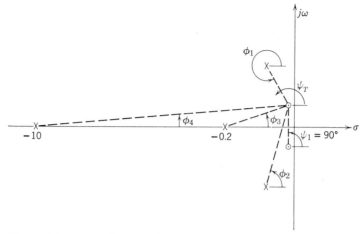

Figure C-10   Determination of the angle of arrival at the complex zero
$s = -0.025 + j0.244$.   *Note*: This figure is not drawn to scale.

7. *Breakaway or break-in point on the real axis:* referring to Figure C-8, it can be seen that there must be a transition point where the root locus leaves the real axis and proceeds toward the asymptotes. This point is referred to as a "breakaway point." There also may be a break-in point, that is, where the root-locus transitions from the complex plane to the real axis. There are several methods for determining these breakaway or break-in points. Two methods are discussed in Ref. 1, pp. 149–152.

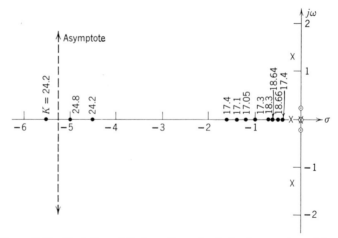

Figure C-11   Variation of K along the real axis for the sample problem.

A third method, which is sometimes referred to as the "hills and dales" method, is explained here. Since K must increase as any pole moves from an open-loop pole to a finite zero or to a fictitious zero at infinity, then if two poles come together on the real axis and break away, the value of K at the breakaway point must be greater than the value of K on either side of the breakaway point. Similarly, if the root locus breaks into the real axis, the K at the break-in point is lower than the value of K on either side of this point. Thus to determine the existence and approximate location of the breakaway or break-in points it is only necessary to check the magnitude condition along the real axis. The results of these calculations are plotted in Figure C-11, where it can be seen that there is a breakaway point near -5 and -0.5 and a break-in point near -1.2. A final check can be made by checking the angle condition about $\frac{1}{4}$ of an inch above the real axis in the vicinity of the maximum and minimum values of K. An observation of the values of K reveals how the name "hills and dales" method originated.

8. *Imaginary axis crossing:* Routh's criteria can be used to determine the values of $j\omega$ at which the root locus crosses the imaginary axis (see Ref. 1, pp. 153–154). However, if the root locus crosses the imaginary axis, the asymptotes also must cross. It is usually easier, unless the problem is very simple, to find the imaginary axis crossing by searching for a point along the $j\omega$ axis that satisfies the angle condition (this point will be near the asymptote crossing). For this problem there is no imaginary axis crossing.

9. *Remaining plot:* after steps 1–8 have been accomplished, as applicable, it is necessary to complete the root locus by locating points that satisfy the angle condition. It is not necessary to find many points in this way. The most important part of the root locus is in the vicinity of the desired closed-loop poles. The final root locus is plotted in Figure C-12 with the points actually determined by use of the angle condition indicated. Notice that by locating only four points (one of these is one of the complex closed-loop poles) the rest of the root locus was determined. Actually it would have been sufficient to determine only the portion of the root locus from the complex poles to the real axis and along the real axis. In checking for points that satisfy the angle condition it has been implied that the angle from each pole and zero must be measured and the results substituted into Eq. C-20. However, this operation is simplified by the use of the "Spirule" made by the Spirule Company of Whittier, California. The use of the Spirule is described in Appendix D of Ref. 1, and operating instructions are furnished with the Spirule.

10. *Location of the closed-loop poles:* after the root locus has been determined it is necessary to determine the location of the closed-loop

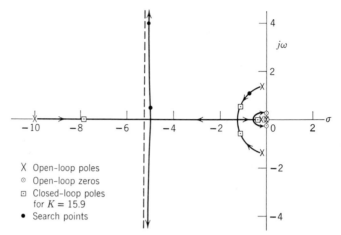

Figure C-12   Root locus for sample problem.

poles.   The location of the dominant pole or poles is usually dictated by the type of response desired.   The dominant poles, either real or imaginary, are those closest to the imaginary axis providing they are not very close to a zero.   After determining the dominant pole or poles, that is, one real pole or a pair of complex poles, the corresponding magnitude condition can be determined.   The rest of the closed-loop poles must satisfy the same magnitude condition.   For the sample problem there is a damping ratio of 0.9 for the complex roots.   Applying the magnitude condition, as given in Eq. C-19, yields

$$K = \frac{(8.86)(1.24)(2.12)(1.1)}{(1.17)(1.38)} = 15.9$$

The remaining closed loop poles must also yield the same value for K. From Figure C-11, the K for $s = -0.4$ is 17.4 and, the K for $s = -0.3$ is 12.8.   Therefore, there is a real axis closed-loop pole between -0.3 and -0.4.   The pole is closer to -0.4 than to -0.3, and is assumed to be at $s = -0.38$.   The pole that started at $s = 0.004$, and is moving toward the origin, is assumed not to have moved.   There is now one more closed-loop pole to determine; this is the pole that started at $s = -10$ and is moving in along the real axis.   From Figure C-11 it can be seen that it is to the left of $s = -5.5$.   This pole can be located by checking the magnitude condition along the real axis between $s = -5.5$ and $s = -10$.   However, there is a simpler way to locate at least the approximate position.   Referring to Table C-2 it can be seen that the sum of the real parts of s_1 and s_2 is a constant and equal to -6; this is not a coincidence, and is always true if the number of open-loop poles of $G(s)H(s)$, minus the number of zeros of $G(s)H(s)$, is equal to or greater than 2. (This characteristic of the root locus was discussed by A. J. Grant of North American Aviation, Inc., in an unpublished paper entitled "The Conservation of the Sum of the System Roots as Applied to the Root Locus Method" dated April 10, 1953, and for this reason is often referred to as Grant's rule.[1]) The initial value of the summation of the poles is -10.576. Thus far the closed-loop poles that have been determined are $s = -1.15 \pm j0.55$; $s = -0.38$; $s = 0.004$. The summation of these poles is -2.676; therefore, the remaining pole should be near -7.9. A check of the magnitude condition for $s = -7.9$ yields a K of 15.8, which is close enough. All the closed loop poles corresponding to a K of 15.9 have been determined. From Eq. C-21 $K = 13.8S_{(yrg)}$. Then, $S_{(yrg)} = K/13.8 = 15.9/13.8 = 1.15$. For simplicity a value of 1 can be used. The remaining step is the writing of the closed-loop transfer function.

C-6 Closed-Loop Transfer Function

The closed-loop transfer function is given by Eq. C-2, which is repeated here for convenience

$$\frac{C(s)}{R(s)} = \frac{G(s)}{1 - [TF]_{OL}} \tag{C-26}$$

and for this problem is

$$\frac{r(s)}{r_c(s)} = \frac{\dfrac{-13.8(s^2 + 0.05s + 0.066)}{(s + 10)(s - 0.004)(s^2 + 0.38s + 1.813)}}{1 + \dfrac{Ks(s^2 + 0.05s + 0.066)}{(s + 10)(s - 0.004)(s + 0.2)(s^2 + 0.38s + 1.813)}} \tag{C-27}$$

Simplifying,

$$\frac{r(s)}{r_c(s)} = \frac{-13.8(s + 0.2)(s^2 + 0.05s + 0.066)}{(s + 10)(s - 0.004)(s + 0.2)(s^2 + 0.38s + 1.813)} \tag{C-28}$$
$$+ Ks(s^2 + 0.05s + 0.066)$$

The root locus has furnished the factors of the denominator of Eq. C-28 for a K of 15.9; thus

$$\frac{r(s)}{r_c(s)} = \frac{-13.8(s + 0.2)(s^2 + 0.05s + 0.066)}{(s + 7.9)(s - 0.004)(s + 0.38)(s^2 + 2.3s + 1.64)} \tag{C-29}$$

As a check on the accuracy of the root locus and the denominator of Eq. C-29, the denominator of Eq. C-28 for $K = 15.9$ is expanded, combined, and compared to the denominator of Eq. C-29 after expanding it.

From Eq. C-28 the denominator is $s^5 + 10.576s^4 + 23.45s^3 + 20.01s^2 + 4.589s - 0.0145$, and from Eq. C-29 the denominator is $s^5 + 10.576s^4 + 23.27s^3 + 20.1s^2 + 4.77s - 0.0194$. There is excellent agreement between the two denominators; in fact, in this case it might be considered exceptional.

Equation C-29 is the closed-loop transfer function for the block diagram given in Figure C-7 for a yaw rate gyro sensitivity of 1.15. Notice that since the variable gain was in the feedback loop, the static loop sensitivity remains unchanged at 13.8. Also the open-loop pole from the washout circuit shows up as a zero in the closed-loop transfer function. Thus any time the feedback loop contains a transfer function the step shown in Eq. C-27 should be used to assure the proper form of the closed-loop transfer function. This can also be accomplished by using Eq. C-30,[1] where $G(s) = N_1(s)/D_1(s)$ and $H(s) = N_2(s)/D_2(s)$.

$$\frac{C(s)}{R(s)} = \frac{N_1 D_2}{D_1 D_2 + N_1 N_2} \tag{C-30}$$

C-7   Zero-Angle Root Locus

In Section C-5 the rules were given for the so-called 180° root locus. That is $1 + G(s)H(s) = 0$ or $G(s)H(s) = 1\underline{/180°}$. In some instances where positive feedback is used, or as in this book for nonminimum phase-angle transfer functions (that is, zeros in the right-half plane), if $[TF]_{OL}$ is positive, the zero-angle root locus must be plotted. If the open-loop transfer function is positive, as defined in Section C-2, then $1 - G(s)H(s) = 0$ and $G(s)H(s) = 1\underline{/0°}$. To plot the zero-angle root locus all the rules given in Section C-5 apply; the only change is the replacement of $(1 + 2m)180°$ in all the angle equations with $2m(180°)$, where $m = 0, 1, 2$, etc. The closed-loop transfer function is obtained as described in Section C-6 for the 180° root locus.

C-8   Frequency Response

Frequency response analysis yields the steady-state behavior of a system to sinusoidal forcing, as indicated in this book, by such figures as 1–9, 1-10, 1-11, etc. This can be accomplished by substituting $j\omega$ for s in the transfer function and then calculating the magnitude and phase angle as a function of ω. (This procedure is covered in detail in Ref. 1, Chapter 8.) At first this sounds time consuming; however, by knowing the characteristic response for a first- and second-order system, the plot for even the

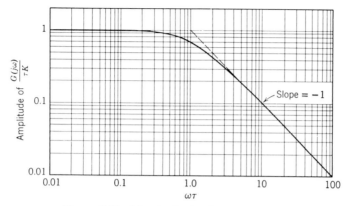

Figure C-13   Magnitude plot for $(j\omega\tau + 1)^{-1}$.

most complicated transfer function can be obtained by the proper combination of the appropriate first- and/or second-order responses.

First-order response:

Given:
$$(G)s = \frac{K}{s + 1/\tau}$$

then
$$G(j\omega) = \frac{K}{j\omega + 1/\tau}$$

Next, the transfer function must be put in the standard form for obtaining the frequency response curves. For this the form is

$$G(j\omega) = \frac{\tau K}{j\omega\tau + 1} \tag{C-31}$$

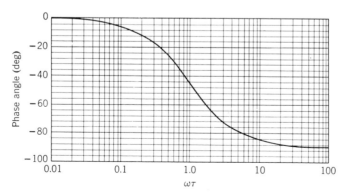

Figure C-14   Phase angle plot for $(j\omega\tau + 1)^{-1}$.

The magnitude and phase angle of Eq. C-31 must now be plotted as a function of ω. For $\omega = 0$, $G(j\omega) = \tau K \underline{/0°}$, and as $\omega \to \infty$, $G(j\omega) \to 0 \underline{/-90°}$. Also for $\omega = 1/\tau$,

$$G(j\omega) = \frac{\tau K}{j1 + 1} = \frac{\tau K}{1.414 \underline{/45°}}$$

Therefore $G(j\omega) = 0.707 \tau K \underline{/-45°}$. The magnitude plot of Eq. C-31 is shown in Figure C-13, and the phase angle plot in Figure C-14. The point where $\omega = 1/\tau$ is called the "break point" and the value of ω is referred to as the "corner frequency." If the first-order term appears in the numerator, the magnitude plot breaks up at the corner frequency and the phase angle goes from 0 to $+90°$.

Second-order response:

Given:
$$G(s) = \frac{K}{s^2 + 2\zeta\omega_n s + \omega_n^2}$$

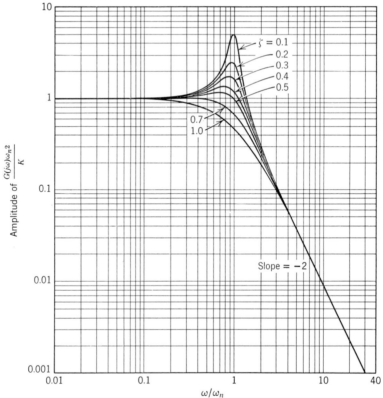

Figure C-15   Magnitude plot for $\left[\left(\dfrac{j\omega}{\omega_n}\right)^2 + \dfrac{j2\zeta\omega}{\omega_n} + 1\right]^{-1}$.

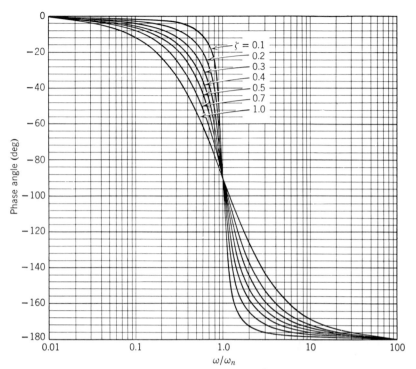

Figure C-16 Phase angle plot for $\left[\left(\dfrac{j\omega}{\omega_n}\right)^2 + \dfrac{j2\zeta\omega}{\omega_n} + 1\right]^{-1}$.

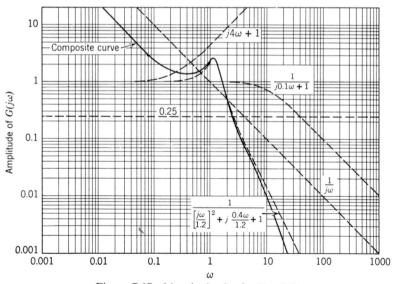

Figure C-17 Magnitude plot for Eq. C-33.

Replacing s by $j\omega$ and going to the standard form yields

$$G(j\omega) = \frac{\dfrac{K}{\omega_n{}^2}}{\left(\dfrac{j\omega}{\omega_n}\right)^2 + \dfrac{j2\zeta\omega}{\omega_n} + 1} \tag{C-32}$$

Again for $\omega = 0$, $G(j\omega) = K/\omega_n{}^2 \underline{/0^\circ}$ and, as $\omega \to \infty$, $G(j\omega) \to 0\underline{/-180^\circ}$. For $\omega = \omega_n$,

$$G(j\omega) = \frac{\dfrac{K}{\omega_n{}^2}}{(j)^2 + j2\zeta + 1} = \frac{\dfrac{K}{\omega_n{}^2}}{j2\zeta}$$

Therefore $G(j\omega) = K/2\zeta\omega_n{}^2 \underline{/-90^\circ}$. Thus the magnitude of $G(j\omega)$ at the corner frequency is dependent upon the damping ratio ζ. The lower ζ,

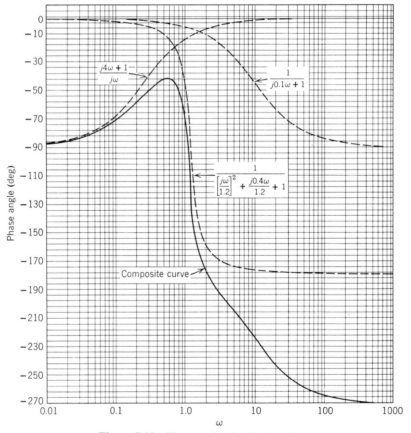

Figure C-18   Phase angle plot for Eq. C-33.

the higher the gain at $\omega = \omega_n$. This phenomenon is known as "resonance." The magnitude and phase plots for various values of ζ are shown in Figures C-15 and C-16. Since the amplitude plots are plotted on a logarithmic scale, the magnitude plot for any transfer function can be obtained by adding the plots of each factor of the transfer function.

Response of a complete transfer function:

Given:

$$G(s) = \frac{14.4(s + 0.25)}{s(s + 10)(s^2 + 0.48s + 1.44)}$$

Then

$$G(j\omega) = \frac{0.25(14.4)(j4\omega + 1)}{14.4j\omega(j0.1\omega + 1)\left[\left(\dfrac{j\omega}{1.2}\right)^2 + \dfrac{j0.4\omega}{1.2} + 1\right]}$$

or

$$G(j\omega) = \frac{0.25(j4\omega + 1)}{j\omega(j0.1\omega + 1)\left[\left(\dfrac{j\omega}{1.2}\right)^2 + \dfrac{j0.4\omega}{1.2} + 1\right]} \tag{C-33}$$

The magnitude and the phase plots are shown in Figures C-17 and C-18. As the $1/j\omega$ term adds a constant phase angle of $-90°$, this term is combined with the numerator factor in plotting Figure C-18.

References

1. J. J. D'Azzo, and C. H. Houpis, *Feedback Control System Analysis and Synthesis*, McGraw-Hill Book Co., New York, 1960.
2. J. G. Truxal, *Automatic Feedback Control System Synthesis*, McGraw-Hill Book Co., New York, 1955.
3. E. O. Doebelin, *Dynamic Analysis and Feedback Control*, McGraw-Hill Book Co., New York, 1962.
4. B. C. Kuo, *Automatic Control Systems*, Prentice-Hall, Englewood Cliffs, New Jersey, 1962.

Appendix *D*
Fundamental Aerodynamic Principles

D-I   Aerodynamic Forces

By use of dimensional analysis the equation relating the aerodynamic force on a body to the parameters of the problem can be derived.[1,2,3] The result is Eq. D-1.

$$F = C_F \frac{\rho}{2} V_T{}^2 S \qquad (D\text{-}1)$$

Where C_F is the force coefficient and is dependent on the Reynolds number $V_T l/v$, ρ is the air density, V_T is the velocity of the air relative to the body, and S is a characteristic area.   In the equation for the Reynolds number l is a characteristic length and

$$v = \frac{\mu}{\rho} \frac{\text{sq ft}}{\text{sec}} = \text{kinematic viscosity}$$

where

$$\mu = \frac{\tau}{\dfrac{dV_T}{dy}} \frac{\text{lb-sec}}{\text{ft}^2} = \text{coefficient of viscosity}$$

and τ is shear stress between layers of the moving fluid (viscosity).   The characteristic area used for the study of aircraft is the wing area, which includes the entire area of wing including that portion of the wing covered by the fuselage and nacelles, but excluding wing fillets.   For missile analysis the S is usually the cross-sectional area of the missile. The aerodynamic force acting on a lifting body may be resolved into two components, which are

$$\text{Lift} = L = C_L \frac{\rho}{2} V_T{}^2 S = C_L q S \text{ lbs} \qquad (D\text{-}2)$$

$$\text{Drag} = D = C_D \frac{\rho}{2} V_T{}^2 S = C_D q S \text{ lbs} \qquad (D\text{-}3)$$

where $q = \frac{\rho}{2} V_T{}^2$ is the dynamic pressure in lb/sq ft.   In resolving the aerodynamic force into the lift and drag components, lift is always taken

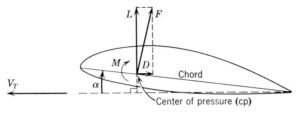

Figure D-1   Forces acting on an airfoil.

normal to V_T and drag along V_T, as shown in Figure D-1.   The lift and drag are shown acting at the "center of pressure" of the airfoil.   The center of pressure is defined as the point on the airfoil through which the total aerodynamic force can be considered to be acting.   The center of pressure in general will not be located at the center of gravity of the airfoil; thus a

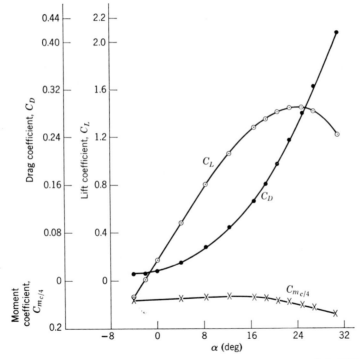

Figure D-2   Data for NACA 4521 airfoil from NACA Report 460, 1933. $R = 3.15 \times 10^6$; size, 5 × 30 in., velocity = 69 ft/sec; pressure (standard atmosphere): 20.8.   (Test V.D.T. 573, at L.M.A.L., Apr. 17, 1931.)   Corrected for tunnel wall effect.

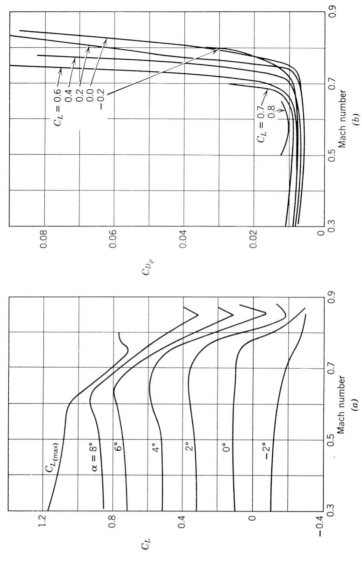

Figure D-3 Variations of lift and drag with Mach number. (a) C_L versus Mach number: BAC 53 airfoil; (b) C_{D_p} versus Mach number: BAC 53 airfoil. By permission from G.S. Schairer, "Systematic Wing Section Development," *Journal of the Aeronautical Sciences*, January 1947.

moment will be produced. The positive direction of the moment is indicated in Figure D-1. To obtain a moment, the force must be multiplied by the proper moment arm; by convention the mean aerodynamic chord, c, is used; thus

$$\text{Moment} = M = C_M q S c \qquad \text{(D-4)}$$

The mean aerodynamic chord (MAC) is defined as the chord of an imaginary airfoil which throughout the flight range has the same force vectors as the actual three-dimensional wing. For a tapered wing

$$c = \frac{2}{3}\left(a + b - \frac{ab}{a + b}\right)$$

where a = wing root chord and b = wing tip chord.

The lift, drag, and moment coefficients are functions of the angle of attack and Mach number. Figure D-2 shows the variation of the lift and drag coefficients and the moment coefficient about the quarter chord point with angle of attack. The lift, drag, and moment coefficients are relatively independent of Mach number for Mach numbers below 0.6 (see Figure D-3 for C_L and C_{D_p}), the widest variation occurring in the transonic region, Mach numbers between 0.6 and 1. Also when the leading edge of the airfoil is supersonic the lift vector is always normal to the chord.

D-2 Induced Drag

An examination of Figure D-2 indicates that the drag coefficient curve is approximately parabolic while the lift coefficient curve is linear up to an angle of attack of 8°. A plot of the drag coefficient versus the square of the lift coefficient is plotted in Figure D-4, as indicated by the dotted line. The solid line is a plot of Eq. D-5.

$$C_D = C_{D_0} + \frac{C_L^2}{\pi R e} \qquad \text{(D-5)}$$

where C_{D_0} is the minimum parasite or profile drag and consists of friction and pressure drag; $R = b^2/S$ and is the aspect ratio; b is the wing span; and e is the "span efficiency factor." The span efficiency factor accounts for the fact that the wing does not have an elliptical lift distribution. The $C_L^2/\pi R e$ term in Eq. D-5 is referred to as the "induced drag" and accounts for the change in parasite drag with angle of attack and the induced drag. From Figure D-4 it can be seen that Eq. D-5 is valid over a wide range of lift coefficients, the only deviations being at extremely low

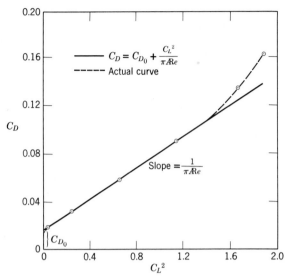

Figure D-4   Comparison of theoretical and actual plot of
C_D versus $C_L{}^2$.

values of C_L and for values of C_L greater than 1.   At the high values of C_L
the wing is approaching the stall, and Eq. D-5 cannot be expected to be
valid.

References

1. C. D. Perkins and R. E. Hage, *Airplane Performance, Stability, and Control*,
   John Wiley and Sons, New York, 1949.
2. D. O. Dommash, S. S. Sharby and T. F. Connolly, *Airplane Aerodynamics*, Pitman
   Publishing Corporation, New York, 1951.
3. C. B. Millikan, *Aerodynamics of the Airplane*, John Wiley and Sons, 1941.

Appendix E

Matrices

E-I Definition

A matrix is an array of numbers used primarily to systemize and simplify the writing and/or manipulation of simultaneous equations. The three basic matrices are as follows:

1. *Rectangular matrix*:

$$[a] = [a_{ij}] = \begin{bmatrix} a_{11} & a_{12} & \cdots & a_{1j} \\ a_{21} & a_{22} & \cdots & a_{2j} \\ a_{31} & a_{32} & \cdots & a_{3j} \\ \vdots & \vdots & \cdots & \vdots \\ a_{i1} & a_{i2} & \cdots & a_{ij} \end{bmatrix} \text{Row}$$

(Column)

where the size of the matrix is designated as a $i \times j$ (read i by j) matrix. If $i = j$, then the rectangular matrix becomes a square matrix.

2. *Column matrix*:

$$\{a_j\} = \begin{bmatrix} a_1 \\ a_2 \\ a_3 \\ \vdots \\ a_j \end{bmatrix}$$

3. *Row matrix*:

$$\lfloor a_i \rfloor = [a_1 \quad a_2 \quad a_3 \quad \cdots \quad a_i]$$

As an example of the use of matrices, the following simultaneous equations are written using matrices.

$$3x + 5y + 2z = 10$$
$$2x - 3y + 4z = 8$$
$$-3x + 2y - 2z = -3$$

In matrix form

$$\begin{bmatrix} 3 & 5 & 2 \\ 2 & -3 & 4 \\ -3 & 2 & -2 \end{bmatrix} \begin{bmatrix} x \\ y \\ z \end{bmatrix} = \begin{bmatrix} 10 \\ 8 \\ -3 \end{bmatrix}$$

or $[A]\{B\} = \{C\}$ where

$$[A] = \begin{bmatrix} 3 & 5 & 2 \\ 2 & -3 & 4 \\ -3 & 2 & -2 \end{bmatrix}$$

$$\{B\} = \begin{bmatrix} x \\ y \\ z \end{bmatrix}, \quad \text{and} \quad \{C\} = \begin{bmatrix} 10 \\ 8 \\ -3 \end{bmatrix}$$

E-2 Matrix Operations

1. *Addition*:   When adding matrices each matrix to be added must be the same size; thus $[a_{ij}] + [b_{ij}] = [c_{ij}]$ where each element of $c_{ij} = a_{ij} + b_{ij}$.

Example:

$$\begin{bmatrix} 1 & 2 & 1 \\ 3 & 2 & 7 \end{bmatrix} + \begin{bmatrix} 1 & -2 & 3 \\ 4 & 1 & 2 \end{bmatrix} = \begin{bmatrix} 2 & 0 & 4 \\ 7 & 3 & 9 \end{bmatrix}$$

2. *Subtraction*:   Again each matrix must be the same size, and each element of $c_{ij} = a_{ij} - b_{ij}$ for $[a_{ij}] - [b_{ij}] = [c_{ij}]$.

3. *Multiplication*:   To multiply two matrices they must be compatible in size; thus the operation $[a_{ij}][b_{jk}] = [c_{ik}]$ can be performed if the number of columns of the *a* matrix is equal to the number of rows of the *b* matrix.   The matrix of the product then has the same number of rows as the *a* matrix and the same number of columns as the *b* matrix.

Example:

$$\begin{bmatrix} 1 & 2 & 4 \\ 3 & 1 & 3 \\ -1 & 2 & 7 \end{bmatrix} \begin{bmatrix} 1 & 3 \\ 2 & 2 \\ 4 & 1 \end{bmatrix} = \begin{bmatrix} 21 & 11 \\ 17 & 14 \\ 31 & 8 \end{bmatrix}$$

The c_{11} term is obtained by adding the products of the terms of the first row of the *a* matrix and the first column of the *b* matrix.   Thus

$$(1)(1) + (2)(2) + (4)(4) = 21$$

In like manner

$$(3)(1) + (1)(2) + (3)(4) = 17$$

and

$$(-1)(1) + (2)(2) + (7)(4) = 31$$

To obtain the second row of the c matrix, this procedure is followed using the second column of the b matrix.   *Note*: $[A][B] \neq [B][A]$

4. *Scalar multiplication*:   Any matrix can be multiplied by a scalar quantity by multiplying each element of the matrix by the scalar quantity.

5. *Division*:   Division can be performed only by inverting the divisor and multiplying.   The inversion can be performed only for square matrices.   The symbol used to indicate the inversion is a superscript -1 outside the bracket; thus the inverse of $[A]$ would be $[A]^{-1}$.   If $[B] = [A]^{-1}$, then

$$[B_{ij}] = \frac{(-1)^{(i+j)} \begin{vmatrix} \text{minor of} \\ A_{ji} \end{vmatrix}}{|A_{ij}|} \tag{E-1}$$

Example:

Let

$$[A] = \begin{bmatrix} 1 & 4 & 6 \\ 3 & 2 & 7 \\ -1 & 4 & 2 \end{bmatrix}$$

Then (see Section E-4 for the method of expanding a determinant),

$$B_{11} = \frac{(-1)^2 \begin{vmatrix} 2 & 7 \\ 4 & 2 \end{vmatrix}}{\begin{vmatrix} 1 & 4 & 6 \\ 3 & 2 & 7 \\ -1 & 4 & 2 \end{vmatrix}} = \frac{4 - 28}{(4 - 28) - 4(6 + 7) + 6(12 + 2)}$$

or

$$B_{11} = \frac{-24}{8} = -3$$

In like manner,

$$B_{12} = \frac{(-1)^3 \begin{vmatrix} 4 & 6 \\ 4 & 2 \end{vmatrix}}{8} = \frac{-(8 - 24)}{8} = 2$$

And

$$B_{13} = \frac{(-1)^4 \begin{vmatrix} 4 & 6 \\ 2 & 7 \end{vmatrix}}{8} = \frac{28 - 12}{8} = 2$$

Finally,

$$[B] = \begin{bmatrix} -3 & 2 & 2 \\ -1.625 & 1 & 1.375 \\ 1.75 & -1 & -1.25 \end{bmatrix}$$

E-3   Special Matrices

There are several special cases of the square matrix.

1. *Symmetric matrix*:   For the symmetric matrix $a_{ij} = a_{ji}$.

Example:

$$[A] = \begin{bmatrix} 1 & 2 & -4 \\ 2 & 3 & 6 \\ -4 & 6 & 5 \end{bmatrix}$$

2. *Diagonal matrix*: For the diagonal matrix only the diagonal terms are nonzero; thus $a_{ij} = 0$, for $i \neq j$; and $a_{ij} \neq 0$, for $i = j$.

Example:

$$\begin{bmatrix} 1 & 0 & 0 \\ 0 & 3 & 0 \\ 0 & 0 & 5 \end{bmatrix}$$

3. *Identity matrix*:   A special case of the diagonal in which each term in the diagonal is equal to one.   The identity matrix is denoted by (I).

Example:

$$[I] = \begin{bmatrix} 1 & 0 & 0 \\ 0 & 1 & 0 \\ 0 & 0 & 1 \end{bmatrix}$$

Any matrix multiplied by the identity matrix is equal to itself.   Also $[A]^{-1}[A] = [I]$.   The proof of this is left as an exercise for the student.

In Chapter 8, in the discussion of the resulting simultaneous equations using matrix techniques, the solution was obtained by multiplying both sides of the equation by the inverse of the coefficient matrix.   The proof of this procedure follows.

If $[A][X] = [C]$, multiply both sides by $[A]^{-1}$.

Then $$[A]^{-1}[A][X] = [A]^{-1}[C]$$

But $$[A]^{-1}[A] = [I] \quad \text{and} \quad [I][X] = [X]$$

Therefore $$[X] = [A]^{-1}[C]$$

E-4 Determinants[1]

In performing the inversion of a matrix use was made of determinants and the minors of determinants. By definition a determinant is a square array of numbers and provides a shorthand form for denoting the sum of all products which can be formed using all the numbers of the determinant. The order of the determinant is determined by the number of rows or columns. Thus the determinant D may be

$$D = \begin{vmatrix} a_{11} & a_{12} & \cdots & a_{1n} \\ a_{21} & a_{22} & \cdots & a_{2n} \\ a_{31} & a_{32} & \cdots & a_{3n} \\ \vdots & \vdots & \cdots & \vdots \\ a_{n1} & a_{n2} & \cdots & a_{nn} \end{vmatrix} \qquad (E\text{-}2)$$

which is an nth order determinant and contains n^2 terms.

Minor of a determinant: The minor of a determinant is the determinant formed by crossing out the row and column containing a given element a_{ij}. Thus for the determinant given in Eq. E-2, the minor of a_{22} is

$$\begin{vmatrix} a_{11} & a_{13} & \cdots & a_{1n} \\ a_{31} & a_{33} & \cdots & a_{3n} \\ \vdots & \vdots & \cdots & \vdots \\ a_{n1} & a_{n3} & \cdots & a_{nn} \end{vmatrix}$$

Expansion of a determinant by minors: The expansion of the determinant given in Eq. E-2 is

$$a_{11} \begin{vmatrix} \text{minor} \\ \text{of } a_{11} \end{vmatrix} - a_{12} \begin{vmatrix} \text{minor} \\ \text{of } a_{12} \end{vmatrix} + a_{13} \begin{vmatrix} \text{minor} \\ \text{of } a_{13} \end{vmatrix} - + - \cdots a_{1n} \begin{vmatrix} \text{minor} \\ \text{of } a_{1n} \end{vmatrix}$$

or

$$a_{11} \begin{vmatrix} \text{minor} \\ \text{of } a_{11} \end{vmatrix} - a_{21} \begin{vmatrix} \text{minor} \\ \text{of } a_{21} \end{vmatrix} + a_{31} \begin{vmatrix} \text{minor} \\ \text{of } a_{31} \end{vmatrix} - + - \cdots a_{n1} \begin{vmatrix} \text{minor} \\ \text{of } a_{n1} \end{vmatrix}$$

$$(E\text{-}3)$$

If the determinants formed by the minors in Eq. E-3 are greater than the second order, the procedure is continued until they are reduced to second-order determinants. The expansion of a second-order determinant is as follows

$$\begin{vmatrix} a_{11} & a_{12} \\ a_{21} & a_{22} \end{vmatrix} = a_{11}a_{22} - a_{12}a_{21}$$

Example:

$$D = \begin{vmatrix} 3 & -1 & 2 & 3 \\ 1 & 0 & 2 & 1 \\ 2 & 3 & 0 & 1 \\ 5 & 2 & 4 & -5 \end{vmatrix}$$

$$= 3 \begin{vmatrix} 0 & 2 & 1 \\ 3 & 0 & 1 \\ 2 & 4 & -5 \end{vmatrix} + 1 \begin{vmatrix} 1 & 2 & 1 \\ 2 & 0 & 1 \\ 5 & 4 & -5 \end{vmatrix} + 2 \begin{vmatrix} 1 & 0 & 1 \\ 2 & 3 & 1 \\ 5 & 2 & -5 \end{vmatrix} - 3 \begin{vmatrix} 1 & 0 & 2 \\ 2 & 3 & 0 \\ 5 & 2 & 4 \end{vmatrix}$$

$$\text{(E-4)}$$

Expanding the first term of Eq. E-4 yields

$$3 \begin{vmatrix} 0 & 2 & 1 \\ 3 & 0 & 1 \\ 2 & 4 & -5 \end{vmatrix} = (3)(0) \begin{vmatrix} 0 & 1 \\ 4 & -5 \end{vmatrix} + (3)(-2) \begin{vmatrix} 3 & 1 \\ 2 & -5 \end{vmatrix} + 3(1) \begin{vmatrix} 3 & 0 \\ 2 & 4 \end{vmatrix}$$

$$= 0 - 6(-15 - 2) + 3(12 - 0)$$

$$= 0 + 102 + 36 = 138$$

The second step in the expansion just given can be eliminated, as will be done in the expansion of the second term; thus

$$1 \begin{vmatrix} 1 & 2 & 1 \\ 2 & 0 & 1 \\ 5 & 4 & -5 \end{vmatrix} = 1[1(0 - 4) - 2(-10 - 5) + 1(8 - 0)]$$

$$= 34$$

In like manner,

$$2 \begin{vmatrix} 1 & 0 & 1 \\ 2 & 3 & 1 \\ 5 & 2 & -5 \end{vmatrix} = -56 \quad \text{and} \quad -3 \begin{vmatrix} 1 & 0 & 2 \\ 2 & 3 & 0 \\ 5 & 2 & 4 \end{vmatrix} = 30$$

Therefore $D = 138 + 34 - 56 + 30 = 146$.

Use of determinants to solve simultaneous equations: Given the simultaneous algebraic equations,

$$x + 2y + 3z = 6$$
$$2x - 2y - z = 3$$
$$3x + 2y + z = 2$$

Solve for x, y, and z

$$x = \frac{\begin{vmatrix} 6 & 2 & 3 \\ 3 & -2 & -1 \\ 2 & 2 & 1 \end{vmatrix}}{\nabla} \tag{E-5}$$

where $\nabla = \begin{vmatrix} 1 & 2 & 3 \\ 2 & -2 & -1 \\ 3 & 2 & 1 \end{vmatrix} = 1(-2 + 2) - 2(2 + 3) + 3(4 + 6)$

Therefore $\nabla = 20$. The determinant of the numerator of Eq. E-5 is obtained by replacing the column of ∇ formed by the coefficient of the unknown being solved for by a column made up of the right-hand side of the equations. Expanding,

$$x = \frac{6(-2 + 2) - 2(3 + 2) + 3(6 + 4)}{20} = 1$$

Similarly,

$$y = \frac{\begin{vmatrix} 1 & 6 & 3 \\ 2 & 3 & -1 \\ 3 & 2 & 1 \end{vmatrix}}{20} = -\frac{40}{20} = -2$$

and $z = 3$.

Reference

1. W. L. Hart, *College Algebra and Trigonometry*, D. C. Heath and Co., Boston, 1959.

Appendix *F*

F-94A Longitudinal and Lateral Aerodynamic Data

S = 239 sq ft   c = 6.4 ft   l_t = 18.5 ft
b = 37.3 ft

		Landing 1	Cruise 2	Top Speed 3	Cruise 4	Cruise 5	High Speed 6	Cruise 7	Cruise 8
				No Tanks			With Tip Tanks		
Longitudinal Data									
h	(ft)	SL	35,000	SL	12,000	15,000	SL	12,000	15,000
V_T	(mph)	135	450	606	288	403	553	288	403
M		...	0.68	0.797	0.396	0.56	0.727	0.396	0.56
WT	(lbs)	12,359	13,614	13,614	13,614	13,614	16,091	16,091	16,091
q	(psf)	46.6	161	940	148	261	781	148	261
I_x	(slug ft²)	7,169	11,029	11,029	11,029	11,029	38,732	38,732	38,732
I_y	(slug ft²)	26,545	26,543	26,543	26,543	26,543	26,723	26,723	26,723
I_z	(slug ft²)	33,009	36,801	36,801	36,801	36,801	64,575	64,575	64,575
J_{xz}	(slug ft²)	−2,557	−660	+335	−790	−200	+450	−770	−150
C.G.	(% MAC)	25.2	26.1	26.1	26.1	26.1	29	29	29
C_L		1.11	0.354	0.0605	0.385	0.219	0.0858	0.454	0.257
C_{L_α}	(per rad)	5.27	5.27	5.27	5.27	5.27	5.27	5.27	5.27
$C_{L_{\delta e}}$	(per rad)	0.43	0.43	0.43	0.43	0.43	0.43	0.43	0.43
C_D		0.172	0.023	0.017	0.025	0.018	0.019	0.036	0.021
C_{D_α}	(per rad)	0.854	0.21	0	0.29	0.0844	0	0.337	0.148
$C_{D_{\delta e}}$	(per rad)	...	...	...	...	...	...	...	...
C_{m_z}	(per rad)	−0.406	−0.44	−0.44	−0.44	−0.44	−0.44	−0.44	−0.44
$C_{m_{\delta e}}$	(per rad)	−0.88	−0.934	−0.944	−0.934	−0.934	−0.94	−0.94	−0.94
$\dfrac{c}{2U} C_{m_{\dot\alpha}}$	(per rad/sec)	−0.069	−0.0213	−0.0159	−0.0324	−0.0232	−0.0173	−0.0324	−0.0232
$\dfrac{c}{2U} C_{m_q}$	(per rad/sec)	−0.132	−0.041	−0.0305	−0.064	−0.046	−0.033	−0.064	−0.046

341

Lateral Data	No Tanks					With Tanks	
	Landing 1	Cruise 2	Top Speed 3	Cruise 4	Cruise 5	High Speed 6	Cruise 7
C_{y_β} (per rad)	−0.86	−0.546	−0.546	−0.546	−0.546	−0.546	−0.546
$C_{y_{\delta_r}}$ (per rad)	0.149	0.149	0.149	0.149	0.149	0.149	0.149
$C_{y_{\delta_a}}$ (per rad)	...	...	...	...	...	...	...
C_{y_p} (per rad)	...	...	...	...	...	...	...
C_{y_r} (per rad)	0.45	0.287	0.287	0.287	0.287	0.291	0.291
C_{l_β} (per rad)	−0.0487	−0.0654	−0.0654	−0.0654	−0.0654	−0.0716	−0.0716
$C_{l_{\delta_a}}$ (per rad)	−0.0916	−0.114	−0.129	−0.114	−0.114	−0.0956	−0.0956
$C_{l_{\delta_r}}$ (per rad)	0.0057	0.0057	0.0057	0.0057	+0.0057	+0.0057	+0.0057
C_{l_r} (per rad)	0.278	0.107	0.043	0.115	0.065	0.0458	0.131
C_{l_p} (per rad)	−0.45	−0.39	−0.39	−0.39	−0.39	−0.39	−0.39
$C_{l_{\dot\delta_a}}$ (per rad/sec)	...	...	...	...	...	...	...
C_{n_β} (per rad)	0.1045	0.115	0.1	0.118	0.106	0.088	0.107
$C_{n_{\delta_a}}$ (per rad)	0.0069	0.0086	0.0057	0.0090	0.0069	0.0057	0.0101
$C_{n_{\delta_r}}$ (per rad)	−0.069	−0.069	−0.069	−0.069	−0.069	−0.069	−0.069
C_{n_p} (per rad)	−0.053	−0.0145	...	−0.016	−0.0077	−0.0007	−0.0197
C_{n_r} (per rad)	−0.210	−0.134	−0.134	−0.134	−0.134	−0.134	−0.134
$C_{n_{\dot\delta_r}}$ (per rad/sec)	−0.053	−0.0145	...	−0.016	−0.0077	−0.0007	−0.0197

Index

Acceleration control system, 71-75
  rate gyro feedback loop, 71
Acceleration effects on pendulum, 255
A-C dither signal, 206-208
Adverse yaw, 114
Aerodynamic coefficients, F-94A, 341-342
  jet transport, 32-33, 116-117
  Vanguard missile, 234
Aerodynamic missiles, axis system, 225
  control, 225-226
  defined, 223
Aileron displacement, defined, 122
  positive direction, 122
Airspeed, automatic control of, 76, 81-83
Altitude hold, 92, 96-98
Angle of attack, definition, 16
  of tail, equations of, 23
  variation, 17
Angular momentum, 9
Apparent mass effect, 22
Apparent vertical, 99
Approach couplers, 86, 169-170
Aspect ratio, 331
Attitude reference, stabilized platform,
  100-105
  vertical gyro, 98-100
Autocorrelation function, 273-277
  equation for, 273
  example of, 274-277
Autopilots; see Flight control systems
Autopilots, self-adaptive, definition, 199
  MH-90, block diagram, 217
    description, 215-220
    effects of low frequency servo on, 218
      with variable compensator, 218-220
    effects of noise, 216
  Minneapolis-Honeywell, 206-209
    A-C dither signal, 206-208
    effect on ideal relay, 208
    bang-bang servo, 206
    block diagram, 207
  MIT model reference, 209-215

Autopilots, MIT model reference, block
  diagram, 210
  error criteria, 211-213
  roll damping loop, 211
  roll stabilization loop, 213
  yaw orientational control loop, 212
  models, use of, 200
  philosophy, 200-201
  Sperry, 201-206
    block diagram, 203
    logic circuit, 204
    Schmitt trigger, 205
Avion Instrument Corp., 81
Axis system, aerodynamic missile, 225
  ballistic missile, 227
  body, 5
  disturbed, 15
  equilibrium, 15
  stability, 15

Ballistic missiles, axis system for, 227
  control system, 230-235, 264-266
  defined, 223
  longitudinal equations for, 228
  transfer function, 230, 252
Beam guidance, 167-180
  block diagram of, 168
  geometry of, 169
  localizer receiver, 167
  Omni tracking, 167
  response of, 178
Block diagrams, 306
  control ratio of, 307
  of adaptive autopilots, 200-201, 203,
    207, 210, 211-213, 216-217
  of lateral control systems, 137-139,
    142-143, 145, 147-149, 158-160,
    162, 164, 165-166, 168, 170-172,
    174-175, 179
  of longitudinal control systems, 56-59,
    62, 65, 71, 73-74, 76-77, 81, 85-86,
    89, 93-94, 96-97

Block diagrams, of missile control systems, 224, 226, 230, 232-233
  summation points, 306
  terminology, 307
  unity feedback, 308

Canards, illustrated, 225
Center of pressure, 329
Closed loop transfer function, 321-322
Coefficient of viscosity, 328
Complementary error function, 271
Control stick steering, discussion of, 64
Control system, definition, 222
Coordinated aircraft, defined, 159
  transfer function of, 159
Coordination techniques, 141-152
  discussion of, 152-157
  use of computed yaw rate, 146-149
  use of lateral acceleration, 144-146
  use of rudder coordination computer, 149-152
  use of sideslip, 141-143
Coriolis acceleration, 297
Coriolis equation, 296-297
Correlation function, auto, 273-277
  cross-, 277-278
Coupler, glide slope, 86
  lateral, 169-170
Cross-correlation function, 277-278
Curtis flying boat, 1

Damping, ratio, 35
  of phugoid mode, 81
  of short-period mode, 60
  time to damp to $\frac{1}{2}$ amplitude, 35
Determinants, 337-339
  expansion of, 337-338
  minor, 337
  solution of simultaneous equations, 339
Distribution function, 270
Downwash, 22-24
  angle, 23
    equation for, 114
  effect on horizontal stabilizer, 23-25
Drag, 328, 331
  induced, 331
  parasite, 331
Dutch roll mode, approximate equation, 126
  approximate transfer function, 125-127

Dutch roll mode, damping of, 138-141
  description of, 118, 125-126
  effects of airspeed and altitude, 128, 131
  equations for, damping ratio, 126
    natural frequency, 126
Dynamic pressure, definition of, 18, 328

Eigenfunctions, 247
Eigenvalues, 247
Elevator displacement, positive direction, defined, 36
End plate effect, 64
Equations of motion, six-degree-of-freedom, 11
  three-degree-of-freedom, lateral, 107
    linearized, 108-109
    nondimensional, 109-112
    solution of, 116-118
  longitudinal, 14
    linearized, 14-16
    nondimensional, 17-21
    solution of, 31-36
Equilibrium, forces, 7
  moments, 7
Erection of vertical gyro, 99-100
Error coefficient, 309
Error criteria, for MIT self-adaptive control system, 211-214
Error integral, 271-272
Euler angles, 12-13

First probability density function, 272
Fixed control neutral point, stick fixed, 28
Flare, automatic control of, 75, 88-92
  definition of, 88
Flight control systems, autopilots, self-adaptive, 199-235; *see also* Autopilots, self-adaptive
  lateral, basic, 137-138
    coordination techniques, 141-157; *see also* Coordination techniques
    beam guidance, 167-180; *see also* Beam guidance
    Dutch roll damper 138-141; *see also* Dutch roll mode
    nonlinear effects on, 180-181
    rate stabilized, 166
    roll angle control system, 164-166
    turn compensation, 166

Flight control systems, yaw orientational
control system, 157-163; *see also*
Yaw orientational control system
longitudinal, acceleration control system, 71-75
altitude hold, 92, 96-98
automatic flare control, 75, 88-92
displacement, 56-62
with rate feedback, 59-62
Mach hold, 92-96
pitch orientational control system,
62-71
missile control systems, 222-235
alternate control system, 231-235
for aerodynamic missiles, 225-226
for Vanguard missile, 230-231,
264-266
roll stabilization, 223-225
Flight path angle, control of, 76-83
definition of, 15
Flight path stabilization, 92
altitude hold, 92, 96-98
Mach hold, 92-96
Forces, applied, 16
equilibrium, 7
Frequency response, 322-327
of complete transfer function,
325-327
of first-order term, 323-324
of lateral transfer function, 119-121,
123-125
of longitudinal transfer function, 38-40,
43
of rudder coordination computer,
150-151
of second-order term, 324-325

Gaussian distribution, 270
Glide slope coupler, 86
effect of beam narrowing, 84
geometry, 84
Gravity, components, 13
Guidance, definition, 222
Gyro, equation for integrating, 303-305
equation for rate gyro, 301-303
integrating, 101
law of, 298-299
rate, 60, 71
vertical, 56, 98-100
erection of, 99-100
Gyro blender, 260-262

Induced drag, 331
Inertia, moment of, 10
product of, 5, 10
Inertial cross-coupling, condition for
stability, 188-189
control system for stabilizing, 191-198
discussion, 183-184
effects of variation of roll rate,
186-189
linearized equations for steady roll rate,
186
mathematical analysis of, 184-189
parameters affecting stability, 189-191
response of basic aircraft, 194-197
with basic control system, 195
with complete control system, 197
without control system, 194
Inertial space, definition, 6, 99
Integrating gyro, 101, 303-305

Kinematic viscosity, 328

Lagrange's equations, 236
application to, double pendulum,
237-239
flexible vehicle, 239-241
mass, spring and damper, 237
Lateral acceleration, 109, 144-146
Lateral autopilot; *see* Flight control
systems
Lift, 328
Limit cycle, 216, 219
Line of nodes, 13
Lin's method, 34
Localizer receiver, 167

Mach hold, 92-96
Matrices, 333-337
addition of, 334
column, 333
diagonal, 336
division of, 335
identity, 336
inversion of, 335-336
multiplication of, 334-335
rectangular, 333
row, 333
subtraction of, 334
symmetric, 336
Mean aerodynamic chord (MAC), 18
definition, 331

Mean-square error, 273
Missiles, guided, 223
    aerodynamic, defined, 223
    axis system, 225, 227
    ballistic, defined, 223
    control system for; *see* Flight control
        system
    transfer function of, 230
        derivation of, 226-230
    Vanguard, aerodynamic coefficients, 234
Moments, applied, 16
    equilibrium, 7
Momentum, moment of, 9
    angular, 9

Natural frequency, 35
Navigation system, definition, 222
Newton's second law of motion, 6

Omni tracker, 167
Open loop transfer function, 73, 307

Parasite drag, 331
Phugoid mode, approximate equation, 44
    approximate transfer functions.
        $'u(s)/\delta_e(s)$, 46
        $\theta(s)/\delta_e(s)$, 46
    damping of, 81; *see* Velocity control
    description of, 35
    effects of airspeed and altitude on,
        49-50
Pitch damper, 59-60
Pitch-up, 64
    automatic system to control, 62-71
    discussion, 64-65
Post, Wiley, 2
Power spectral density, 278-280
    equation for, 278-279
    example of, 279-280
    of gusts, 282
Probability, 269
Probability density, 270
    first function, 272
    second function, 272
Probability density function, 271
Propellant sloshing, 254-259
    effects on control system, 258-259
    equation of the effects of, 257
    mechanical analog of, 255
    model for one tank, 256
    transfer function for, 258

Quartic equation, factoring, 34
Quasisteady flow, 22

Random process, 272
    ergodic hypothesis, 272
    stationary, 272
Random variable, 269-270
Rate gyro, 60, 71, 301-303
Response; *see also* Transient response
    to atmospheric turbulence, 281-285
Roll angle control system, 164-166
    control stick steering for, 165
Roll damping loop, use of, 211
Roll stabilization loop, use of, 213-214
Roll stabilization of missiles, control,
    223-225
    sensors, 223, 225
    system, 224
Roll subsidence, description, 118, 127
    effects of airspeed and altitude, 128, 132
    equation, 127
    one-degree-of-freedom rolling mode,
        127-128
    time constant of, 127
    transfer function, 127-128
Root locus, 310-322
    construction of, 314-321
        angle, condition, 314
        of arrival, 317
        of departure, 316
        asymptotes, 316
        breakaway point, 318-319
        break-in point, 318-319
        closed loop poles, 319
        imaginary axis crossing, 319
        magnitude condition, 314
        real axis intercept, 315
        real axis loci, 315
        Spirule, 319
    of adaptive autopilots, 202, 211-214,
        218
    of lateral control systems, 139-140, 148,
        159-160, 164-165, 173-176
    of longitudinal control systems, 57-58,
        60-62, 66-69, 75-77, 82, 86, 90,
        93-94, 97
    of missile control systems, 225, 231, 233
    theory of, 310, 314
    zero angle, 73, 322
Rudder displacement, positive direction,
    defined, 118

Sample point, 269
Sample space, 269
Schmitt trigger, 205
Second probability density function, 272
Servo, elevator, 56
Set-point, 216
Short-period mode, approximate equation, 40
  approximate transfer functions,
    $'\alpha(s)/\delta e(s)$, 42
    $\theta(s)/\delta e(s)$, 43
  damping of, 60
  description, 35
  effects of airspeed and altitude, 48, 50
  equations for ballistic missiles, 228
Sideslip angle, defined, 107
  measurement, 143
Sideslip sensor, 142
Space integrator, 101-105
  base motion isolation mode, 102-103
  commanded angular velocity input
    mode, 103-105
Span efficiency factor, 331
Sperry Aeroplane Stabilizer, 1
Sperry, Dr. E. A., gyropilot, 2
Spiral mode, description, 118
  effects of airspeed and altitude on,
    133
  equation for time constant, 135
Stability derivatives, lateral, table,
    112
  values, 112, 116-117, 122
  longitudinal, table, 19
  values, 19, 26-33
Standard deviation, 270
Static loop sensitivity, 308
Static margin, stick fixed, 28
Stick, force, 63, 71
  maneuver, 63
Stochastic variable, 269
Structural filter, 266
Structural flexibility, compensation required,
    259-266
  gyro blender, 260-262
  derivation of equations, 239-241
  mode frequencies, 242-247
  mode shapes, 242-247
  normal coordinates, 247-249
    eigenfunctions, 247
    eigenvalues, 247
  "tail-wags-dog" zero, 253-254

Structural flexibility, transfer function, including derivation of, 249-253
  structural filter, 266

"Tail-wags-dog" zero, 253-254
Tangential velocity, 9
Time constant, roll subsidence, 127
  spiral divergence, 135
Transfer functions, closed loop, 321-322
  definition, 3
  including, propellant sloshing, 258
    structural flexibility, 249-253
  lateral, aileron displacement, $\beta(s)/\delta_a(s)$,
    124
    $\phi(s)/\delta_a(s)$, 122
    $\psi(s)/\delta_a(s)$, 123
  approximate transfer function, one-
    degree-of-freedom Dutch roll mode,
    125-127
  one-degree-of-freedom rolling mode,
    127-128
    $\beta(s)/r(s)$, 142
  rudder coordination computer,
    150-151
  rudder displacement, $\beta(s)/\delta_r(s)$, 121
    $\phi(s)/\delta_r(s)$, 119
    $\psi(s)/\delta_r(s)$, 120
  longitudinal, $a_z(s)/\delta_e(s)$, 72
    $'\alpha(s)/\delta_e(s)$, 37
    $\gamma(s)/\theta(s)$, 85
    $h(s)/\theta(s)$, 90
    $u(s)/\delta$ (rpm) (s), 81
    $'u(s)/\delta_e(s)$, 36
    $\theta(s)/\delta_e(s)$, 38
  phugoid approximation, $'u(s)/\delta_e(s)$,
    46
    $\theta(s)/\delta_e(s)$, 46
  short-period approximation,
    $'\alpha(s)/\delta_e(s)$, 42
    $\theta(s)/\delta_e(s)$, 43
  missiles, guided, 230
  nonminimum phase angle, 72
  open loop, 73, 307
  Vanguard missile, 230
Transient response, effects of airspeed
    and altitude on, Dutch roll, 128,
    131
  phugoid mode, 49
  roll subsidence, 128, 132
  short-period mode, 48
  spiral mode, 133

Transient response, effects of stability
  derivative variation on, lateral
  modes, description of, 133-136
    table, 134
  longitudinal modes, 53-55
  effects of velocity control on, 83
  of adaptive autopilot, 219
  of lateral control systems, 141, 153-157,
    161, 178
  of longitudinal control systems, 70, 78-80,
    83, 87-88, 91-92, 95, 98
  of yaw orientational control system, 161
  three-degree-of-freedom,
    lateral for δ_a input, 130
    lateral for δ_r input, 129
    longitudinal for δ_e input, 47, 78
Type 0 system, 56, 308-309
Type 1 system, 309-310
Type 2 system, 310
Type 3 system, 310

Unit vectors, 8, 291

Vanguard missile, aerodynamic coefficients,
    234
  control systems for, 230-235, 264-266
  transfer function of, 230
Variance, 270
Vectors, addition of, 290
  components of, 291
  cross-product of, 292
  definition of, 289-290
  differentiation of, 8, 293-296
  dot product of, 292

Vectors, multiplication of, 291
  triple cross product of, 293
  triple scalar product of, 293
  unit, 8, 291
Velocity, angular, 6
  components, 6
  linear, 6
  nondimensional variation, 17, 21
  tangential, 9
  total angular, 8
    linear, 6
Velocity control, effects on transient
    response, 83
Viscosity, 328
  coefficient of, 328
  kinematic, 328

Washout circuit, 139
White noise, 279
Winnie May, 2
Wing area, 18
  definition, 328
Wing span, 331
Wing tip vortices, 22-23
Wright brothers, 1

Yaw angle, defined, 107
Yaw damper, 138-141
Yaw orientational control loop, 212, 214
Yaw orientational control system, 157-163
  block diagram of, 158
  step response of, 161
  yaw rate limiting, 162-163